The Biologist's Imagination

The symphony of scientific research, discovery and technological innovation. © Lynn Fellman,
Fellman Studio — All rights reserved

The Biologist's Imagination

INNOVATION IN THE BIOSCIENCES

William Hoffman
and
Leo Furcht

OXFORD
UNIVERSITY PRESS

OXFORD
UNIVERSITY PRESS

Oxford University Press is a department of the University of Oxford.
It furthers the University's objective of excellence in research, scholarship,
and education by publishing worldwide.

Oxford   New York
Auckland   Cape Town   Dar es Salaam   Hong Kong   Karachi
Kuala Lumpur   Madrid   Melbourne   Mexico City   Nairobi
New Delhi   Shanghai   Taipei   Toronto

With offices in
Argentina   Austria   Brazil   Chile   Czech Republic   France   Greece
Guatemala   Hungary   Italy   Japan   Poland   Portugal   Singapore
South Korea   Switzerland   Thailand   Turkey   Ukraine   Vietnam

Oxford is a registered trademark of Oxford University Press
in the UK and certain other countries.

Published in the United States of America by
Oxford University Press
198 Madison Avenue, New York, NY 10016

Library of Congress Cataloging-in-Publication Data
Hoffman, William R., author.
The biologist's imagination : innovation in the biosciences / William Hoffman & Leo Furcht.
pages cm
Summary: "Discusses the history of technological innovation in the biosciences"—Provided
by publisher.
Includes bibliographical references and index.
ISBN 978–0–19–997459–7 (hardback)
1. Medical innovations.   2. Biology—Technological innovations.   3. Life sciences—
Technological innovations.   I. Furcht, Leo, author.   II. Title.
RA418.5.M4H64 2014
610.28—dc23
2013040842

9 8 7 6 5 4 3 2 1
Printed in the United States of America
on acid-free paper

CONTENTS

ACKNOWLEDGMENTS

Many people gave us helpful information and insights during this book's long gestation. Among them are former US Senator David Durenberger, economists Michael Mandel and Louis Johnston, science journalist Clive Cookson, synthetic biologist and entrepreneur Robert Carlson, geostrategist Parag Khanna, technology forecaster Anthony Townsend, regional historian and author Joseph Amato, patent attorney Kevin Noonan, biotech consultant William Johnson, and IT consultant James Hudak. We acknowledge the excellent staff at the US Bureau of Labor Statistics for assistance in tracking US bioscience employment trends. The late scientist and Nobel economist Robert W. Fogel graciously allowed us to modify his timeline of technological innovation and world population growth in a way that reflects advances in biological technologies. Lastly, we thank Jean Kurata of the Department of Laboratory Medicine and Pathology, University of Minnesota, and Jeremy Lewis and Erik Hane of Oxford University Press for their assistance in bringing our manuscript to print.

INTRODUCTION

The ability of human societies to modify and transform
biological systems will increase more in this century than it has
in the hundred centuries since the dawn of agriculture . . . .

—Nature

Without deviation from the norm, progress is not possible.

—Frank Zappa

During an interview in 1929, Albert Einstein was asked if he trusted more to his imagination than to his knowledge. His answer has echoed through the decades since. "I am enough of an artist to draw freely upon my imagination," he said in "What Life Means to Einstein," published in *The Saturday Evening Post*. "Imagination is more important than knowledge. Knowledge is limited. Imagination encircles the world."

Imagination is a mysterious thing. Springing from the provinces and principalities of the unconscious, imagination takes us from what we know to the infinite realm of what we do not know, a poet once said. Imagination is the life force of our literature, which has its ancient roots in the oral tradition of storytelling; our art, which continues to keep archeologists and anthropologists busy; and our music, which probably filled the air long before recorded history when the acoustics of some prehistoric cave chambers are taken into account. Imagination is what has brought human beings from the earliest days of cave painting and tool making to the days of moon shots and iPhones and genomes decoded to reveal how life works as an information system, a bioinformation system. Imagination is the wellspring of creativity and the key to health, wealth, and well-being. The twentieth century established what Einstein called "the democracy of the intellect," a new epoch in which intellect, imagination, and creativity can blossom anywhere in the world, not just in elite enclaves among a privileged few.

It was imagination that drove James Watson and Francis Crick in their search for the structure of DNA in the middle of the twentieth century. Their discovery was one of the most socially and economically valuable in the history of science, "ranking alongside the transistor and laser," reflected a veteran science journalist on its sixtieth anniversary. Less than a quarter century after the discovery, venture capitalist Robert Swanson and biochemist Herbert Boyer imagined and then founded Genentech, giving birth to a new industry based on the idea that value could be created by moving DNA from one organism to another. Their dream was to harness recombinant DNA technology to design and manufacture new types of drugs. Today those drugs, called biologics, are used to treat millions of patients, enabling many to live normal lives. Swanson and Boyer located their new company in the technological field of dreams south of San Francisco: Silicon Valley, the *über*-cluster of innovation.

This book is about the history and current state of innovation in the biosciences in our changing world. It is about how imagination pushes back the frontiers of what we understand about living things. It is about how imagination challenges conventional thinking and norms by bringing together the sciences of life, creative enterprise, and a restless entrepreneurial drive. And it is about the places where these interactions tend to occur, the fertile soil where ideas take root.

What we know about biology figures into the medications we take, the vaccines we use to immunize our children, the food we eat, the water we drink, the air we breathe, and the way we dispose of our waste. Biology gives us fabrics that clothe us and ecosystems that sustain us. From biological knowledge we create genetic tests that enable us to learn about our health and risks for developing disease, about where we came from and possibly about where we are headed. Detailed understanding of biological components and processes plus converging domains of knowledge and the explosion in sharing information are accelerating the pace of discovery. The technology of biology has already entered our economic affairs in a big way. Today our ability to make changes in the DNA molecule, shuttle DNA from one life form to another, manipulate biological processes, alter biological endowment, and test biological response accounts for more than 2 percent of what the US economy produces every year in revenues. Those revenues are growing 15 percent annually by some estimates (see Figure I.1).

The tools of bioscience innovation are also telling us a lot about the living systems of which we are part. As the biologist Rachel Carson was writing and publishing her 1962 environmental classic *Silent Spring*, Marshall Nirenberg and Har Gobind Khorana and other geneticists and biochemists were deciphering the genetic code and protein synthesis and explaining the chemistry of life. Today the tools of bioscience and genetics are showing that healthy ecosystems are critical for economic development. They help us understand that clean air and water are vital for human health and productivity, that biodiversity may ease the burden of infectious disease, and that forests are important carbon storehouses for greenhouse gas emissions. They also help us connect ocean acidification to declining coral reefs and commercial fish stocks and empower us to develop new food crop lines that are better suited to thrive in a changing climate, yielding more food from less land using less water, fertilizers, and pesticides.

We have grown accustomed to taking the remarkable advances in the biosciences fairly in stride. That was highlighted by what happened in just one week in the spring of 2013. First, geneticists confirmed the hybrid ancestry of New World cattle such as Texas Longhorns with their rich genetic diversity and ability to adapt

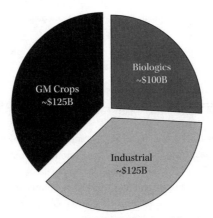

FIGURE I.1 United States biotech revenue in billions of US dollars (B). Genetically modified products in the US Bioeconomy (2012 estimate): 350B or equivalent to 2.5 percent of GDP.

Source: Robert Carlson, Biodesic, synthesis.cc blog, January 1, 2014, with permission.

to drought. Their DNA is derived from cattle that spanned the globe, from the Iberian cattle that came with Columbus to the ancestral aurochs of the Middle East and India, telling a story of human and cattle co-migration. Then scientists reported the genetic sequence of the mountain pine beetle, a notorious pest that has destroyed more than 15 million hectares of forest in western North America and made them vulnerable to wildfires. Unusually dry conditions in the pine forests, a consequence of climate change, enable the beetle to expand its range northward. In a pernicious feedback loop, the beetle itself contributes to a warming planet by destroying forests. Shortly thereafter, bioengineers reported eyebrow-raising experimental results. They constructed functional transistor-like logic gates inside cells, paving the way for living computers. Cancer researchers then reported that mutated-driver genes responsible for most if not all malignancies operate through as few as twelve cell-signaling pathways, giving innovative drug designers a clearer pathway to developing safe and effective therapies.

To innovate stems from the root Latin word *innovare*, to renew or change, which in turn stems from *novare,* to make new. *Innovation* is a word that gained currency in published materials beginning in the eighteenth century with the Industrial Revolution (Figure I.2), although the word *innovation* does not appear in Adam Smith's *The Wealth of Nations* (1776). *Innovation* is a word inextricably bound with the American democratic experiment. Indeed, America's form of democracy has itself proved to be an unparalleled innovation. James Madison, the father of the Constitution and a student of the "science of Government," employed the term in the *Federalist Papers* to describe the way critics saw the Framers' proposed structure of a national government. "And it is asked by what authority this bold and radical innovation was undertaken."

To Madison, innovation in the structure of republican government was both desirable and necessary for the American experiment to work. To his *Federalist*

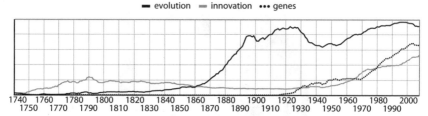

FIGURE I.2 Innovation (light line), evolution (dark line), and genes (broken line) 1700–2008. The graph illustrates the frequency trend line of each word appearing in 5.2 million books published in American English, British English, French, German, Spanish, Russian, and Chinese scanned by Google Books in an exercise that has been called cultural genomics or culturomics. Jean-Baptiste Michel et al., "Quantitative Analysis of Culture Using Millions of Digitized Books." *Science* 331 (2011):176–182.
Source: Google Books Ngram Viewer, http://books.google.com/ngrams.

Papers coauthor Alexander Hamilton, innovation could be a creative and beneficial action or a mischief-making nuisance when it came to administration and governance. Yet it was Hamilton more than any of his contemporary nation-builders who understood that innovation in economic affairs—the mix of diverse ideas, experimentation, and capital—was the dynamo of the capitalist system and the secret to material affluence. The role of government is to ensure that the dynamo performs its task as efficiently as possible consistent with the public good. Hamilton wrote in the *Federalist Papers* that "*in the usual progress of things, the necessities of a nation, in every stage of its existence, will be found at least equal to its resources*" (emphasis in original). His idea of energetic government and Madison's idea of individual liberty were in some tension with each other in 1787. That tension still exists. Will today's *necessities* for long-term economic growth find the *resources* to support federally funded research and development essential for that growth? Or will the current and projected decline in federal research funding as a percentage of the nation's overall economic output—headlined as the "The Coming R&D Crash"—put Hamilton's emphatic belief at risk?

Living things respond to changes in their ecosystems through experimentation, adaptation, and natural selection. When viewed as dynamic systems, economies also seem to respond organically, from the ground up, to changing environments. Entrepreneurship is the source of capitalism's "creative destruction," an ongoing process that economist Joseph Schumpeter likened to evolution. Innovation can be thought of as the means by which dynamic economies adapt to new realities, new necessities, new forces in the economic ecosystem, producing economic growth and social change. Today human ingenuity, the application of imagination, is expressed in the economic arena principally through technology. New tools, devices, products, processes, methods, and practices that arise from the adoption and use of technologies generate positive feedbacks, stimulating further technology development and economic activity. We reorchestrate the parts and processes that constitute technology to better serve our needs. We adapt to new economic realities by creating new technologies and then find ways to expedite their entry into the marketplace and society.

Innovation systems in the biosciences build wealth through brainpower using micro-amounts of elements and natural resources: DNA, RNA, proteins, enzymes, carbohydrates, cells, silicon, carbon, gallium, platinum, phosphorous, and a host of other elements, molecules, compounds, substances, fibers, and materials. Innovation is shrinking industrial machines and medical devices to the scale of atoms and molecules and building microbial factories. Among the biological technologies flowing into our economic ecosystem are tools to read (sequence), write (synthesize), and edit with exquisite precision the code of life—DNA—and engineer the fundamental unit of life—the cell. New instruments have driven down the cost of reading and writing DNA dramatically in the past decade, at a much faster pace than the drop in the cost of computing power over the past five decades. These tools and instruments are or will be available to more and more people—creative individuals, research teams, entrepreneurs, college and high-school students, even do-it-yourself hobbyists working at home, in garages, or in community laboratories.

Our story about the history and current state of innovation in the biosciences is divided into seven chapters, all bound by recurrent themes woven throughout the narrative. We cover subjects as seemingly disparate as the history of technology, economics, molecular biology and genetics, neuroscience, geography, evolution, education, globalization, clinical trials, technology transfer, the digital revolution, patent law, and public policy.

To put the remarkable world of technological change we entered more than two centuries ago into context, we reflect on the history of technology in chapter 1. Technological innovation has is roots both in myth and in the observations of pre-Socratic natural philosophers. It came to the fore in late Middle Ages and exploded beginning with the Industrial Revolution. The Lunar Society of Birmingham linked science, industry, and society for the first time, setting the stage for the technology-driven modern economy. What science historian Horace Judson calls "the great endeavor of biology" beginning in the nineteenth century provided the foundation for the rise of molecular biology and the biotechnology industry in the twentieth century and completion of the Human Genome Project early in the twenty-first century.

In chapter 2, we describe how the pharmaceutical industry came to account for 90 percent of economic output in today's trillion-dollar-plus global life sciences sector. Its small-molecule drugs still dominate the market, though biopharmaceutical drugs based on advances in molecular biology constitute a growing proportion of innovation in the industry and of drug sales. U.S. Food and Drug Administration approvals of new molecular entities have grown in recent years after a period of stagnation, yet R&D efficiency in pharmaceutical industry has declined with increases in investment. High-throughput screening of candidate molecules together with advances in "omics" technologies (genomics, transcriptomics, proteomics, metabolomics) and biomarker identification may help reverse this trend, as may patient selection for early-stage trials based on their genetic profile. Meanwhile, manufacturing of biopharmaceutical drugs has become highly efficient. We explain how the offshoring of pharmaceutical R&D, biomanufacturing, and clinical trials gives host countries greater opportunity to innovate themselves with domestic talent and explore differences in the way their ethnic groups respond to drugs based on their genetic makeup (pharmacogenomics). Regulatory bodies and public health agencies are hard pressed to keep up with these rapid, technology-driven developments.

The European "Age of Discovery" five centuries ago opened up global trade, including the exchange of animal, plant, and microbial species between the Old and New Worlds. Today genomic sequences of many species are transported over high-speed data networks rather than in organisms aboard ships, as in Columbus's day. Yet even in an era of global networks, bioscience startups continue to show a strong tendency to cluster in specific regions with strong research assets, which is the focus of chapter 3. Bioscience tends to catalyze itself in these urban regions, making the landscape of bioscience innovation spiky rather than flat on the global map. Technology clusters that join entrepreneurship with finance, support services, research, and education tend to emerge spontaneously. Around the world, nations and regions are trying to seed clusters by building incubators, accelerators, and science parks, typically near research facilities. Innovation and location like to march together, especially in drug, diagnostic testing, and medical device entrepreneurship. The global distribution of entrepreneurial bioscience will depend on the forces of technological innovation, urbanization, globalization, and research investment.

The Augustinian monk Gregor Mendel's pea experiments laid the foundation for modern genetics. Mendel's aptitude for math gives us license to track his legacy from pea genetics to petabytes of genomic information in chapter 4. Yet Mendel's heavy use of statistics made his famous paper practically impenetrable to the botanists of his day. Today's communications technologies, search engines, and open access make specialized scientific information readily available to scientists and innovators alike. The biostatistical analysis Mendel did in his head is done now with powerful computers and bioinformatics software that generate cumulative genomic data measured in petabytes. Biological and digital technologies will be critical for mitigating the effects of greenhouse gas emissions on biogeochemical cycles and food production. They are beginning to revolutionize how we think about human health and disease. Though direct-to-consumer (DTC) genetic testing companies marketing disease-risk information face regulatory hurdles, the market for DTC testing services is expected to grow because many people want to know what their disease risks may be. In the laboratory, next-generation DNA sequencing is developing at a faster pace than Moore's Law in computing. The expected deluge of whole genome sequence information will challenge our ability to interpret it accurately so that clinical genomics can realize its promise of improving health outcomes and reducing costs. Meanwhile, mobile devices that carry applications for biomedical research, community health reporting, and clinical genomics have made their debut.

The story of universities and technology transfer, the subject of chapter 5, begins at the University of Bologna, the first university in the West, located in a dynamic, entrepreneurial region of Italy. Since their first appearance a millennium ago, universities have been in the business of knowledge transfer and, for the past two centuries, of knowledge generation through research. Today universities, government, and industry are viewed as the triple helix of knowledge production and innovation. Enactment of the Morrill Act in 1862 laid the foundation for US federal involvement in knowledge transfer from academia to the economy. Private investment and government initiatives to spur technology transfer in the twentieth century culminated with enactment of the Bayh–Dole Act in 1980, which enabled university faculty to patent technologies and processes they invented. As we describe at length, the Bayh–Dole Act was key to the rise of the biotechnology industry and transformed academic bioscience and innovation in the United States. Subsequent federal legislation sought to enhance technology transfer from federal agencies, national laboratories, and universities to the private sector, particularly to small businesses and entrepreneurial startups. Today, universities are exploring new models for technology transfer through venture centers, partnerships, and innovation networks.

Intellectual property, information, and incentives are well understood to be keys to innovation, as we show in chapter 6. Although much has changed since enactment of the US Patent Act of 1790, "the basic idea that inventors have the right to patent their inventions has not," wrote John Roberts, chief justice of the US Supreme Court. Yet the patent system has been burdened with administrative backlogs and legal disputes. Congress sought to update the system in 2011 when it passed and President Barack Obama signed the America Invents Act, which replaces the first-to-invent with a first-to-file system and implements a post-grant review process for issued patents. In the biosciences, the public debate over whether genes can be patented went before the US Supreme Court in 2013. The gene patenting issue reveals the tension between proprietary interests important for innovation and the practice of medicine and public health uninhibited by limits imposed by proprietary interests. Successful innovation depends on a system that aligns incentives in a way that maximizes value both to the innovator and to society. Intellectual property, open-source and networked science, and precompetitive collaboration all contribute to innovation in drug discovery and development and in agriculture as well as in emerging technological fields like synthetic biology and genomic medicine.

In the twenty-first century, innovation will be broadly framed by exploration of evolution—the evolution of the universe, evolution by natural selection on Earth and perhaps beyond, and ongoing technological and cultural evolution. As we describe in chapter 7, we will see the evolution of institutions, organizations, and innovation networks; the evolution of public health agencies in response to pandemic influenza and other biothreats; and the evolution of the global middle class. The systematic practice of innovation in the industrial era—the application of ideas, inventions and technology to markets, trade, and social systems—is now being joined with the code of life. Our ability to read, write, and edit DNA and re-engineer the cell empowers new processes for discovery and development in medicine, agriculture, industry, energy production, and environmental health. It also opens a window to our evolutionary past and will influence how we live, how well we live out our years, and even how we evolve.

For the first three decades of its existence as an industrial concern, biotechnology required us "to indulge in a mass suspension of disbelief, to relax our critical facilities to allow ourselves to believe—despite all the reasons to the contrary—that we can create something that simply did not exist before." That was the way a critic described the industry's persistent lure and regular supply of gaffes and shortfalls. But biotechnology is moving beyond the drug development model that has dominated the field. Whether it is through harnessing the computational power of the code of life, the regenerative power of the cell, or the manufacturing and biomass conversion powers of microorganisms, innovators are already are already making their presence felt. Their productive handiwork is reflected in the growth of regional, national, and international collaborations in the field and in global initiatives to address the changing circumstances of life and living systems. Over time, it will be reflected in national economic growth statistics.

Finding the right combination of policies, processes, incentives, and initiatives to spur innovation from the swelling biological knowledge bank will take what the application of new knowledge normally takes and what Einstein and Hamilton understood implicitly: imagination and experimentation. It will also take new forms of organization from the self-organizing creatures that we are, as evidenced by the symphonic orchestration of our genes, our brain's ability to reorganize its neural pathways in response to new conditions and experiences, and our complex social systems.

In the late eighteenth century, scientists, philosophers, mechanics, entrepreneurs, and investors applied their individual ingenuity in a systematic way to usher in a world of labor-saving and time-liberating machines, devices, and processes. In the early twenty-first century, we imagine and experiment at will. Ingenuity is every bit as valuable in the big-data era as it was in the machine age. We gather around good ideas arising from what we know and what we learn about the codes, elements, mechanisms, and mazes of biology to build valuable and useful things. Nature has supplied innovators in the field with billions of years of experimental results. Technology has equipped them with powerful new tools, enabling them to undertake their own experiments in laboratories, *in silico*, in garages, and in the marketplace. And imagination, as Einstein said nearly a century ago, encircles the world.

References

Listed sequentially based on their order in the Introduction.

George Sylvester Viereck, "What Life Means to Einstein," *The Saturday Evening Post*, October 26, 1929, pp. 17, 110–117.

Clive Cookson, "The Lucrative Allure of the Double Helix," *Financial Times*, April 19, 2013, accessed April 23, 2013, http://www.ft.com/intl/cms/s/0/4269758c-a776-11e2-9fbe-00144feabdc0.html.

"The Benefits and Costs of the Clean Air Act from 1990 to 2020," US Environmental Protection Agency Office of Air and Radiation, March 2011, accessed December 28, 2012, http://www.epa.gov/oar/sect812/feb11/fullreport.pdf.

Matthew H. Bonds, Andrew P. Dobson, and Donald C. Keenan, "Disease Ecology, Biodiversity, and the Latitudinal Gradient in Income," *PLOS Biology* 10.12 (2012): e1001456, accessed December 28, 2012, doi:10.1371/journal.pbio.1001456.

Yude Pan et al., "A Large and Persistent Carbon Sink in the World's Forests," *Science* 333 (2011):988–993, accessed December 28, 2012, doi: 10.1126/science.1201609.

Sarah R. Cooley and Scott C. Doney, "Anticipating Ocean Acidification's Economic Consequences for Commercial Fisheries," *Environmental Research Letters* 4.2 (2009): 024007, accessed December 28, 2012, doi:10.1088/1748-9326/4/2/024007.

Scott Malcolm. et al., "Agricultural Adaptation to a Changing Climate," Economic Research Service, Economic Research Report No. 136, U. S. Department of Agriculture, July 2012, accessed December 28, 2012, http://www.ers.usda.gov/media/848748/err136.pdf.

Lewis H. Ziska et al., "Food Security and Climate Change: On the Potential to Adapt Global Crop Production by Active Selection to Rising Atmospheric Carbon Dioxide," *Proceedings of the Royal Society B* 279.1745 (2012): 4097–4105, accessed December 28, 2012, doi: 10.1098/rspb.2012.1005.

"Genome Research: Discovery as an Everyday Event," *The Guardian*, March 29, 2013, accessed April 3, 2013, http://www.guardian.co.uk/science/2013/mar/29/genome-research-adventure.

Emily J. McTavish. et al., "New World Cattle Show Ancestry from Multiple Independent Domestication Events," *Proceedings of the National Academy of Science USA* (2013), accessed April 3, 2013, doi: 10.1073/pnas.1303367110.

Christopher I. Keeling et al., "Draft Genome of the Mountain Pine Beetle, Dendroctonus ponderosae Hopkins, a Major Forest Pest," *Genome Biology* 14 (2013):R27, accessed April 3, 2013, http://genomebiology.com/2013/14/3/R27/abstract.

Holly Maness, Paul J. Kushner, and Inez Fung, "Summertime Climate Response to Mountain Pine Beetle Disturbance in British Columbia," *Nature Geoscience* 6 (2013):65–70.

Jerome Bonnet et al., "Amplifying Genetic Logic Gates," *Science* 340 (2013):599–603.

Bert Vogelstein et al., "Cancer Genomics Landscapes," *Science* 339 (2013):1546–1558.

James Madison, "The Conformity of the Plan to Republican Principles," Federalist Paper No. 39 (n.d.), accessed March 4, 2013, http://thomas.loc.gov/home/histdox/fedpapers.html.

Alexander Hamilton, "Concerning the General Power of Taxation," Federalist Paper No. 30, December 28, 1787, accessed March 4, 2013, http://thomas.loc.gov/home/histdox/fedpapers.html.

Brad Plummer, "The Coming R&D Crash," Wonkblog, *The Washington Post*, February 26, 2013, accessed March 4, 2013, http://www.washingtonpost.com/blogs/wonkblog/wp/2013/02/26/the-coming-rd-crash/.

Justin Hicks and Robert D. Atkinson, "Eroding Our Foundation: Sequestration, R&D, Innovation and U.S. Economic Growth," Information Technology & Innovation Foundation, September 2012, accessed March 4, 2013, http://www2.itif.org/2012-eroding-foundation.pdf.

Joseph A Schumpeter, *The Theory of Economic Development: An Inquiry into Profits, Capital, Credit, Interest, and the Business Cycle* (Cambridge, MA: Harvard University Press, 1934).

John Hodgson, "Ten Years of Biotech Gaffes," *Nature Biotechnology* 24 (2006):270–273.

The Biologist's Imagination

Ideas, Progress, Wealth, and the Biological Revolution

The rise of western economies based on economic growth and
technological progress "is the central event of modern history.
Nothing else even comes close."

—Joel Mokyr, *The Gifts of Athena*

One day long ago, a group of people in East Africa set out on the longest terrestrial
journey the world has ever known, a journey that continues to this day. It was the
first leg of a trek out of Africa that took their descendants to the farthest reaches
of the globe—to Eurasia, Oceania, East Asia, and the Americas. These migrating
humans, members of our earliest ancestral population whose genetic markers we
bear in our own life script, left few tangible markers of their travels across the land.
Indeed, we have little evidence to explain why they left Africa in the first place.

What we do know is that, for whatever reason, they felt it necessary to risk the
venture, to leave the familiar and go where they had never gone, taking with them
what little they had. The savannah of East Africa was first to witness a dimension
of the human spirit that remains very much with us. Innovation is and always has
been the realization of a successful search. Today we venture forth to explore the
unknown in research laboratories as well as across the earth's surface, deep in its
oceans, and beyond its atmosphere. We plumb the secrets of the sciences of life and
seek to capture that knowledge and put it to work. We may have new ideas and new
tools, but we are infused by the same spirit that prompted our distant forbearers to
leave the African savannah.

That spirit to venture out, to seek the unfamiliar, arises from what the American
poet W. S. Merwin calls the shadow, the unknown and ultimately unknowable side
of human existence that guides us and gives depth and dimension to life. "The
imagination, which goes beyond knowledge, moves closer to the unknown, and
of course comes out of the unknown," Merwin says. Deep history shrouds our
earliest days as a species in the unknown. We will never know what motivated

our prehistoric forbearers to do what they did, to leave the familiar behind. Was it a search for new sources of food? an escape from danger posed by enemies? a response to a changing climate? the culmination of pent-up curiosity to see what lay over the horizon? All the evidence from archeological or genetic digs doesn't really tell us. We can only imagine.

When prehistoric people began to leave behind a pictorial record on the walls of caves, they gave us a glimpse of the mysterious realm of early human culture and belief. One theory holds that shamans would go into trances and convey their visions by painting images of wild animals on the stone, perhaps dreaming that their number would thereby increase for the next hunt for the early hunter-gatherers. The brains of prehistoric people were much the same as our own thoroughly modern brains. These people used their brains to catalog the knowledge of important plants and animals within their sphere of activity, their home range, their universe. Elders and shamans, the tribal encyclopedists, shared that critical knowledge with their open, egalitarian, and cooperative communities through successive generations. That exercise was an early instance of organized "knowledge sharing" that is so much a part of how innovation occurs today.

Knowledge sharing from the flood of biological data arising from genomics, integrated systems biology, imaging, and other information-intensive fields has set the stage for major innovations in health and well-being, food and nutrition, renewable energy, novel materials, and the design of products, processes, and living spaces. The tools of bioscience give us the power to influence our own evolution and to delve into the human brain to visualize the seeds of imagination and creativity, the neural networks that bring art and tools into existence. Thus our capacity to innovate in the biosciences brings us full circle. "In the end, we self-perceiving, self-inventing, locked-in mirages are little miracles of self-reference," wrote cognitive scientist and author Douglas Hofstadter in his book *I Am a Strange Loop*. Our quest to illuminate the "shadow" of our existence, to expand our comprehension of the natural world and improve our circumstances, is dogged and unrelenting, strange loops that we are.

Ancient Gifts Get Things Started

Innovation is what happens when new thinking is successfully introduced and valued by organizations, write Mark Dodgson and David Gann in their book *Innovation: A Very Short Introduction* (2010). It is an arena where the creation and application of new ideas are formally organized and managed, a theater where the excitement of experimentation and learning meets organizational realities and constraints.

The theater was "a place for viewing" in ancient Greece. Audiences witnessed innovations in the dramatic arts never before seen, including a shift from the world of *mythos* transmitted by the solitary poet to the world of human interaction and plot. The contemporary theater of innovation that Dodgson and Gann describe is a collective affair, an outward, interactive realm beyond what neurologist and author Oliver Sacks calls the imaginative space of our own minds, "our own private theaters." That is especially true in the biosciences. Innovation in the biosciences, biopharmaceuticals, cellular therapies, diagnostics, and biomedical devices tends to occur more often in certain places where ideas, investment, and talent come together. It is in these places that the emerging field's economic output, now worth several hundred billion dollars annually in the United States alone, is concentrated. Think of these places as arenas of discovery, creativity, and interaction among the innovators and the end-users of innovations—among the players and the audience. They are often called clusters, hubs, or hotbeds of bioscience where exponential growth in knowledge spurs innovation. Their evolution on the global stage, as we will see, is inextricably tied both to personal and to economic health and well-being in the twenty-first century, the century of biology.

The exercise of imagination that describes biological research and innovation as well as classical Greek drama can be traced to the pre-Socratic philosophers. Thanks to these philosophers and their descendants—and to Byzantium and Islam for transmitting ancient Greek learning to Renaissance Europe—the effort to understand the natural world through *mythos* was complemented with a new view based on reason rather than solely on the supernatural. The key principle governing the cosmos, these philosophers discerned, was *logos*, the universal divine reason and the power of reasoning resident in the human soul. The key figure in introducing *logos* was Thales of Miletus, regarded by Plato as one of the Seven Sages of ancient Greece and by Aristotle as the founder of Greek natural philosophy, the father of science.

The earliest of the pre-Socratic natural philosophers, Thales was not satisfied with what *mythos* had to offer when it came to understanding the natural world. So he tried to find naturalistic explanations using reason for how the world worked rather than relying on *mythos*—on stories about how anthropomorphic gods caused storms and lightening as well as birth, disease, and death. He distrusted the truth of the world as told by myth. Based on study and reasoning, including the abstract reasoning at the heart of mathematics, Thales concluded that the world originated from water, not from the creative inclinations of gods. His explanation for earthquakes, for example, was that the world sits on the element that gave it birth. Thales put Western thought on a course that elevated the use of *logos*, of reason for comprehending the world and how it works. That meant acquiring knowledge systematically, what became the scientific method with the rise of modern science in the seventeenth century. "The scientific method has released us, intellectually, from the absolutes of the mythological ages," wrote the celebrated mythologist Joseph Campbell.

In addition to *mythos* and *logos*, the ancient Greeks gave us another pair of words that are fundamental to modern life and to our story: *episteme* and *technê*. *Episteme* is literally "science" but is translated as "knowledge." *Technê*, literally

"craftsmanship," is usually translated as "craft" or "art." It is using reason to produce an object or accomplish a goal or objective. The means of doing it was understood by Plato to be art. But *techné* is also the root of the word "technology." Today
we understand *techné* to be the application of science in everyday life through the
design, development, and production of instruments and tools that raise our living
standard and enrich our lives.

Just as imagination operates full throttle in the world of myth, so it does in
the world of technology, including biotechnology. Neither just the measurable
nor just the immeasurable but deeply felt can tell the whole story of how we
understand our world. Otherwise we would have no need to see science in mythic
terms, as we do in the stories of Prometheus and Pandora and in terms like
"miracle cure," "magic bullet," and the "Holy Grail." We would not name bioscience and technology companies and research projects Amazon, Apollo, Artemis,
Athena, Beowulf, Chiron, Genesis, Hydra, Isis, Jupiter, Minerva, Odyssey, Odin,
Orion, Osiris, Pegasus, Saturn, and Zeus. We would not name a master stem cell
gene Nanog after the Celtic legend of Tír na nÓg, the "land of forever young,"
a therapeutic island far across the western sea. We would not name so-called
jumping genes Sleeping Beauty and Frog Prince. We would not call the fruit fly
the Cinderella of genetics.

Horses and Horsepower

After the decline of the Roman Empire and the Dark Ages in the West, China possessed a higher level of social development measured by the key factors of energy
use (coal), urbanization, military capacity, and information technology (woodblock
printing, paper, the magnetic compass). Europe was in a position, however, to create the conditions for general well-being and prosperity. It had a mild climate and
even rainfall, good soil for growing crops, and superb river drainage systems. What
was holding peoples of Western Europe back was a rigid social order reinforced by
their own myths, beliefs, and legends. They were surrounded by dense forest that
stoked fear of the unknown captured in fairy tales like *Little Red Riding Hood,
Hansel and Gretel, Tom Thumb*, and other stories of woods, wolves, and witches.

It was Europe's biological endowment coupled with its soil and climate that put
it on the path to growth. Europe had something other parts of the world did not
have—an abundance of large mammals, especially the horse. Evolution and species
migration had given Europe a comparative advantage. Large horses could do work
no other draft animal could do as well. Draft horses could plow the soil better than
oxen and were best for moving fresh crops to urban markets, wrote David Landes in
The Wealth and Poverty of Nations, his account of Europe's explosive growth after
the Middle Ages. Plus they generated abundant fertilizer for the next years' crops.
This virtuous cycle led to intensive and expanding cultivation and larger yields,
providing Europeans with a diet rich in animal protein.

The power of horses reigned supreme, giving Europeans a standard of living the rest of the world did not have. It also gave them the opportunity to fight wars and invent useful things, such as the sextant and the astrolabe. These tools, in turn, gave them liberty to travel over the seas on ships built from the abundant supply of wood from their forests to seek treasure and conquer other peoples. The horse supplied power for the surge in agricultural productivity from new farming implements and techniques that spurred population growth in Europe beginning around 1750. That surge meant fewer hands were needed to farm and more could engage in practices associated with civic society and the rapidly expanding marketplace, a marketplace for innovation.

As the Industrial Age made its debut in the second half of the century, the horse began to face competition from the mobile machine even as it lent the machine its name as a source of power. "Horsepower" came to mean work done by a machine over time. The concept of horsepower originated with none other than James Watt, the Scottish inventor of the steam engine with a critical separate condenser design. Watt's name is inextricably linked with the beginning of the Industrial Age but also with the "watt," the unit of energy associated with electricity that powered an energy revolution that he did not live to see. Watt needed to show the power advantage of his machines over horses, so he invented a formula for horsepower. We still use the term today to gauge engine power but also metaphorically, as in computer processing horsepower or DNA sequencing horsepower.

The genesis of efficient machines owes much to ingenious inventors and entrepreneurs like Watt but also to the nature of the times, themselves the product of a long evolutionary process. The stage was set first in England by the social reaction against mercantilism, monopolies, regulation, guilds, and other forms of institutionalized protections that economic historian Joel Mokyr calls "rent-seeking," an activity that inhibited economic activity and trade, as it does today. Rent-seeking is the attempt to generate income by manipulating the social or political environment in which economic activities occur rather than by creating new wealth. One of the rising forces counteracting rent-seeking behavior was governmental establishment of intellectual property rights for inventors. Changing cultural values meant that solving practical problems of the day took on greater urgency once minds were unshackled from social, political, economic, and religious restraints and practices and material self-improvement was given a freer hand.

The rise of capitalism also coincided with the emergence of idea ecosystems, "the propitious linkage of technological curiosity to economic opportunities and a supportive social environment," Joyce Appleby wrote in her history of capitalism. These local economic ecosystems arose initially because of transportation cost-savings yet thrive today in spite of the comparative ease of moving goods and information over the face of the earth. Such local and regional economic ecosystems, if economists and forecasters are right, will drive an increasing portion of the global economy as the twenty-first century unfolds. Relentless urbanization across the globe is one reason. Another is that they foster the adoption and diffusion of new ideas and generate positive "spillover effects." They are at the heart of techno-logical innovation.

The Industrial Revolution got its geographical start in one such economic eco-system—the West Midlands of England, a region of gently rolling hills, rustic beauty, and abundant iron ore and coal deposits. Birmingham became the revolu-tion's powerhouse, "the workshop of the world" and "city of a thousand trades," harboring a vast array of small craft shops and metal works. Birmingham boasted an extensive network of canals within the city and linking it to nearby deposits of coal, the energy from compressed plants of the Carboniferous period that pow-ered the new engines and, with its burning at an industrial scale, began to change the atmospheric concentration of carbon dioxide. Later, the railroad picked up where the canals left off, linking Birmingham with all of Britain. It was in one of Birmingham's instrument shops, in 1769, that Watt began his journey into history with a patent on an engine he had invented.

"Everything flows" wrote the pre-Socratic philosopher Heraclitus, arguing that fundamental reality is not about being but *becoming,* because the world is in a state of perpetual change. For the first time in history both the economy and knowl-edge were growing fast enough to generate "a continuing flow of improvements" in Landes's account, leading to sustained improvements in living standards as the mechanization of industry penetrated deep into the culture. The continuing flow of improvements, beginning with steam-powered mechanization, has been expo-nential, a burst in the evolution of technology perhaps not unlike the Cambrian explosion that produced a rich diversity of species over a relatively short period. Indeed, technological evolution has enough in common with biological evolution that economic historians like Mokyr and Nobel Laureate Robert Fogel employ the language of evolutionary biology in their narratives whether describing the history of technological change (Mokyr) or the effect of that change on how healthy we are and how long we live (Fogel).

The "Lunaticks" Link Science, Industry, and Society

The continuing flow of improvements in biological technology today is mostly the result of innovation and collaboration, not the genius of the heroic inventor of the Industrial Age like Watt. It is the consequence of people getting together, talk-ing, corresponding, and exchanging information and ideas in person or virtually. Research advances in biotechnology occur on the shoulders and backs of discover-ies made by an earlier generation. The next time you hear about a breakthrough in genetic sequencing or biological microchips or the formation of a collaborative brain science network or science park, remember the "Lunaticks." That is what the men of the Lunar Society of Birmingham liked to call themselves. They were a group of scientists, manufacturers, and entrepreneurs in the England's West Midlands who began to meet sometime around 1765 to discuss what they called "natural philosophy," the study of nature and the physical universe that Thales of Miletus set in motion, and figure out ways to put that knowledge to good use. They put aside the customary disdain that practical men often held for scholars with their book learning, lectures, and colloquies and the indifference scholars typically showed toward inventors, traders, and businessmen. Never numbering more than fourteen, some of them are legend. Besides James Watt they included the chemists and inventors Joseph Priestley and James Keir, the agrarian reformer Richard Edgeworth, the manufacturing potter and factory organizer Josiah Wedgwood who was Charles Darwin's maternal grandfather, and William Murdoch, the inventor of gas lighting. Because the society was formed just prior to the era of gas lighting, members decided to meet each month on the Sunday nearest to the full moon so that they had enough light to find their way home (Figure 1.1).

Of all the society members, Josiah Wedgwood stood out for his acumen about business organization, management, and innovation. When he installed steam engines into his pottery works, the modern ceramics industry was born. The new machines turned pottery factory lathes that once were turned by power generated by animals,

FIGURE 1.1 Meeting of the Lunar Society of Birmingham. The Lunar Society was a kind of Silicon Valley networking association at the dawn of the Industrial Revolution, linking the world of science with the world of practical affairs. Members met to debate scientific questions and the application of science to manufacturing, mining, transportation, education, and medicine. Meetings took place under a full moon so that members could travel at night. This meeting took place at the home of James Watt, inventor of the steam engine with a separate condenser in 1765. Watt's business partner, the entrepreneur Matthew Boulton, was also a member of the group. © Science Museum/Science & Society Picture Library–All rights reserved.

watermills, or human arms and legs. Wedgwood's friendship with Watt put a production system that had its roots in the Neolithic age on a new innovation pathway. For our purposes, Wedgwood's legacy is reflected in two contemporary phenomena that go to the heart of bioscience innovation and economic development today: innovation clusters and open innovation. Wedgwood's Staffordshire factory, Etruria, freed up workers to concentrate on the artistic and design side of craftsmanship. By concentrating production processes in one place and facilitating distribution through a network of roads and canals he built, Wedgwood laid the foundation for today's industrial cluster. Dodgson and Gann describe how Wedgwood also encouraged collaborative research within the cluster, proposing perhaps the first industrial cooperative research program to solve a technical problem. Thus he was a proponent of what today is called "open innovation," a key feature of dynamic industry clusters today.

The Lunar Society had affiliates in other countries. Benjamin Franklin in colonial Philadelphia was a society correspondent, as was the father of modern chemistry in Paris, Antoine-Laurent de Lavoisier. The queen bee of the club, the first scientific collaborative circle of the Industrial Age, was Erasmus Darwin. Darwin was a physician, botanist, poet, and author of *Zoönomia* or "The Organic Laws of Life," a treatise on the generation of living things. The book was a foreshadowing of the theory of evolution that his grandson, Charles, gave to the world in *The Origin of Species*. It was Wedgwood who sent Erasmus Darwin down the pathway of evolution when he unearthed curious bones from a canal tunnel he was excavating for his pottery transport. The bones put the good doctor on an intellectual journey. At the journey's end, Erasmus Darwin concluded that all living creatures are descended from a single microscopic ancestor. As the story goes, Charles read his grandfather's book when he was a boy. Toward the end of his life he acknowledged that becoming acquainted with his grandfather's views early in his life "may have favoured my upholding them under a different form."

What may appear to be just a curious anecdote about bones and evolution carries significance well beyond innocent coincidence. In our time, exploring the role of genes in the evolution of life and development of organisms constitutes an expanding research front for describing disease pathways, regenerating tissues and organs, and constructing novel microorganisms with potentially valuable economic function. Around the time Wedgwood discovered the bones in the canal tunnel he was excavating and Erasmus Darwin was inspired to write his book, we entered a period of what economic historian and scientist Robert Fogel terms "technophysio evolution." This evolution constitutes the synergistic interplay between rapid technological change and the improvement in human physiology reflected in body size, shape, and longevity. Technophysio evolution holds that the economic output of a given generation is partly determined by its inheritance from past generations through maternal and childhood health and nutrition, with the standard of living mediating the process. The standard of living, in turn, is influenced by investment in technologies, including those that ease the burden of disease and disability and extend lifespan.

More than any other group of the day, the Lunar Society of Birmingham represented the forces of change in late-eighteenth-century England, what culture historian Maureen McNeil calls "a brilliant microcosm of that scattered community of provincial manufacturers and professional men who found England a rural society with an agricultural economy and left it urban and industrial." The Lunar Men were interested in everything—science, technology, entrepreneurship, education—and in agricultural innovations to which they made contributions. Fogel credits such interdisciplinary interests and innovations as responsible for the surge in population growth even as the economist and cleric Thomas Malthus warned that such growth was not sustainable. Lunar Society members would meet for dinner and drink and would then burn the midnight oil in their discussions and debates. The society could lay claim to representing perhaps the first melding of science and technology with an emerging economic system called capitalism. The conversation of the "friends who made the future," the subtitle of Jenny Uglow's book about the Lunar Men, ran from exploring new theories and technologies to discussions of fields ranging from astronomy, biology, botany, chemistry, geology and mineralogy, optics, physics, and zoology to electricity, hydraulics, mechanics, and political economy. They typically gathered at Darwin's home or entrepreneur Matthew Boulton's Soho House next to his Soho Manufactory and Foundry, both in the Birmingham area, the Silicon Valley of its day.

The legacy of the Lunar Men is found in innovation and entrepreneurship initiatives, community collaboratives, scientific and trade associations, business incubators and accelerators, and science and technology prizes and fairs. Their legacy is also found in online communities, creative commons, open-source networks, virtual clusters, science and technology blogs, RSS feeds of science news, wikis like Wikipedia, web-based mapping tools like Google Earth, and, of course, social media like Facebook and Twitter. The gathering of bright and enterprising individuals with disparate backgrounds and interests was a hallmark of the new industrial order, as it is today in the bioscience companies and startups with their PhDs and technicians from biology, chemistry, engineering, and information technology. The new colloquies of which the Lunar Society was most famous were not Enlightenment salons. They were oriented around the dual purpose of both understanding the natural world and shaping it to the material improvement of people living in it.

If Birmingham was the Industrial Revolution's Silicon Valley just as it picked up steam, the hub was Matthew Boulton's Soho Manufactory and Foundry. As its reputation grew, people from all over the world traveled to Birmingham to see for themselves the new coin presses, steel jewelry, silver-plated articles, and innumerable other inventions and manufactured items, and then the Boulton-Watt steam engine. Today the Soho Foundry is a museum, a tribute to the day when Brum, the shortened version of the city's ancient name of Brummagem, ruled the world. On the nearby University of Birmingham campus, another foundry is seeking to make its name in a new industry. Alta Bioscience is a manufacturing laboratory that synthesizes DNA, protein peptides, and other biochemical molecules, like Boulton's manufactures and the Boulton-Watt engine, for worldwide distribution.

The Generative Economy: From Machines to Bioscience

When the wealth-producing ideas and organization that took root in Birmingham, England, and altered the world's economic landscape, the United States was just coming into being as an independent republic. Today the United States, with less than 5 percent of the world's population, produces about one-fifth of the world's total economic output. But its position in overall manufacturing, including in textiles, consumer electronics, and automobiles as well as its production of commodities like steel, has been slipping steadily for decades. Many of these activities are now carried on outside the United States, particularly in low-cost high-production centers in China. Indeed, the global economy's center of gravity, which has been moving eastward from the Atlantic since 1980, is projected to be positioned between India and China by mid-century. The financial crisis of 2008 and its aftermath accelerated the shift.

The stage is set for new thinking about how economies foster growth. Economist and technology historian W. Brian Arthur describes the "generative economy" as an organic rather than mechanical process. Such an economy is driven by connected individuals and nimble, interlinked enterprises drawing on deep knowledge derived from the research base. It is an expression of know-how often embedded in regions, new organizational structures and processes, and novel configurations of technology that are reconfigurations of nature itself—of materials and molecules and forces that we have retrieved and molded and channeled to perform useful functions. Consider the rise of digitized, additive, and continuous manufacturing processes, layer-by-layer 3D printing (including with cells and biomolecules), cheaper and smarter factory robots, and crowdsourced production, all of which are combining to bring about what the *Economist* calls "a third industrial revolution."

No field of entrepreneurial endeavor is more deeply rooted in basic science and technology than the biosciences. Even in an era when the term "death of distance" is bandied about and technology, talent, and money can be sourced from anywhere in the world, it is no surprise that the great majority of biotechnology firms tend to be located in certain places on the globe, places with something "in the air" as Cambridge historian Alfred Marshall described the English factory and mill towns as the Industrial Revolution picked up momentum. Places like Birmingham, where the Boulton-Watt steam engine made its debut; New Bedford, Massachusetts, where whaling and wealth were once synonymous; and Detroit, Michigan, where the automobile was once king.

Even in an era of globalization, place matters for innovation. The world's cities and surrounding regions are where the lion's share of new scientific knowledge is produced and technical innovation is spawned. That has been true ever since the scientific revolution in Western Europe during the seventeenth century. Adam Smith himself was a product of the "Edinburgh science" of the Scottish Enlightenment. Smith published *The Wealth of Nations* in 1776, the same year Thomas Jefferson wrote the "Declaration of Independence" in the American colonies and Smith's fellow Scotsman James Watt put his fascination with the elasticity of steam to work in an engine he designed. All three 1776 events were quintessential expressions of liberty, a reminder that liberty of scientific exploration and business entrepreneurship is an essential ingredient for growth in a knowledge-based economy as well as for liberal democracy to flourish.

For two centuries following publication of *The Wealth of Nations*, land, labor, and capital were the compelling and unchallenged inputs that economists took into account in their calculations for predicting economic output. Not until the second half of the twentieth century did that tried-and-true construct begin to give way. Massachusetts of Institute of Technology economist and Nobel Laureate Robert Solow introduced the idea of technological progress as an additional factor in economic output, the "Solow residual." We translate what we discover and reconfigure what we have on hand into useful things: *technê*. New technology spurs innovation in products, processes, services, and organizational structure and behavior. That leads to higher levels of total factor productivity, a variable that accounts for effects in total economic output not caused by traditionally measured inputs such as labor and capital. Technological innovation is responsible for more than half of the growth in advanced economies by most accounts. Countries that readily adopt new technologies and into which they deeply penetrate see the greatest productivity gains.

Stanford economist Paul Romer brought technology inside the economic growth equation with his writings about new growth theory in the 1980s. "Economic growth occurs whenever people take resources and rearrange them in ways that are more valuable," Romer argued. Unlike other economic inputs, ideas have a special character. They do not obey the law of diminishing returns. To grow the economy, we constantly need new ideas and new recipes. Research, development, and technological innovation enhance the raw materials, the ingredients. People supply the recipes with their novel ideas, with their knowledge turbocharged with their imagination. Simply put, more comes from more knowledge about how to do things, not just from more conventional inputs. The more we learn about the science of DNA, the science of materials, or quantum mechanics—"the more we learn about this stuff, the better we get at finding new, ever more valuable mixtures." It is the underlying rate of technological change that determines the growth rate. Nothing is more central to continuing technological change and a generative economy than the institutions of science and education and the institutions of the market and how effectively they are linked.

It is no accident that Romer's theory of ideas-driven growth fueled by science and technology took shape at the academic research institution most closely connected with Silicon Valley, the world's largest high-technology cluster. It is in

Silicon Valley where ideas flow freely and the law of increasing returns to scale characteristic of successful clustering is enhanced by the highly networked structure of the regional economy. "Geography matters," Romer said in describing the value of face-to-face encounters for successful innovation. That was just before the global financial crisis struck in 2008, precipitating what economist Tyler Cowen termed the "Great Stagnation." In Cowen's telling, we face the prospect of many years of poor growth because the technological "low-hanging fruit" has been picked. Mass effect innovations and efficiencies associated with the likes of James Watt, Thomas Edison, Henry Bessemer, Henry Ford, Frank Whittle, Alfred P. Sloan, George Merck, and countless others are harder to repeat. In his survey of health, nutrition, and technology from 1700 to 2000, economist Robert Fogel found that the era of the household accumulation of consumer durables that fueled the growth of many manufacturing industries during the decades following World War II "is largely over in the United States." General-purpose technologies such as labor-saving machines and follow-up inventions from the second Industrial Revolution were largely in place by the 1970s, in economist Robert J. Gordon's bleak scenario of future innovation and economic growth. Innovating with broad effects across society that also produces large numbers of jobs in the US has become more difficult. Globalization pushes down the cost of labor available to multinational firms while technology and automation allow companies to produce more with fewer workers. Thus relatively slow rates of job-producing technological progress will be with us for a while, Cowen says, noting that the rate of technological progress has never been even or predictable. Does that mean Romer's new growth theory no longer applies? Or is it more relevant than ever? How does a generative economy square with a "great stagnation?"

Technological innovation is the growth recipe for modern economies. As Joel Mokyr observes, nothing else comes close. But innovation comes at a price. Everyone is in favor of growth, but no one wants change, Romer would tell his economics students. Automated systems disrupt industries and cost jobs. Time will tell whether new technologies will replace those jobs or whether the share of the adult population in the United States that is employed will proceed on a downward slope. Smart system control loops are routinely replacing what macroeconomist, economic historian and blogger Brad DeLong calls the "human eye-brain-hand loops" that made and ran things for more than two centuries. As DeLong reads the future, the control loops of cellular growth are next to be conscripted. The interface of different industries with biotechnology and healthcare is "where the money's going to be. That's what people are going to care about."

In the biosciences, change is coming, and so is opportunity. New institutional arrangements and configurations for converting basic biological research into commercial value are being designed and tested. New tools are making biology more accessible to creative people wherever they are, not just in research laboratories and bioscience startups. More people than ever before are working on the recipes.

Scientists, bioengineers, and entrepreneurs are dreaming them up with the ingredients of DNA, RNA, genes, proteins, signaling pathways, cells, silicon, and software. They are borrowing ideas from biological evolution itself with its unrivaled 3.5 billion-year history of experimentation.

The Galápagos Islands: A Tale of Two Proteins

At the time molecular biology was coming into its own, Horace Judson published his classic chronicle of its history, *The Eighth Day of Creation* (1979). He announced what he called the approaching culmination "of the great endeavor in biology that has swept on for a century and a quarter—an achievement of imagination that rivals the parallel, junior enterprise in physics that began with relativity and quantum mechanics."

By Judson's timeline, the "great endeavor in biology" had its origin in the 1850s. That was the decade Rudolf Virchow promoted his version of cell theory whereby all cells arise from preexisting cells, one of the foundations of modern biology. Alfred Russel Wallace, while doing field studies in Malaysia and Indonesia, established biogeography, the study of the distribution of species in geographic space and through geological time and today a key measure of species and food crop migration in response to climate change. As Wallace labored in Southeast Asia, Charles Darwin reflected on his experience as a passenger on the *HMS Beagle* two decades earlier and began to construct his theory of evolution by natural selection. Darwin's observations on the Galápagos Islands and the collection of native species he brought back to England served as the platform for his "dangerous" idea.

Evolution by natural selection is the living edifice of the life sciences. The geneticist Theodosius Dobzhansky famously titled an essay "Nothing in Biology Makes Sense Except in Light of Evolution." Nature is the great experimenter, the ultimate innovator. Creating social and market value from biological knowledge, whether in the form of a biopharmaceutical drug or cellular therapy, a drought-tolerant seed, a biosynthetic microorganism, a cloned mammal, a regenerated organ, or a revitalized ecosystem, requires taking mechanisms such as mutation, variation, adaptation, selection, and immunity into account. Indeed, to understand how complex social and market systems work we borrow from principles of evolutionary biology all the time. The same applies to phenomena as disparate as the spread of contagious disease and artificial intelligence. IBM's supercomputer Watson, made famous in 2011 by winning the quiz show *Jeopardy!*, employs what are called genetic algorithms to frame a context, a more complex task than linear mathematical logic is equipped to fulfill. Watson is interning in oncology by tracking the genetic evolution of tumors in experiments designed to test its potential to provide clinical support for diagnosing disease and recommending treatment. In searching for evidence of life on Mars, the NASA rover *Curiosity* employs genetic algorithms in CheMin, a chemical and mineralogical instrument designed to detect biosignatures,

FIGURE 1.2 A self-portrait of NASA's rover Curiosity on the surface of Mars. Curiosity employs genetic algorithms inspired by evolutionary biology in CheMin, a chemical and mineralogical instrument designed to detect biosignatures, the fingerprints of life. NASA reported in 2013 that Curiosity had identified carbon, nitrogen, hydrogen, oxygen, phosphorus, and sulfur in Martian rocks, key elements of life as we know it.

Source: NASA/JPL-Caltech/MSSS.

the fingerprints of life, on the Red Planet (Figure 1.2). NASA reported in 2013 that *Curiosity* had identified sulfur, nitrogen, hydrogen, oxygen, phosphorus, and carbon, the key elements of life as we know it, though it noted the absence of methane that many microbes produce.

The Galápagos Islands bear witness to Darwin's theory as an unparalleled laboratory of nature where the ability to adapt to changing conditions can spell the difference between evolutionary success and extinction. Evolutionary ecosystems and economic ecosystems like technology clusters are crucibles of experimentation and innovation. They share the characteristics of competition, cooperation and altruism, innovation, adaptation, and geographic space (niche). Galápagos niches are where species variation and adaptation take place. The reason bioscience firms tend to locate near each other is evolutionary in origin, involving the economics of search and selection practices of firms in contexts where variety acts as "evolutionary fuel" as described by Philip Cooke, who studies the global bioeconomy and bioregions. The greater the variety and experimentation, the greater the opportunity for innovation arising from interactions with other actors in the ecosystem.

While visiting the islands as a crew member of the *HMS Beagle*, Darwin observed birds closely and collected specimens. Later these specimens became known as Darwin's finches. Though the birds possessed quite different beak sizes and shapes and had varied eating habits, they were all finches, something Darwin was surprised to learn upon his return to England. Today Darwin's finches are a model for studying how a species adapts to the varied ecological niches of the isolated islands—niches best suited for seed eaters, insect eaters, fruit eaters, and cactus eaters—and the corresponding effects of these niches on beak shape and structure.

Adaptation to environmental circumstances is mediated through genes. The genes responsible for regulating skeletal and cartilage growth in the Galápagos Island finches have already yielded clues to solving the puzzle of how beaks develop such an amazing display of evolutionary diversification. DNA microarray analysis revealed that the gene for calmodulin, a calcium-binding protein whose DNA sequence and folding characteristics have been well studied, was more highly expressed in finches with long and pointed beaks than in those with short, deep, and broad beaks. Calmodulin together with bone morphogenic protein (Bmp) control the genetic signals that influence the behavior of cells responsible for beak and jaw sculpturing in birds, fish, and mammals. Since the shape of a finch's beak influences the sounds it emits, calmodulin is in a sense the tuner for the finch's song.

We are learning a lot about proteins, how they are structured, how they fold, which is the key to their function, and the genes that code for them. Integrating information from "omics" studies—genomics, transcriptomics, proteomics, and metabolomics—with that from molecular biology and protein chemistry is revealing the versatility of proteins like calmodulin. It turns out that calmodulin and its associated calcium signaling pathway are prime suspects in a variety of health conditions including our response to inflammation by signaling white blood cells to move to the inflamed or infected site in the body. Chronic inflammation is believed to be involved in heart disease, diabetes, Alzheimer's, stroke, and cancer. Anti-inflammatory drugs constituted a $65 billion worldwide market in 2010.

Most interesting is that the calmodulin-calcium signaling pathway has been proposed as a mechanism for how we store our memories via a latticework of microtubules—what has been described as "a real-time biomolecular information code akin to the genetic code." That would make calmodulin, by implication, a potential factor in Alzheimer's disease, which strips victims of their memories and leaves their families with memories of the most painful kind. Is it too far-fetched to imagine Darwin in his Down House study constructing his theory while summoning memories of the Galápagos encoded in his brain—memories encoded with the aid of the molecule that produces such amazing variation in the beaks of the finches that caught his eye?

Besides being a laboratory for finch beak variation, the Galápagos Islands are a favorite feeding and breeding grounds for whales, including the sperm whale, the largest toothed predator on the planet. This prime habitat for the endangered sperm whale is the focus of international efforts to restore whale populations from the days when whalers in search of whale oil biofuel to illuminate lamps decimated populations. Anyone who has read Herman Melville's *Moby Dick* or watched the smashing and sinking of the *Pequod* in the film versions can appreciate the strength of the sperm whale. Its muscles are saturated with myoglobin, a protein found in muscle fibers. A relative of hemoglobin, myoglobin is an iron-containing molecule that supplies oxygen to muscle and other tissues where it is stored. Sperm whales dive deep into the sea. They typically have lots of myoglobin, ten times as much as we have in our muscle tissue. In sperm whales, myoglobin actually carries more oxygen than its cousin hemoglobin, an amazing evolutionary feat. They need it, for oxygen is scarce in the ocean depths where sperm whales search for squid, their favorite food.

Sperm whales are among the species that have made major contributions to our understanding of how critical proteins like myoglobin are put together, how they fold, and how these molecules have evolved over time. Indeed, myoglobin from a sperm whale was the first protein whose molecular structure was sorted out. That was reported in 1958, a landmark in the field of protein chemistry. Since then, sperm whale myoglobin has become an important model system to help researchers make comparisons to the amino acid sequence of related molecules, including human hemoglobin, and understand the important function of binding or carrying oxygen.

Diving mammals from whales to seals and otters to beavers and muskrats have evolved positively charged oxygen-binding myoglobin enabling them to store lots of oxygen, a valuable innovation. Evolutionary biologists see this as an instance of convergent evolution in which species of different lineages living in similar environments evolve similar features. Myoglobin is proving to be a versatile molecule. It packs oxygen for deep sea diving and, in a different capacity, captures solar energy in biosensitized solar cells. In this capacity myoglobin is converted from an oxygen storage molecule to an enzyme with electron transfer properties and chemical reactivity, a promising avenue for developing artificial photosynthetic systems.

Our ability to visualize the intricacies of how proteins fold and unfold, bind to other molecules, and participate in biochemical signaling pathways has advanced enormously since the molecular structure of sperm whale myoglobin was first described. That information is essential for designing novel drugs, because how proteins fold affects how they bind to receptor molecules and carry out their functions. The market for what are called therapeutic proteins rests in large part on knowledge of how proteins fold and our ability to induce them to fold properly in the laboratory and in biomanufacturing facilities. Therapeutic proteins are the breakthrough drugs that have come from biotechnology R&D in the past three decades. Monoclonal antibody-based drugs (mAbs) commanded a $44 billion global market in 2011, the leading segment of a protein therapeutics market that is projected to reach more than $140 billion in 2017.

The early days of protein structure and folding research coincided with the rise of a new field of science in the 1950s: molecular biology. It is molecular biology—the study of the structure, function, and makeup of biologically important molecules—and its successor biological technologies, that will make our century, in the eyes of many, the century of biology. The story of how the atoms in the myoglobin molecule are arranged is a seminal chapter of the story of how molecular biology got its start in Cambridge, England, an unparalleled ideas hub in the years following World War II. How curious it is that the muscle of the sperm whale, a creature whose fat once illuminated house, shop, and street lamps of the newly industrialized world, was the raw material that gave the world its first accurate glimpse of the design of a protein, the workhorse molecule of the cell and the key building block of life itself.

Designs for Life: The Rise of Molecular Biology

The Lunar Society of Birmingham was a beacon for the spirit of inquiry and imagination in the early years of the first Industrial Revolution. Another extraordinary association of luminaries began to take shape in the years leading up to and immediately following World War II in a few scientific laboratories in England and the United States. The objects of the scientists' fascination were not machines, engines, and chemical processes but the molecules of life. The tools they used for their explorations were instruments provided by advances in electromagnetism, chemistry, and physics. Their association was not through informal gatherings at private homes but in the rarified setting of the scientific meeting.

Horace Judson observes in *The Eighth Day of Creation* that it is imagination that made molecular biology what it became and what it is today. He tells the story of how Nobel Laureate Max Perutz, a scientific mind of the first order and one of the founders of the field, paid James Watson a compliment for his role with Francis Crick in elucidating the structure of DNA—for finding an "elegant solution" to a problem, the design of DNA, the molecule of life. "He [Watson] found it partly because he never made the mistake of confusing hard work with hard thinking; he always refused to substitute the one for the other," Perutz said, reacting to the widely held perception that Watson was as fond of play as of work, a perception Watson himself later earned with his 1968 book *The Double Helix*. In brief, success in biological research may result from being smart and working hard, as it was in the case of Perutz, who spent much of his career disclosing the secrets of the hemoglobin molecule with its 10,000 atoms, or it may result from an innate ability to see into a difficult problem, create a mental picture of possible solutions, and follow the evidence to the solution that works, as in the case of Watson.

In its early days, molecular biology possessed the advantage of having on board plenty of people characterized by dogged determination or intuitive grasp and everything in between. Their names are legend, as are the institutions with which they were affiliated. The Cavendish Laboratory at Cambridge University in England under the direction of the biochemist Sir Lawrence Bragg, heir as director to the "father" of nuclear physics Ernest Rutherford, was ground zero. Bragg had on staff, among others, Perutz, John Kendrew, Francis Crick, and James Watson, all versed in chemistry and biophysical research, with Watson harboring a fascination with the secrets of the gene. Before the war the laboratory had received funding from the Rockefeller Foundation. Warren Weaver, the foundation's director of research grants, coined the term "molecular biology" in a report to the Foundation's president in 1938. With explorations in the borderline areas where chemistry and physics merge with biology, Weaver wrote, "gradually there is coming into being a new branch of science—molecular biology—which is beginning to uncover many of the secrets concerning the ultimate units of the living cell." In a letter to Weaver in 1967, Crick wrote that Oswald Avery, Colin MacLeod, and Maclyn McCarthy's experiment at the Rockefeller Institute showing that DNA and not protein is the "transforming factor" in bacteria, which they published in 1944, "is usually taken to be the beginning of that part of molecular biology which led to the genetic code."

The Cavendish Laboratory was not the only crucible for hot ideas in the new field. Molecular X-ray experts Maurice Wilkins and Rosalind Franklin worked at nearby King's College in London. Across the pond, Erwin Chargaff, a biochemist at Columbia University, and Salvador Luria, a bacterial geneticist at Indiana University in Bloomington and Watson's doctoral advisor, were making seminal discoveries. The Western node of the emerging field was the California Institute of Technology in Pasadena, where two contemporary giants of their respective fields held court: the incomparable Linus Pauling in chemistry and physicist-cum -biologist Max Delbrück, one of the founders of molecular genetics. As Judson notes, the flight of top-notch physicists like Albert Einstein, Erwin Schrödinger, Leó Szilárd, Hans Bethe, and Enrico Fermi from the rise of fascism in Europe had its counterpart in biology. Perutz fled his native Austria for England in 1936, two years after his fellow Austrian Chargaff migrated to New York City. Delbruck, a biological newcomer from the realm of physics, left his native Germany for the United States in 1937 as a Rockefeller Fellow and stayed on, first at the California Institute of Technology and then Vanderbilt University. His journey from physics to biology was followed by, among others, that of Szilárd, Crick, and Wilkins, so strong was the pull and the excitement of the emerging science of life. Luria left fascist Italy in 1938, evading the German army in France before emigrating to the United States. Schrödinger fled Germany in 1934 eventually settling in Dublin where he wrote *What is Life?* (1944), his speculations about the physical basis of biology and the gene.

Perutz's research partner, John Kendrew, was not an émigré but an Englishman to the core. After a stint studying sheep hemoglobin, Kendrew set about sorting out the structure of myoglobin. The myoglobin molecule has only a quarter of the molecular weight of hemoglobin, the molecule Perutz was studying. Thus it seemed like a good candidate molecule for X-ray study. Like Perutz and many fellow biochemists, Kendrew used the tools of physics to see if he could solve the molecular structure of the myoglobin molecule in three dimensions. Cavendish had some of the best such tools then in existence, including X-ray crystallography, a technique in which proteins in the form of crystals are exposed to X-rays to reveal their atomic structure. The pattern of black dots and streaks produced by the diffraction or scattering of X-rays through the closely spaced lattice of atoms in the crystal is recorded and analyzed. Models were constructed based on the data.

Following World War II, model building became increasingly important in many fields of the physical sciences as well as in the social sciences, including economics. By the 1950s, model building took off as electrons were recruited and put to work by digital computers running ever-more sophisticated software programs. Today it is impossible to exaggerate the importance of model building for simulating the behavior of complex systems, the systems that underlie climate change, epidemics and pandemics, urbanization, investment decisions, shopping behaviors, and global terrorism. Linus Pauling was the greatest molecular visualizer of his day. Pauling recognized the limitations of hand-drawn two-dimensional molecular models. So he set about to develop a "tinker-toy" modeling device to help visualize and comprehend the complicated molecules under investigation. Judson describes these simple toys as "one of Pauling's most remarkable contributions to molecular biology." The most famous biomolecular model is the one showing the double helical structure of DNA. It was built by Watson and Crick in their laboratory at Cavendish, the veritable "in the beginning" moment of the life sciences revolution, the "Eighth Day of Creation" in Judson's authoritative account.

After working with myoglobin from horse heart that refused to yield crystals large enough for X-ray analysis, Kendrew realized that diving mammals and birds offered a better prospect for success than land mammals because of the much greater quantity of myoglobin in their muscle. The ever-resourceful Perutz set about to obtain for Kendrew a large chunk of sperm whale from the Peruvian whaling grounds, which included the Galápagos Islands. To the delight of all involved, "its myoglobin yielded large sapphire-like crystals which gave beautiful X-ray diffraction diagrams."

But the X-ray diffraction diagrams provided information sufficient to tell only half the story of the structure of myoglobin. It took Kendrew many years of sleuthing, until the late 1950s, to finally come up with enough data to begin building an atomic model of myoglobin (Figure 1.3). Data processing from his experiments was done not by computers but by a team of women calculators. At the time, the mathematics needed to process the some 25,000 spots on film, the typical output from an experiment in which an X-ray beam passed through a rotating crystal, was beyond the capacity of any existing computational machine. In time, Kendrew and

FIGURE 1.3 British biochemist John Kendrew with his "forest of rods" model used as a basis for determining the atomic structure of myoglobin in 1958 from X-ray crystallography diffraction data, a first in protein chemistry. Kendrew and his MRC Laboratory colleague Max Perutz were jointly awarded the Nobel Prize in Chemistry in 1962 for solving, respectively, the atomic structures of myoglobin and hemoglobin.

Source: MRC Laboratory of Molecular Biology, Cambridge, England, with permission.

colleague John Bennett developed a computer program for analyzing spots in an X-ray diffraction photograph.

Progress in the biosciences today is inextricably bound to the vast computational power that supercomputers, computer workstations, and research institutes bring to the task. Analysis of laboratory and animal studies depends on computers. Information and communications technologies are increasingly the tools of scientific analysis and communication as well as public information about scientific and medical research. The intricacies of Francis Crick's famous "central dogma" of molecular biology—that genetic information moves from nucleic acid to protein and not from protein to nucleic acid—are mapped in exquisite detail by bits and bytes. The scientific descendants of Perutz and Kendrew model their proteins in dazzling three-dimensional virtual reality environments in which protein folding and receptor configurations for drug discovery are simulated.

Fifty years after Kendrew's feat, Harvard scientists reported that they had developed a computer model that could map and predict how small proteins fold into three-dimensional, biologically active shapes. The phenomenon of protein folding is the most complex dance in the chemical ballet of life. "One of the great problems in science has been deciphering how they amino acid sequence—a protein's primary structure—also determines its three-dimensional structure, and through that its biological function," said Harvard's Eugene Shakhnovich. The ability to predict how proteins fold not only would pave the way for the development of novel drugs targeted to cell receptor molecules; it could shed light on the formation of abnormal protein found in some devastating conditions like Alzheimer's disease.

Perutz and Kendrew shared the Nobel Prize in 1962 for their work on the molecular structures of hemoglobin and myoglobin. In the decades following, hemoglobin and myoglobin sequences were determined for many different species. These sequences were so clearly related that they could be compared with confidence to the three-dimensional structures of two selected standards—horse hemoglobin and whale myoglobin. The differences between sequences from different organisms were used to construct a family tree of hemoglobin and myoglobin variation among organisms, a tree that illustrates the process of molecular evolution.

Genentech and Amgen: Biotechnology Is Born

Matthew Boulton's Soho Manufactory helped make Birmingham become what the English statesman Edmund Burke called the "Toyshop of Europe." It pioneered the mass production of "toys," not child playthings but small items and accessories such as coins, buttons, buckles, boxes, and other manufactured metal-plated goods. Mass production of what we today think of as toys came about in due course. The Tinkertoys Linus Pauling used to build his models of protein molecules were invented a century ago by stonemason Charles Pajeau and partner Robert Petit. Legos, the interlocking stud-and-tube plastic blocks, were invented by the Dane Godtfred Christiansen in the 1950s. To mark the fiftieth anniversary of the discovery of the structure of DNA by Watson and Crick, educator John Schollar created a giant Lego model of the DNA double helix at Experimentarium, Denmark's Science Center. Today Legos are widely used not only to build models of biological molecules but also to illustrate engineered biological designs using standardized biological parts, so-called BioBricks.

Biotechnology's "toyshop" officially came into being with the founding of Genentech by Herb Boyer and Robert Swanson in 1976. In two centuries, the source of power had shrunk by many orders of magnitude. The powerful parts of the new industry—the DNA Watson and Crick modeled with wood, plastic, steel rods, clamps and fasteners, and the proteins Pauling modeled with Tinkertoys—are invisible to the naked eye. But they can be manipulated with the right tools in the right hands. The "Golden Days" of molecular biology, Judson wrote, awaited the arrival of such tools before the vision of the field's founders could be realized. The tools that made biomolecular and genetic manipulation possible debuted in the 1970s with the arrival of recombinant DNA technology, monoclonal antibodies, RNA reverse transcriptase, DNA sequencing techniques, and, in the 1980s, polymerase chain reaction (PCR) technology.

More than any other new laboratory method for manipulating molecules, it was recombinant DNA technology that launched a new industry founded on the life sciences. The story began at Stanford University and what became Silicon Valley where engineers and entrepreneurs began to cluster following the establishment of an electronics firm in a garage by the name of Hewlett Packard in the 1930s. In 1958, the year Kendrew described the first molecular structure of a protein, the "Traitorous Eight" physicists and engineers abruptly left Shockley Semiconductor Laboratory and established Fairchild Semiconductor in Palo Alto to manufacture advanced silicon transistors. The culture of innovation that characterized the electronics industry would in time launch a new industry based on biological technologies.

Stanford biochemist Paul Berg was the first scientist to construct a recombinant DNA molecule. Working with viruses, he succeeded in splicing genetic information taken from one virus into the genetic material of a different virus. To do the job, Berg used restriction enzymes, the chemical tools that organisms like bacteria use to cut DNA apart and stitch it back together again. Hamilton Smith and his graduate student Kent Wilcox, and Daniel Nathans and his graduate student Kathleen Danna, all of Johns Hopkins University, had just described how these unique molecular scissors that would become the workhorses of molecular biology cut DNA at specific sequences. Berg and colleague John Morrow discovered that the cancer-causing SV40 monkey virus DNA contained a single site where the restriction enzyme called EcoRI cut cleanly and reliably. They used a cut-and-stitch approach to combine the DNA from SV40 with that of a bacterial virus called lambda, reporting their success in 1972. It was a tour de force. But it was with viruses, not cellular organisms.

Herbert Boyer and Stanley Cohen soon came up with the even better idea than Berg's for making recombinant DNA. As the story goes, Boyer of the University of California in San Francisco and Cohen at Stanford were chatting over sandwiches late one night in 1972 in a delicatessen during a scientific conference in Hawaii. Both were interested in moving pieces of DNA between bacteria. In his laboratory at Stanford, Cohen was commandeering plasmids, free-floating strings of bacterial DNA, to transfer genes between organisms. Boyer had isolated a certain restriction enzyme that not only cut DNA at specific, predetermined sites—it also left behind "sticky" ends that enabled strands of DNA to be linked together in a complementary manner. This end-to-end complementarity would make it fairly easy to recombine DNA molecules and enable any DNA to be cloned into fast-growing *E. coli* bacteria.

Thus *E. coli* could be recruited and modified to mass produce desirable proteins. Molecular biologists had a method for growing large quantities of one or a few known, identical genes, through the process of cloning (Figure 1.4). By the time they finished their late-night snack, Cohen and Boyer understood they had the makings of a powerful new technology, as told in *Time*: "Using restriction enzymes and plasmids, they could take just about any gene—say, one for making human insulin—insert it into a plasmid, and transfer it into the genome of a bacterium like *E. coli*. Acting like little factories, the rapidly multiplying microbes could crank out highly pure stores of human insulin."

FIGURE 1.4 A scientist studies bacterial colonies growing on an agar plate. From the plate, a recombinant clone containing a gene of interest is identified, grown, and harvested. The DNA is then extracted and used for studying genes including cancer genes. Recombinant DNA technology was the foundational technology of the biotechnology industry.

Source: Dr. Stuart Aronson, Laboratory of Cellular And Molecular Biology, National Cancer Institute, 1985. Wikimedia Commons.

From these early tools of the molecular biologists, a multibillion-dollar biotechnology industry arose, and restriction enzymes became what one chronicler called "the workhorses of molecular biology." With these enzymes, then, biologists are able to alter the genetic instructions in living organisms. The Soho Manufactory of biology began to take shape. Cohen and Boyer took out a patent on their process, some say reluctantly because at the time the academic world of bioscience, secure within its olive groves and ivory towers, had little truck with the business world. Berg, Cohen, Nathans, and Smith all won Nobel Prizes for their pioneering work, but not Boyer. In his case, it may be that the Nobel committee was annoyed by "his willingness to turn genes into gold" when he left academia to found Genentech, surmised *Time*. Be that as it may, their universities were the clear winners. They would never be the same. A local business columnist wrote: "When we spliced the profit gene into academic culture, we created a new organism—the recombinant university." The Cohen-Boyer cloning patent—USPTO Patent 4,237,224—has been worth more than $300 million to their respective institutions, the biggest payout in the University of California's history for a technology Cohen initially thought was not patentable and had no commercial value.

A phone call was perhaps biotechnology's equivalent event to the introduction of inventor James Watt to entrepreneur and financier Matthew Boulton by the English-American scientist and fellow Lunar Society member William Small, one of Thomas Jefferson's teachers at the College of William and Mary. Twenty-seven-year-old venture capitalist Robert A. Swanson from the marquee Silicon Valley investment firm Kleiner Perkins called up Boyer on the telephone and asked for a meeting. Boyer replied he had ten minutes to spare for a discussion. But Swanson wasn't a successful entrepreneur by accident. True, he was a greenhorn when it came to biotechnology, but he was on a fast learning curve. He had heard about recombinant DNA, and he had an idea. The ten-minute meeting with Boyer became a three-hour meeting. "Everybody said I was too early," Swanson said. Everyone told him it would take ten years to turn out a human hormone by inserting a gene into a bacteria and twenty years to come up with a commercial product. Everyone, that is, except Boyer. In contrast to Boyer's research partner Cohen, who dismissed the commercial potential of recombinant DNA technology, Boyer's enthusiasm for it and faith in its commercial potential was evident from the beginning. It was Boyer who named their venture Genentech, an acronym for genetic engineering technology.

Just seven months later, Genentech announced its first success, a genetically engineered human brain hormone, somatostatin. Genentech scientists had one goal, its former president and CEO Arthur Levinson said: "Clone a human protein that was relatively easy, if not therapeutically relevant, so they could prove its scientific feasibility and obtain funding to grow the company." The proof of concept worked. Genentech's cloning of somatostatin proved that human proteins could be produced in and harvested from microorganisms. Swanson said that they accomplished ten years of development in seven months.

The initial public offering of Genentech stock in 1980 set a Wall Street record for the fastest increase of price per share at the time. Two years later genetically engineered insulin became the first marketed product of biotechnology, the first application of recombinant DNA technology put into industrial production and wide clinical use. Genentech scientists gave special strains of *E. coli* bacteria or yeast a copy of the insulin gene to produce human insulin. And produce it they did, in large quantity. The firm licensed the technology to the pharmaceutical giant Eli Lilly. For decades Lilly was a leading US manufacturer of animal insulin, which it extracted from the pancreatic tissue of cows and pigs. In 2006 Lilly stopped selling animal-derived insulin and turned exclusively to its biosynthetic human insulin—Humulin and Humalog—produced by recombinant DNA technology.

Today Genentech is a major force in endocrinology, cardiovascular medicine, and especially oncology, the field in which the firm has had the least experience historically. Yet it has had major success stories in cancer treatment: the humanized monoclonal antibody Herceptin for treating breast cancer and the monoclonal antibody Rituxan (with IDEC Pharmaceuticals) for treating certain kinds of lymphoma. Avastin, another humanized monoclonal antibody, was approved by the FDA in 2004 for treatment of metastatic colorectal cancer and was subsequently approved for treatment of lung, kidney, and brain cancer. Avastin is the first FDA-approved therapy designed to inhibit angiogenesis, the process by which new blood vessels develop and carry essential nutrients to a tumor.

The Swiss pharmaceutical giant Roche acquired Genentech for $46.8 billion in 2009. The sale of Genentech marked the end of an era for the biotechnology industry. It will take years to see whether the firm's entrepreneurial culture can retain its reputation for Silicon Valley–style innovation as the property of a foreign-owned pharmaceutical giant. Indeed, after it acquired Genentech, Roche's three best-selling drugs were the cancer-fighting biopharmaceuticals Avastin, Herceptin, and Rituxan—all developed by Genentech. Roche's pipeline very quickly was populated with Genentech drugs in late-stage clinical trials. Three years after Roche purchased Genentech, Reuters found that the "jeans-wearing scientists...have proven they are the ones driving the drugs pipeline of the 116-year-old Basel-based pharma giant, casting a shadow over Roche's own research operations." Genentech accounted for 55 percent of Roche's $46 billion in sales in 2011.

Amgen is the world's largest independent biotechnology company with more than 17,000 employees. Annual sales surpassed $15 billion in 2011. Headquartered in Thousand Oaks, California, just up the coast from Hollywood, Amgen had a market capitalization of $65 billion in 2011, ranking it in the top fifty American corporations. George Rathmann, chairman and CEO of Amgen from its inception in 1980 to 1988, is regarded as one of the founding fathers of the biotechnology industry, along with Paul Berg, Stanley Cohen, Robert Swanson, and Herb Boyer. Rathmann was a PhD chemist from a family of chemists. His first job was at 3M Company, a firm that would become practically synonymous with innovation under its long-time chairman William L. McKnight. At 3M Rathmann helped developed several diagnostic tests, X-ray film, and also Scotchgard, among 3M's most successful products. After two decades at 3M he left for Litton Industries in 1972 and then moved to Abbott Laboratories, where he served as vice president of R&D. At Abbott he developed a highly successful immunoassay diagnostic test based on fluorescence, a technology he became familiar with at 3M.

But Rathmann had the entrepreneurial bug. He surrounded himself with a coterie of dedicated researchers and scientists who were risk takers and very smart. That was the way Gordon Binder, his successor at Amgen, described him in honoring Rathmann at a meeting of the Newcomen Society, an educational foundation named after the entrepreneur Thomas Newcomen, who invented the first practical steam engine decades before Watt's much improved machine. Binder credited Rathmann's shrewdness in locating Amgen where he did, near such major research universities as UCLA, UC Santa Barbara, and the California Institute of Technology. Location near academic research resources would constitute a key factor in the biotech growth culture—lower-risk places for high-risk innovation.

As a glamour industry in its early days, biotechnology had more than its share of hucksters. In the fashion of a genuine entrepreneur within the firm, an intrapreneur, Rathmann made no apologies for being one of them. He said he learned to be a huckster on the cutting edge at 3M Company, where McKnight instilled a culture of innovation that has lasted more than half a century. The huckster "was the guy that got you keyed up," Rathmann told an interviewer for a program on the history of the biological sciences and biotechnology at UC Berkeley. "He was the guy that predicted early success, frequently when there wasn't one. But without the huckster, you didn't do anything. So this was not viewed as a bad thing by 3M. If somebody got behind something, and they really pushed it too hard or too soon, that's far better than being too late." To Rathmann, being a huckster meant being visionary, not just someone who imagines what is possible and asks probing questions of nature but someone who is also nagged by an irresistible urge to do something about it, to "make it happen," if not inside the firm then on the outside.

Soon after recombinant DNA technology and monoclonal antibodies made their industrial debut in the late 1970s, Rathmann, then at Abbott Laboratories, took a six-month sabbatical in the UCLA laboratory of Winston Salser. The sabbatical persuaded him of the commercial potential of the new biological technologies. He agreed to be CEO of Salser's startup, Applied Molecular Genetics, Amgen, in the fall of 1980. Abbott invested $5 million in the startup after failing in a bid to keep Rathmann with the company. The entrepreneur set up shop in a concrete building in a business park taking over space that, perhaps fortuitously, had been occupied by an evangelical choir.

One of Salser's postdoctoral researchers worked with University of Chicago biochemist Eugene Goldwasser, who had been doing research on the molecule in blood that signals bone marrow to replace red blood cells, erythropoietin (EPO). After identifying and isolating EPO, Goldwasser tried unsuccessfully to interest a number of firms, including Abbott, in developing the molecule as a drug to treat patients with anemia seen in chronic kidney disease dialysis patients. These patients were typically treated with blood transfusions, which can come with complications. So it was no surprise that EPO was among Amgen's lead projects at the outset. By 1983 Amgen scientist Fu-Kuen Lin had sequenced the EPO protein and made multiple copies of the EPO gene by cloning it. At year's end he filed a patent for recombinant EPO. In 1985 Amgen applied to the FDA to begin clinical trials for Epogen, the first genetically engineered or recombinant version of human EPO.

Epogen is a therapeutic protein taken by injection or infusion typically once to three times a week. The first patient was injected in December 1985. Soon the drug virtually eliminated the need for dialysis patients who experienced anemia to require blood transfusions. "Within a matter of a few weeks, we had patients who could run a 10K [race]," Rathmann said. "I mean, it was just an astonishing fact. And [patients] could hardly believe it, because they thought their energy was being sapped by the dialysis itself. Actually, that was happening because their kidneys were not producing erythropoietin." The FDA approved the drug for treating kidney failure patients in 1989. Since it was approved, Epogen has earned Amgen more than $40 billion, constituting what a commentator called "one of the most lucrative monopolies of all time."

Because of Epogen, more than a million patients enjoy a higher quality of life, including Rathmann himself before his death in 2012. In 1991, the year after Rathmann left the company to launch ICOS Corporation, Amgen won FDA approval for Neupogen, a genetically engineered protein that stimulates formation of white blood cells that can help fight infections in cancer patients undergoing chemotherapy. The one-two market punch of these engineered proteins manufactured using recombinant DNA and advanced cell culture technologies gave Amgen a virtual monopoly in its respective therapeutic fields through the 1990s.

When Kevin Sharer took over Amgen leadership from Gordon Binder in 2000, he began to focus the firm's R&D efforts to combine its expertise with bioengineered proteins with more conventional drug manufacturing. He also made several strategic acquisitions, including the purchase of Seattle-based Immunex for $11 billion, at the time the biggest biotech purchase ever. Immunex made Enbrel, a breakthrough drug for treating rheumatoid arthritis. Under Amgen, sales of Enbrel more than doubled in just two years—to nearly $2 billion—and nearly doubled again by 2011. On the R&D side, Amgen moved well beyond its one-two punch in the fields of anemia and cancer care treatment to diabetes, arthritis, lupus, obesity, cancer and other fields expected to loom large as the baby boomers retire.

Biotechnology, Sequencing Technology and the HGP

The first wave of biotechnology firms in the generative economy of technology enterprises sprouted from the fertile soil of entrepreneurial opportunity with names like Amgen, Biogen, Cetus, Chiron, Genentech, Genzyme, and Hybritech. These firms specialized in making genetically engineered proteins and enzymes, monoclonal antibodies, vaccines, and advanced diagnostic tests. Their new approach, harnessing the molecules of life to make their products, was the first major advance drug manufacturing since the rise of chemically synthesized drugs half a century earlier. Small-molecule drugs have remained the gold standard for the major pharmaceutical firms, but biopharma is making major inroads into the multi-billion dollar pharmaceutical market and providing much-needed innovation in an industry that has lacked it for more than a decade.

Yet it was not until the mid-1990s, with the arrival of genomics two decades after the founding of Genentech, that biotech R&D truly took on the character of an industrialized system, in the view of Harvard University's Gary Pisano. "Software engineering, computational horsepower, and equipment design became as important as, if not more important than, the hand of the skilled bench scientist." The seeds for the new R&D system were sown in the late 1970s with dramatic improvements in DNA sequencing technologies developed by Fred Sanger of Cambridge University in England and Walter Gilbert of Harvard, a founder of Biogen, together with Gilbert's graduate student Allan Maxam. The sequencing techniques of Sanger, Gilbert, and Maxam were ingenious but laborious. If a way could be found to multiply DNA fragments and then automate the process, that would open the avenue to rapid sequencing.

The American biochemist Arthur Kornberg was first to identify and isolate DNA polymerase, the enzyme responsible for adding nucleotides to the new DNA strand during DNA replication. It earned him the Nobel Prize in Physiology or Medicine in 1959. In 1983, while working for Cetus, biochemist Kary Mullis came up with the idea of using a pair of primers to bracket the desired DNA sequence and to copy it using DNA polymerase, a technique that would allow a small strand of DNA to be copied a practically limitless number of times. Because the DNA polymerase used in his process was destroyed by the heat at the start of each replication cycle and had to be replaced, in 1986 Mullis started to use a heat-resistant enzyme from the bacterium *Thermus aquaticus* (Taq) found in and around geysers. Taq DNA polymerase can amplify segments of DNA and is thermostable, making the polymerase chain reaction (PCR) technique dramatically more affordable and subject to automation. The first commercial PCR enzyme and thermal cycler systems were introduced the following year. The PCR technique was transformational and soon became an essential tool for doing molecular biology, an accomplished for which Mullis was awarded the Nobel Prize in Chemistry in 1993.

As PCR was catching fire, robotic assembly line-like production systems became available for amplifying and sequencing DNA on a "factory" scale, thereby revolutionizing molecular biology and genetics, forensics, paleontology, and other fields. In 1986 Leroy Hood and Lloyd Smith of the California Institute of Technology and their colleagues described a fluorescence-based automated DNA sequencing machine they had developed. Hood was the co-founder of Applied Biosystems, Inc., an instrumentation company based in Foster City, California. Soon ABI shipped the very first commercial automated DNA sequencer. Later ABI introduced models incorporating real-time PCR, thereby providing speed and accuracy to sequencing that was scarcely imaginable in the days when Sanger and Gilbert were heralded for their sequencing breakthroughs. Today, automated DNA sequencers are standard equipment in virtually every major genetics research lab and many clinical labs around the world. Price-performance improvements over the past decade have been mind-boggling. Whole genome sequencing and interpretation as a standard for patient care are no longer the pipedreams they were when Fred Sanger introduced his DNA sequencing method in 1977.

As PCR and DNA sequencing were being automated in the mid-1980s, Robert Sinsheimer, then chancellor at the University of California in Santa Cruz, hosted a group of leaders in the genetics field to discuss the feasibility of sequencing the entire human genome. "It had occurred to me to wonder whether, were we missing something in biology, was there something we weren't doing in biology, because we didn't think in terms of big science?" Sinsheimer reflected. "It seemed so successful in other fields. We didn't need a big machine like an accelerator or like a telescope. But it did occur to me that maybe we use a big database, like a database of genomes, and sequences of genomes." Sinsheimer's historic workshop generated no funds but did create buzz in scientific circles. It was spurred on by an article published by Salk Institute cancer researcher and 1975 Nobel Laureate Renato Dulbecco, "A Turning Point in Cancer Research: Sequencing the Human Genome," published in *Science* in 1986.

That year, Charles DeLisi of the U.S. Department of Energy (DOE) gave the go-ahead for funding research into genome mapping and sequencing. Since the rise of the nuclear age following World War II, the DOE had been involved in funding research into the mechanisms of mutagenesis and the effects of radiation on genes, including the long-term effects of the Hiroshima and Nagasaki bombs on the people of those cities and particularly their offspring. Direct sequencing of DNA was considered to be a logical method for tracking disease and genetic mutation in these populations, particularly with the new sequencing methods then available. Government clearly had a major role to play in the emerging field of large-scale biology, and the DOE wanted to be first agency through the door.

Two years later, in 1988, the Human Genome Project (HGP) got off the ground. A special committee of the U.S. National Research Council recommended launching a fifteen-year project funded at $200 million a year, $3 billion in all. It was a modest "big science" program compared to Superconducting Super Collider ($11 billion) and Space Station Freedom ($30 billion). Unlike those projects, the HGP would be done once, with a theoretically infinite life span. James Watson himself agreed to lead the National Institutes of Health (NIH) part of what soon became a joint NIH–DOE project, giving it a high-profile figure to help bring the public along. As head of NIH's National Center for Human Genome Research (NHGRI), Watson assailed the growing number of skeptics that the project was worth the money or would even succeed: "It's essentially immoral not to get it done as fast as possible."

The project had its share of critics from the very beginning. One of them was none other than Horace Judson. In the epilogue to the expanded edition of his monumental *The Eighth Day of Creation* published in 1996, just as the project was picking up technological steam, and in *Nature*, Judson weighed in on "*the* project." Despite the promise of federal funding agencies of its being a one-shot project that could be wrapped up soon after the turn of the millennium, "in fact and for excellent reasons the research has no foreseeable terminus," Judson wrote, arguing that extending and exploiting it would occupy scientists and drain funds for more than half a century. "Try titling it accurately—for example, 'the genomes business.' Falls flat; won't sell." A true golden age of science, in Judson's view, is more than about small size, modest needs, and big ideas. It is also an age of innocence that thrives in a fiercely competitive ocean as an island of idealism devoted to intellectual enthusiasm, excitement, and openness.

By the time of Judson's indictment, the HGP train had already left the station well behind. The public "selling" of the project that chagrined Judson and others who lamented the arrival of big science in the realm of biology was beginning to bear fruit. Though a big science project, HGP managers were determined to see that the process was driven by the creativity of individual scientists and groups of scientists, which was a long-standing mark of the biological research field. Writing on the fiftieth anniversary of the discovery of the structure of DNA in 2003, then director of the NHGRI Francis Collins and colleagues Michael Morgan of the United Kingdom's Wellcome Trust and Aristides Patrinos of DOE described the HGP decision-making process as intentionally "bottom-up." That took "a managerial leap of faith." The project met its deadline and came in under budget despite a complex production system, the requirements of quality assurance and peer review, and the involvement of international laboratories in Britain, France, Germany, Japan, and China.

Even though HGP's managers tried to avoid a top-down approach that had characterized the Manhattan Project during World War II and big science projects since, federal funding meant the process and its results had to be fully accountable to the public. Bold managerial "leaps of faith" did not necessarily accommodate bold "leaps of imagination." In 1981 both the NIH and the National Science Foundation rejected a request from Leroy Hood and colleagues at Caltech for research funding support to automate DNA sequencing. By the end of the decade, Hood's nascent technologies, midwifed with private funding, dominated the scientific tool market. Even then, the NHGRI rejected novel proposals by Hood and NIH's J. Craig Venter to test large-scale DNA sequencing in a peer-reviewed competition, opting for older, tried-and-true technology. In assessing gene-sequencing technologies, a federal review council reported in 2003 that it took about a year for the NIH to develop and announce the competition and another year to review proposals and make funding decisions, "but two years is a long time in a fast moving field."

The public–private "clash of civilizations" over the best tools to sequence the human genome began in earnest in the mid-1990s. At a meeting near Oxford University organized by the Wellcome Trust in April 1995, J. Craig Venter announced that he and his team had succeeded in sequencing the entire genome of the pathogenic bacterium *Haemophilus influenzae*, the first free-living organism to be fully sequenced. In addition, Venter announced that his team had nearly completed the sequence of *Mycoplasma genitalium*, the smallest known genome that can constitute a cell. Venter had left the NIH in 1992, frustrated with what he thought was its slow pace of adopting new technologies. He had it in mind to set up a nimble, responsive organization to take advantage of the emerging opportunities in the sequencing arena. Instead, he and Harvard faculty member William Haseltine created two. Venter ran the nonprofit organization, The Institute for Genome Research (TIGR) near Washington, while Haseltine, with his dreams of genome-based medicines, ran the commercial organization, Human Genome Sciences, which had first intellectual property claims to TIGR's work. Thus TIGR would benefit not only from advances of funds from Human Genome Sciences' capital but also from the contracts it entered into with the NIH and the DOE.

A gifted scientist and inventor but also a showman and dreamer at heart, Venter believed that genomics would become "the foundation of research in biology in the 21st century." Thus the field required scale-up of industrial proportions to succeed. Venter also understood that working with public bodies involved a long and difficult struggle with red tape and that if he wanted rapid success that route was out of the question. Venter would prove Horace Judson right. It was indeed "the genomes business." In July 2000, four months after President Bill Clinton in Washington, D.C., and Prime Minister Tony Blair in London celebrated completion of the first draft of the HGP, *Scientific American* ran a special industry report on "The Business of the Human Genome." As the magazine editorialized: "What a difference a decade makes."

Venter realized early on that he needed a massive computer infrastructure to support TIGR's industrial-scale sequencing activities. He also took advantage of nearby expertise, specifically that of Hamilton Smith of Johns Hopkins University in Baltimore. Smith had shared the 1978 Nobel Prize in Physiology or Medicine with Werner Arber and Daniel Nathans for the discovery of the cut-and-paste power of restriction enzymes, the foundation of genetic engineering, and the key technology for early biotech firms like Genentech. It was Smith's familiarity with the bacterium *Haemophilus influenzae*, which itself produces restriction enzymes, that gave Venter the idea to "get on the board," first by recruiting Smith's expertise and then sequencing the complete genome of that organism. To do it, Venter developed a random fragmentation procedure called the "shotgun" technique, the process of breaking a long DNA sequence or a whole genome into many small pieces, sequencing the pieces, and then reassembling the fragments with the aid of ingenious software and powerful computers.

In August 1998 Venter shocked the scientific world by launching Celera Genomics. His partner in the joint venture was the Perkin-Elmer Corporation, known best for commercializing the PCR thermocycler. Their goal was to sequence the human genome and "to build an information business to provide researchers in industry and academia with an integrated information and discovery system for genomic information." The sequencing would be done using several hundred of Perkin-Elmer's state-of-the-art capillary sequencers. For accuracy and quality assurance, the planned "coverage" of the genome was ten times the estimated 3 billion bases in the complete haploid human genome, or 30 billion bases, putting the horses of sequencing power at that time to the supreme test.

Of course, Celera wanted to sell some of the information it derived through its efforts, just as James Watt and Matthew Boulton had it in mind to sell the steam engines they manufactured to the coal-mining industry. The genome business had arrived much to the consternation of many in the scientific community who regarded Celera as a renegade operation. The post–World War II scientific establishment, created through the vision of Vannevar Bush and his 1945 "Science, the Endless Frontier" manifesto, was butting heads with one of those agents of "creative destruction" of the capitalistic creed. The "Genome War"—the very public rivalry between the private-sector initiative by Celera and the federally funded HGP, and their respective proponents, Venter and Francis Collins—was underway, with the "genome business" technologically in the driver's seat, as history would show.

Venter declared victory when he and nearly 300 colleagues at Celera Genomics published "The Sequence of the Human Genome" in the February 16, 2001, edition of *Science*. The international consortium published "Initial Sequencing and Analysis of the Human Genome" in the February 15, 2001, edition of *Nature*. The efforts of the publicly funded project and Celera were, in the end, complementary. In assembling its sequences, Celera made extensive use of sequences put into the public domain by the public consortium. Celera's whole genome shotgun technique, developed by Venter, Hamilton Smith, Robert Holt, and many others, relied on sequencing DNA libraries. Though the public project used clone-based sequencing as opposed to Venter's shotgun method, both approaches required construction of DNA libraries and databases. The technology of constructing DNA libraries had been advanced considerably by the publicly funded project when Celera threw its hat into the competition. DNA libraries were key to Venter's approach, as he himself described it. "With whole genome shotgun sequencing, if you don't have DNA libraries—these clones that are derived from the breaking the genome down into smaller pieces that really represent the genome in a random fashion—it doesn't matter how good the rest of the techniques are of sequencing, mathematics, or assembly, it doesn't matter how big your computer is, you can't regenerate the genome." By Venter's account, the Celera team worked with pieces of DNA that were 2,000 letters long, pieces that were 10,000 letters long, and, thanks to a new cloning technique developed by Smith and Holt at Celera, pieces that were as many as 150,000 letters long. The various DNA pieces of known, predetermined length were then stitched together by computation power and clever algorithms developed by Celera's software engineers.

Principal among them was Gene Meyers. Meyers set out to design algorithms that identified the nonrepeating parts of the genome, which are comparatively easy to reconstruct. He then linked those parts to pairs that bookend a repeat sequence. "This self-checking system was built into the algorithm so the probability of reassembling any piece incorrectly was virtually zero," Meyers said. "They all said we were nuts, but nobody could give me a logical reason why it wouldn't work. That's when I knew we had a winner." It took less than a year for Meyers' team to write the 500,000 letters of code necessary to sequence the genome of the *Drosophila* fruit fly, for a century the geneticist's favorite model organism. Results of that study, a collaboration between Celera scientists and Gerald Rubin's group at Berkeley, were published in *Science* in 2000 and earned recognition as the journal's best paper of the year.

Amgen founder George Rathmann has been described by any number of commentators as the Henry Ford of the biotechnology industry. But there is an argument that Venter is a better candidate in terms of pure mass assembly. "You know, this is the most futuristic manufacturing plant on the planet right now," Venter said in describing his genomics center while inadvertently revealing the toll automation takes on the scientific workforce. "You're seeing Henry Ford's first assembly plant. What don't you see? People, right? There are three people working in this room.

A year ago, this work would have taken one thousand to two thousand scientists. With this technology, we are literally coming out of the dark ages of biology."

Genomics Goes for the Gold

The technology of biology and especially genomics rode the dot.com boom in the late 1990s. Genomics companies like Celera, Human Genome Sciences, Incyte Genomics, Millennium Pharmaceuticals, Myriad Genetics, Large Scale Biology, DeCode Genetics, Rosetta Inpharmatics, Galápagos Genomics, and Gene Logic inhaled the ethereal vapors of the promise of big profits. The number of publicly traded genomics firms jumped from eight in 1994 to seventy-three in 2000; total market capitalization for these firms jumped nearly a hundred-fold during the same years, to $96 billion. For those in the business of forecasts, pricing their stock was a little like trying to map the human genome with little understanding of genetics. "Whether you're an investor or a scientist, the going is slow, and the finds are often incredibly difficult to decipher, but the end promises great riches," *Business Week* wrote in 2000 as part of a cover story on "The Genome Gold Rush." "The problem is you'll have to follow a long, winding helix to get there."

If gold rushes are rare, gold rushes sprouting from the sanctum of modern basic bioscience laboratories are rarer still if the early experience of genomics is any guide. In a 2001 article "Genomania Meets the Bottom Line" published in *Science*, analysts attempted to lay out the spectrum of where the gold would come from. At one end were the toolmakers, like the peddlers who sold shovels and blankets during the gold rush, the first to make a profit. Chief among these were two California companies: the sequencing machine-maker Applied Biosystems and Affymetrix, manufacturer of DNA microarrays or gene chips that give researchers the ability to screen thousands of genes at a time, today even whole genomes. The second category were service sector companies such as Incyte Genomics and Celera Genomics that made their names "as gene discoverers and information brokers, selling up-to-date information on genes and their products to companies searching for drugs and diagnostic tests." Profits in this sector were far less certain than among the toolmakers because public databases compete with privately held information. These information service companies found themselves hedging against the perils of genomic "freeware" by applying for patents on genes that had potentially a beeline to drug development. The third category of companies consisted of the drug discoverers like Millennium Pharmaceuticals and Human Genome Sciences. They were interested in designing drugs that could target receptors on the surfaces of cells that are involved in disease and arrest the disease process. These firms set out "to use concepts borrowed from the steel, computer, and other established industries to scale up and speed drug discovery."

Genomics is the "new Internet," Microsoft's chief technical officer Nathan Myhrvold told the group of journalists and industry leaders gathered by Rosetta

Inpharmatics president Stephen Friend in March 2000 as the dot.com bubble was about to burst. "It will rival anything the computer or communication industries gave done up until now. It has the potential to reshape the entire world economy in exactly the same way information technology has." Friend stayed with Myhrvold's vision of genomics, co-founding the nonprofit Sage Bionetworks with its goal of using open-source and computational tools to transform to preclinical drug development. What was at play during the genomics gold rush, based on a quantitative analysis of HGP development during the years 1986 to 2003, appeared to be an instance of "exuberant innovation," a tech bubble but one that can drive future innovation in a way a financial or housing bubble cannot. High-speed backbone networks were overbuilt during the 1990s dot.com bubble. In the 2000s that excess capacity came in handy. It was put to work transmitting data, genomic sequence data included.

Leonardo da Vinci envisioned the body as a machine. But knowledge derived from the study of living systems is more complex than bodies of knowledge *techné* had previously brought to market. The hyperglandular excitement of "Genomania," riding the wave of the dot-com boom, was soon tempered by the inescapable reality of what propels Wall Street and private investors: the prospect of near-term profit. For entrepreneurs, there are no quick-fix bioscience counterparts to Google. In 2001, the NASDAQ Composite Index of 4,000 technology companies went into a free fall and was down some 20 percent by the end of the year. The Burrill Genomics Index comprised of ten biotech companies with technologies spanning the field of genomics fell 20 percent. Genomics companies like Celera, Millennium, Incyte, and Curagen were off by more than that.

When Bill Clinton and Tony Blair announced that the human genome sequence should be made freely available to all researchers and not patented just as the dot.com bubble popped, the entire biotech sector nosedived, losing $50 billion in market capitalization in just two days. With Venter's departure in 2002, Celera abandoned its business model of generating revenue from genomic data subscription services and set out on a quest to reposition itself as a "fully integrated pharmaceutical company." Genomania was put under sedation. The values of genomics companies on Wall Street went under the knife. A decade passed before speculative enthusiasm for biotechnology stocks returned in the United States, reaching an all-time high in 2013.

In the years following genomania, the new scientific tools proved to be invaluable in sorting out the genetics of development, disease, and aging; relationships in the web of life and within diverse ecosystems; the rise of hominids in Africa; and the great migrations of *Homo sapiens* out of Africa and across the globe. After a decade, scientists used genetic sequencing to map the human microbiome, our personal microbial ecosystem that accounts for nine of every ten cells we carry around with us, about 100 trillion all together. "Like 15th century explorers describing the outline of a new continent, Human Microbiome Project researchers employed a new technological strategy to define, for the first time, the normal microbial makeup of the human body," said NIH director Francis Collins. It was a vivid reminder of how the most complex and the simplest organisms on Earth have co-evolved, a concept Darwin himself described. Human-microbe co-evolution joins the estimated 23,000 protein-coding human genes with 3 to 5 million microbial genes, constituting a human superorganism, a vastly expanded genetic universe for probing the secrets of human health and disease. Genomic medicine based on human genes alone generates tens to hundreds of billions of dollars of economic activity in the United States, depending on how it is defined and measured. Human microbiomic medicine, barely out of the birth canal, is destined to add to that genome-based economic output.

Lessons from Lazarus in the Sequencing Era

Modern bioscience has been brought to Darwin's inspirational garden in hopes of preserving an astonishing evolutionary experience. The Galápagos Islands take their name from the Spanish word *galápago* or "saddle"—after the shells of the giant saddlebacked Galápagos tortoises. Herman Melville described in his story "The Encantadas" how whaling ships were lured to the Galápagos Islands because of the nearby prime habitat for sperm whales, the source of coveted oil for superior illumination and for lubricating machines in the early Industrial Age. The whalers hunted the tortoises almost to extinction.

"Lonesome George" was the last member of one of the eleven species of tortoise that the islands sustain (Figure 1.5). Scientists at the Charles Darwin Research Station were excited in 2008 when female tortoises closely related to Lonesome George's species laid eggs in his vicinity, possibly because Lonesome George might have mated with them. None hatched, and none showed evidence of Lonesome George's DNA. Alas, Lonesome George died in 2012 at the estimated age of 100 years. Geneticists from the San Diego Zoo raced to the Galápagos to help preserve Lonesome George's tissues, which someday could be used in an attempt to clone his species. Meanwhile, through DNA analysis of Lonesome George and his genetic relatives, evidence of what was thought to be a vanished species, *Chelonoidis elephantopus*, turned up in the DNA of its hybrid descendants. Would it be possible to resurrect that species through several generations of selective breeding? The answer is possibly yes, though no one today would be alive to see the results. "Theoretically, we can rescue a species that has gone extinct," said Yale University's Adalgisa Caccone. "Our lab calls it the Lazarus project." The project advanced considerably in 2012 when Caccone's team reported finding hybrid tortoises of purebred *Chelonoidis elephantopus* ancestry on a distant island in the Galápagos chain.

While scientists were doing their utmost to induce a giant tortoise to perpetuate his species, half a world away a member of an organizational species teetered on the edge and finally went bankrupt. Iceland's DeCode Genetics is one of the

FIGURE 1.5 Lonesome George (*Chelonoidis abingdoni*) Pinta Island Tortoise, Galápagos Islands. Through analysis of his DNA and DNA from his genetic relatives, scientists have found evidence of a vanished species of tortoise hunted to extinction by whalers in the nineteenth century. Like his observations of the island's finches, Charles Darwin's close study of Galápagos tortoises contributed to the development of his theory of evolution by natural selection.

Source: Wikipedia, Creative Commons.

iconic companies of the genomics revolution of the late 1990s. The "gene hunting" company was founded in 1996 by Harvard University neuropathologist and Iceland native Kari Stefansson. One biotech writer described DeCode as "a subset of the grand experiment of biotechnology itself, which over thirty-five years has seen scientists sink tens of billions of dollars into mostly early-stage projects that in some cases has provided breathtaking cures but more often has not," a player in an industry that "has suffered losses in every year except one." The financial meltdown in the fall of the 2008, the ensuing global recession, and the collapse of Iceland's banks combined to deliver the final blow. The company was resurrected in early 2010 by an alliance of life sciences investment companies.

DeCode Genetics is anything but the Lonesome George of the genomics revolution. For more than a decade the company generated numerous scientific papers identifying genetic risk factors and published its findings in the top journals. Its massive DNA biobank contains a wealth of genetic data on some 140,000 Icelanders and is widely regarded as a treasure trove for tracking down genetic variations underlying common diseases. *Science* described DeCode's biobank in 2009 as "a boon for genome-wide association studies" and credited company researchers with identifying many of the reliable biomarkers found up to that time for diseases such as diabetes, heart disease, and cancer. More than 60 percent of adults in Iceland have donated DNA to the biobank. Britain, Canada, Norway, and Sweden are among the countries expanding their national biobanks.

Like its genomics forerunners, and indeed reflective of the biotechnology industry as a whole, DeCode Genetics never found a commercial model that enabled it
to turn a profit. Its efforts to develop drugs from the abundance of data it produced
fell short. At the time its own financial crisis hit, it was retooling to develop diagnostic genetic tests based on its genetic discoveries. That move is proving to be a
wise one. When DeCode was slipping into bankruptcy, Stefansson was described as
"a survivor and an unrepentant advocate of the power of genomics as both science
and a source of potential profit, even if it remains more promise than reality." But
the promise was closer to reality than most analysts realized. The pharmaceutical giant Pfizer formed a partnership with DeCode in 2011. A year later Amgen
acquired the firm, demonstrating that large biopharma firms are moving to integrate human genetics research into their discovery and clinical programs, reported
Nature Biotechnology. "The big question for Amgen—and for the drug discovery
industry as a whole—is whether the field has reached the point where the intriguing
findings emanating from genome mining can be captured and translated into clinically useful and commercially feasible therapies and diagnostics."

Stefansson is the Lazarus of the genomics revolution. He proved it in 2012 when
he and his research team led an international collaboration that reported a gene with
a strong protective effect against the development of Alzheimer's disease and against
cognitive decline in people who already have it. In so doing, they also provided confirming evidence for the reigning theory about what causes Alzheimer's disease with
its personal and familial toll and its staggering economic costs—more than $600
billion worldwide and more than $300 billion in the United States in 2010, costs
expected to grow rapidly. Drug discovery can now focus on a very specific Achilles
heel in the pathogenesis of the disease. The feat was accomplished not through
hypothesis-driven science but through a complex systematic analysis of every new
DNA variant revealed by the whole genome sequencing of 1,800 Icelanders, redeem-
ing the inherent value of DeCode's vast biobank. Having completed whole genome
sequencing on 2,500 Icelanders and having genotyped 120,000, Stefansson said his
team had discerned "the whole genome sequence of an entire nation."

Despite residing a continent and ocean away from each other, Iceland and
the Galápagos Islands have something in common: Both sit on the edges of the
tectonic plates that created them, seams of the supercontinent Pangaea. The
natural ecosystem of the Galápagos Islands harbors genes important for understanding plant and animal growth and development, adaptation, speciation, and
regeneration. Its tortoises presumably possess genes for longevity, living upward
of 200 years. The human ecosystem of Iceland, with its ethnic homogeneity,
harbors genes important in genetic and chronic disease and also longevity and
memory, a special interest of Stefansson's given the many Icelanders who live
into their nineties. Finding genes involved in development, disease, longevity,
and memory and making practical use of their organic power through innovation and entrepreneurship are tasks the century of biology is well equipped to
undertake.

References

Listed sequentially based on their order in the chapter.

"W. S. Merwin with Naomi Shihab Nye, October 18, 2000," Lannan Podcasts, accessed January 20, 2013, http://podcast.lannan.org/2007/08/13/ws-merwin-with-naomi-shihab-nye/

Ancient Gifts Get Things Started

Douglas Hofstadter, *I Am a Strange Loop* (New York: Basic Books, 2007), 363.
Mark Dodgson and David Gann, *Innovation: A Very Short History* (Oxford: Oxford University Press, 2010).
Oliver Sacks, *Hallucinations* (New York: Knopf, 2012), 242.
Karen Armstrong, *A Short History of Myth* (Edinburgh, UK: Canongate Books, 2005).
Joseph Campbell, *The Flight of the Wild Gander* (Novato, CA: New World Library, 2002), 154.

Horses and Horsepower

Ian Morris, *Why the West Rules—for Now* (New York: Farrar, Straus, and Giroux, 2010).
David Landes, *The Wealth and Poverty of Nations: Why Some Are So Rich and Others So Poor* (New York: W.W. Norton, 1998).
Joel Mokyr, *The Gifts of Athena: Historical Origins of the Knowledge Economy* (Princeton, NJ: Princeton University Press, 2002).
Joyce Appleby, *Relentless Revolution: A History of Capitalism* (New York: W.W. Norton, 2010).
Joel Mokyr, "Punctuated Equilibria and Technological Progress," *American Economic Review* 80.2 (1990):350–354.
Robert W. Fogel, "Catching Up with the Economy," *American Economic Review* 89.1 (1999):1–21.

The "Lunaticks" Link Science, Industry, and Society

Adam Hart-David, "James Watt and the Lunaticks of Birmingham," *Science*: 292 (2001):55–56.
Dodgson and Gann, *Innovation*.
Robert W. Fogel, *The Escape from Hunger and Premature Death, 1700–2100* (Cambridge, UK: Cambridge University Press, 2004).
Roderick Floud, Robert W. Fogel, Bernard Harris, and Sok Chui Hong, *The Changing Body: Health, Nutrition, and Human Development in the Western World since 1700* (Cambridge, UK: Cambridge University Press, 2011).
Maureen McNeil, *Under the Banner of Science: Erasmus Darwin and His Age* (Manchester, UK: Manchester University Press, 1987).
R. E. Schofield, *The Lunar Society of Birmingham: A Social History of Provincial Science and Industry in Eighteenth-Century England* (Oxford: Oxford University Press, 1963).
Jenny Uglow, *The Lunar Men: Five Friends Whose Curiosity Changed the World* (New York: Farrar, Straus and Giroux, 2002).
Desmond King-Hele, "The 1997 Wilkins Lecture: Erasmus Darwin, The Lunaticks, and Evolution," *Notes and Records of the Royal Society of London* 52.1 (1998):153–180.

Charles Darwin and Francis Darwin, eds., *The Life and Letters of Charles Darwin, Including an Autobiographical Chapter* (London: John Murray, 1887).

The Generative Economy: From Machines to Bioscience

Danny Quah, "The Global Economy's Shifting Centre of Gravity," *Global Policy* 2.1 (5 January 2011):3–9, accessed July 30, 2012, doi: 10.1111/j.1758-5899.2010.00066.x.

W. Brian Arthur, *The Nature of Technology: What It Is and How It Evolves* (New York: Free Press, 2009).

"Special Report: Manufacturing and Innovation," *The Economist* (April 21, 2012):3–20, accessed January 6, 2012, http://www.economist.com/node/21552901

Nathan Rosenberg, "Innovation and Economic Growth," Paris: Organisation for Economic Co-operation and Development, 2004, accessed July 30, 2012, http://www.oecd.org.

Robert M. Solow, "Technical Change and the Aggregate Production Function," *Review of Economics and Statistics* 39.3 (1957):312–320.

Diego Comin and Martí Mestieri Ferrer, "If Technology Has Arrived Everywhere, Why Has Income Diverged?" NBER Working Paper 19010 (Cambridge, MA: National Bureau of Economic Research, 2012), accessed May 12, 2013, http://www.nber.org/papers/w19010.

Paul M. Romer, "Economic Growth," in *The Concise Encyclopedia of Economics* (New York: Warner Books, 1993), accessed February 2010, http://www.econlib.org.

Arnold Kling and Nick Schulz, *From Poverty to Prosperity: Intangible Assets, Hidden Liabilities and the Lasting Triumph over Scarcity* (New York: Encounter Books, 2009), 83–84.

G. Pascal Zachary, "When It Comes to Innovation, Geography is Destiny," *The New York Times,* February 11, 2007.

Tyler Cowen, *The Great Stagnation: How America Ate All The Low-Hanging Fruit of Modern History, Got Sick, and Will (Eventually) Feel Better* (New York: Dutton, 2011), 86.

Fogel, *The Escape from Hunger and Premature Death,* 71.

Robert J. Gordon, "Is U.S. Economic Growth Over? Faltering Innovation Confronts the Six Headwinds," NBER Working Paper No. 18315 (Cambridge, MA: National Bureau of Economic Research, 2012), accessed December 5, 2012, http://www.nber.org/papers/w18315.

J. Bradford DeLong, "Where the Money Is Going to Be: Infotech, Health Care, and the Future of 'Industry Studies,'" Paper presented at the 2013 Industry Studies Association Conference, Kansas City, Missouri, June 4, 2013, accessed June 4, 2013, http://delong.typepad.com.

The Galápagos Islands: A Tale of Two Proteins

Horace Judson, *The Eighth Day of Creation: Makers of the Revolution in Biology* (New York: Simon & Schuster, 1979).

Alfred Russel Wallace, *The Malay Archipelago* (New York: Dover Publications, 1962).

Theodosius Dobzhansky, "Nothing in Biology Makes Sense Except in the Light of Evolution," *American Biology Teacher* 35 (1973):125–129.

Jonathan Cohn, "The Robot Will See You Now," *The Atlantic* (March 2013):59–67.

David Blake et al., "Characterization and Calibration of the CheMin Mineralogical Instrument on Mars Science Laboratory," *Space Science Reviews* 170 (2012):341–399.

Kenneth Chang, "Mars Could Have Supported Life Long Ago, NASA Says," *The New York Times,* March 12, 2013.

Philip Cooke, *Growth Cultures: The Global Bioeconomy and its Bioregions* (London: Routledge, 2007).

Arhat Abzhanov et al., "The Calmodulin Pathway and Evolution of Elongated Beak Morphology in Darwin's Finches," *Nature* 442 (2006):563–567.

"Anti-Inflammatory Therapeutics Market to Grow to $85.9 Billion in 2017," *The Pharmaletter,* November 6, 2011.

Travis J. A. Craddock, Jack A. Tuszynski, and Stuart Hameroff, "Cytoskeletal Signaling: Is Memory Encoded in Microtubule Lattices by CaMKII Phosphorylation?" *PLOS Computational Biology* 8.3 (2012): e1002421, accessed July 24, 2014, doi:10.1371/journal.pcbi.1002421.

William E. Royer et al., "Allosteric Hemoglobin Assembly: Diversity and Similarity," *Journal of Biological Chemistry* 280.30 (2005):27477–27480.

Scott Mirceta et al., "Evolution of Mammalian Diving Capacity Traced by Myoglobin Net Surface Charge," *Science* 340 (2013), accessed December 3, 2013, doi: 10.1126/science.1234192.

C.-H. Chang et al., "Photoactivation Studies of Zinc Porphyrin-Myoglobin System and its Application for Light-Chemical Energy Conversion," *International Journal of Biological Science* 7 (2011):1203–1213.

"Therapeutic Proteins Market to 2017—High Demand for Monoclonal Antibodies Will Drive the Market" (Rockville, MD: GBI Research, 2011).

Designs for Life: The Rise of Molecular Biology

Judson, *The Eighth Day of Creation.*

James D. Watson, *The Double Helix: A Personal Account of the Discovery of the Structure of DNA* (New York: Scribner, 1968).

Soraya De Chadarevian, *Designs for Life: Molecular Biology After World War II* (Cambridge, UK: Cambridge University Press: 2002).

Warren Weaver, *"The Natural Sciences Report"* (New York: Rockefeller Foundation, 1938), 203–204.

Francis Crick, Letter to Warren Weaver, May 16, 1967, Wellcome Library for the History and Understanding of Medicine. Francis Harry Compton Crick Papers, accessed November 30, 2013, http://archives.wellcome.ac.uk.

De Chadarevian, *Designs for Life.*

Judson, *The Eighth Day of Creation.*

Max Perutz, "Obituary of John Kendrew," *MRC Newsletter,* Autumn 1997.

"Comprehensive Model First to Map Protein Folding at Atomic Level," *Harvard Gazette* November 9, 2006.

A. Hubner, E. J. Deeds, and E. I. Shakhnovich, "Understanding Ensemble Protein Folding at Atomic Detail," *Proceedings of the National Academy of Sciences USA* 103 (2006):17747–17752.

Ross Hardison, "Hemoglobins From Bacteria to Man: Evolution of Different Patterns of Gene Expression." *The Journal of Experimental Biology* 201 (1998):1099–1117.

Genentech and Amgen: Biotechnology Is Born

Judson, *The Eighth Day of Creation.*

David A. Jackson, Robert H. Symons, and Paul Berg, "Biochemical Method for Inserting New Genetic Information into DNA of Simian Virus 40: Circular SV40 DNA Molecules Containing Lambda Phage Genes and the Galactose Operon of *Escherichia coli,*" *Proceedings of the National Academy of Sciences USA* 69.10 (1972):2904–2909.

Richard J. Roberts, "Perspective: How Restriction Enzymes Became the Workhorses of Molecular Biology," *Proceedings of the National Academy of Sciences USA* 102.17 (2005):5905–5008.

Frederic Golden, "The Pioneers of Molecular Biology: Herb Boyer," *Time,* February 9, 2002.

Tom Abate, "Scientists' 'Publish or Perish' Credo Now 'Patent and Profit,'" *San Francisco Chronicle,* August 13, 2001.

Rajendra K. Berra, "The Story of the Cohen-Boyer Patents," *Current Science* 96.6 (2009): 761–762.

Maryann P. Feldman, Alessandra Colaianni, and Connie Kang Liu, "Lessons from the Commercialization of the Cohen-Boyer Patents: The Stanford University Licensing Program," in *Putting Intellectual Property to Work: Experiences from Around the World,* A. Krattiger, ed. (Oxford: PIPRA/MIHR, 2007).

"Bertram Rowland and the Cohen/Boyer Cloning Patent" (Washington, DC: The George Washington University Law School, 2012), accessed July 31, 2012, http://www.law.gwu.edu/Academics/FocusAreas/IP/Pages/Cloning.aspx.

William D. Bygrave and Andrew Zackarakis, eds. *The Portable MBA in Entrepreneurship* (Hoboken, NJ: John Wiley & Sons, 2004), 8–9.

Bernard R. Glick and Jack J. Pasternak, *Molecular Biotechnology: Principles and Applications of Recombinant DNA,* 2nd ed. (Washington, DC: American Society for Microbiology, 1998).

Thomas M. Doerflinger and Jack L. Rivkin, *Risk and Reward: Venture Capital and the Making of America's Great Industries.* (New York: Random House, 1987).

Arthur Levinson, "For Success, Focus on Your Strengths," *Nature Biotechnology* 16 (Suppl.) (1999):45–46.

New Developments in Biotechnology: Patenting Life (Washington, DC: US Office of Technology Assessment, 1989).

Michael Butler, ed., *Cell Culture and Upstream Processing* (New York: Taylor & Francis, 2007).

Caroline Copley, "After Roche Merger, Biotech Tail Wags Big Pharma Dog," Reuters, July 3, 2012.

George B. Rathman, PhD., Chairman, CEO and President of Amgen 1980–1988: Oral History Transcript / George B. Rathmann, BANC MSS 2005/108, The Bancroft Library, University of California Berkeley.

Gordon Binder, *Amgen,* Newcomen Pub. 1518 (New York, Newcomen Society of the United States, 1998).

George B. Rathmann, PhD., Chairman, CEO, and President of Amgen, 1980–1988.

Paul Francuch, "Biotech Builder," *Northwestern,* Fall 2000.

"Biotech Drugs Go Blockbuster," *Biotechnology Healthcare,* April 2006.

Andrew Pollack, "FDA Approves New Anemia Drug," *The New York Times,* March 27, 2012.

Biotechnology, Sequencing Technology and the HGP

Gary P. Pisano, *Science Business: the Promise, the Reality, and the Future of Biotech* (Cambridge, MA: Harvard Business School Press, 2006).

R. Saiki et al., "Primer-directed Enzymatic Amplification of DNA with a Thermostable DNA Polymerase," *Science* 239 (1988):487–491.

Lloyd M. Smith et al., "Fluorescence Detection in Automated DNA Sequence Analysis," *Nature* 321 (1986):674–679.

Robert Sinsheimer, "A Database of Genomes," Interview for DNA Interactive (DNAi), Dolan DNA Learning Center, Cold Spring Harbor, New York, accessed January 21, 2012, http://www.dnalc.org/view/15334-A-database-of-genomes-Robert-Sinsheimer.html.

Renato Delbecco, "A Turning Point in Cancer Research: Sequencing the Genome," *Science* 231 (1986):1055–1056.

Renato Delbecco, "From the Molecular Biology of Oncogenic DNA Viruses to Cancer," Nobel Lecture, December 12, 1975, accessed January 21, 2010, http://nobelprize.org/nobel_prizes/medicine/laureates/1975/dulbecco-lecture.html.

Natalie Angier, "Great 15-Year Project to Decipher Genes Stirs Opposition," *The New York Times*, June 5, 1990.

Horace Judson, *The Eighth Day of Creation: Makers of the Revolution in Biology* (commemorative edition) (New York: Cold Spring Harbor Laboratory Press, 1996).

Horace Freeland Judson, "The Genomes Business," *Nature* 371 (1994):753–754.

Francis S. Collins, Michael Morgan, and Aristides Patrinos, "The Human Genome Project: Lessons from Large-Scale Biology," *Science* 300 (2003):286–290.

National Research Council, Committee on the Organizational Structure of the National Institutes of Health, *Enhancing the Vitality of the National Institutes of Health* (Washington, DC: National Academies Press, 2003).

J. Craig Venter, "The National Institutes of Health: Decoding our Federal Investment in Genomic Research," in *Hearing before the Subcommittee on Health of the Committee on Energy and Commerce, US House of Representatives* (Washington, DC: US Government Printing Office, 2003).

"Special Industry Report: The Business of the Human Genome," Introduction, *Scientific American* (July 2000):48–49.

J. Craig Venter, Written statement on Behalf of the Biotechnology Industry Organization Before the Subcommittee on Consumer Protection, US House Committee on Energy and Commerce, July 11, 2001.

Robert Cook-Deegan, "The Urge to Commercialize: Interactions between Public and Private Research and Development," in *The Role of Scientific and Technical Data and Information in the Public Domain: Proceedings of a Symposium* (Washington, DC: National Academies Press, 2003), 87–94.

J. Craig Venter et al., "The Sequence of the Human Genome," *Science* 291 (2001):1304–1351.

Eric S. Lander et al., "Initial Sequencing and Analysis of the Human Genome," *Nature* 401 (2001):860–921.

J. Craig Venter, "Whole Genome Shotgun," Interview for DNA Interactive (DNAi), Dolan DNA Learning Center, Cold Spring Harbor, New York, January 21, 2010, accessed July 31, http://www.dnalc.org/view/15365-Whole-genome-shotgun-Craig-Venter.html.

"Computer Algorithm Pioneer to Join UC Berkeley Faculty," UC Newsroom, University of California, October 29, 2002.

M.D. Adams et al., "The Genome Sequence of Drosophila Melanogaster," *Science* 287 (2000):2185–2195.
Richard Preston, "The Genome Warrior," *New Yorker,* June 12, 2000.

Genomics Goes for the Gold

"The Genome Gold Rush," *BusinessWeek,* June 12, 2000.
David Malakoff and Robert F. Service, "Genomania Meets the Bottom Line," *Science* 291 (2001):1193–1203.
Kristen Philipkoski, "Myhrvold: Genomics Will Rule," *Wired*, December 8, 2000.
Philip Ball, "Bursting the Genomics Bubble," *Nature*, March 21, 2010, accessed July 31, 2012, doi:10.1038/news.2010.145.
James Macintosh, "Beware Side-Effects of Biotech Boom," *Financial Times*, May 13, 2013.
"NIH Human Microbiome Project Defines Normal Bacterial Makeup of the Body," *NIH News,* June 13, 2012.
"Economic Impact of the Human Genome Project," Battelle Technology Partnership Practice, Battelle Memorial Institute, May 2011, accessed January 9, 2012, http://www.battelle.org.
"The Impact of Genomics on the U.S. Economy," Battelle Technology Partnership Practice, Battelle Memorial Institute, June 2013, accessed June 12, 2013, http://www.unitedformedicalresearch.com.

Lessons from Lazarus in the Sequencing Era

Adalgisa Caccone et al., "Origin and Evolutionary Relationships of Giant Galápagos Tortoises," *Proceedings of the National Academy of Sciences USA* 96 (1999):13223–13228.
Jennifer Viegas, "Should We Have Cloned Rare Tortoise Lonesome George?" NBCNews.com, July 6, 2012.
"Genetic Analysis Gives Hope that Extinct Tortoise Species May Live Again," Yale University, January 15, 2010.
Michael A. Russello et al., "DNA from the Past Informs *Ex Situ* Conservation for the Future: An 'Extinct' Species of Galápagos Tortoise Identified in Captivity," *PLoS ONE* 5.1 (2010):e8683, accessed July 13, 2012, doi: 10.1371/journal.pone.0008683.
Ryan C. Garrick et al., "Genetic Rediscovery of an 'Extinct' Galápagos Giant Tortoise Species," *Current Biology* 22.1 (2012): R10–R11, accessed December 9, 2013, doi:10.1016/j.cub.2011.12.004.
David Ewing Duncan, "Decoding the Profit Gene: Is the Last Pure Genomics Company from the 1990s About To Go Bust?" *Technology Review,* August 19, 2009.
Jocelyn Kaiser, "Cash-Starved deCODE Is Looking for a Rescuer for Its Biobank," *Science* 325 (2009):1054.
Thorlakur Jonsson et al., "A Mutation in APP Protects Against Alzheimer's Disease and Age-related Cognitive Decline," *Nature* 488 (2012):96–99.
Anders Wimo and Martin Prince, "World Alzheimer Report 2010: The Global Economic Impact of Dementia" (London: Alzheimer's Disease International, September 21, 2010).
Cormac Sheridan, "Amgen Punts on deCODE's Genetics Know-how," *Nature Biotechnology* 31 (2013):87–88.
Kevin Davies, "Kari Stefansson on deCODE's Alzheimer's Discovery, Future Plans," *Bio-IT World*, July 11, 2012.

2

Drugs, Biomolecules, Brainpower, and the Shifting Currents of Innovation

The system itself seems to be changing, becoming more global in character.

—Richard Florida, *New Geography of Science*

Life sciences commercial production is a trillion-dollar-plus global industry. The heavily concentrated pharmaceutical sector alone accounts for 90 percent of life sciences output by some measures. With a projected 3–6 percent compound annual growth rate, drug-making is projected to be worth $1.2 trillion globally by 2016. Its largest markets remain the United States and Europe, but its fastest growing markets are "pharmerging" countries like Brazil, China, India, Russia, Mexico, Turkey, Argentina, Indonesia, South Africa, Egypt, Pakistan, and Vietnam. A rising middle class in these countries will consume a growing proportion of the world's prescription drugs. It will also provide brainpower for future innovation.

The pharmaceutical industry has long been one of the most profitable businesses of the industrial era. It is also one with the highest investment in research and development as a share of revenue, around 15 percent annually among US firms compared to 4 percent for US industry in general. Today the active ingredients of the prescription small-molecule drugs it synthesizes course through the veins of some 80 million adult Americans every day and hundreds of millions of people worldwide. These active ingredients carry with them indisputable health benefits for many patients. Yet our experience at ingesting these drugs, beginning with aspirin synthesized by Felix Hoffmann at Bayer & Co., is little more than a century old, making them newcomers to our evolutionary history.

We have consumed synthesized drugs to counter infections and inflammatory and chronic diseases during a period of rapid technological change and overall improvement in human physiology and longevity. Better nutrition and public health measures such as vaccination, sanitation, and water treatment account for these improvements but so do drug therapies. The creative chemistry behind small-molecule drug development is inextricably linked to our understanding and treatment of disease. Chemically synthesized drugs are a cornerstone of modern medicine. Yet in no other industry is the process of innovation under as much scrutiny and pressure as in pharmaceuticals, because in no other industry is the performance from massive R&D investment falling so precipitously. How that problem is resolved will shape the contours of biomedical research, health care, and the quality of life around the world in the decades to come.

Chemistry helped to launch the industrial era. The organization and application of the knowledge of chemistry that underpins the global economy followed a major expansion of that knowledge in the late eighteenth and early nineteenth centuries. That surge in the knowledge of chemistry arose from the theories and experiments of John Dalton, Henry Cavendish, Antoine Lavoisier, Jöns Jacob Berzelius, Carl Wilhelm Scheele, Alessandro Volta, and Humphry Davy with contributions from Lunar Society chemists Joseph Priestley, James Keir, and Joseph Black. It was an era when mechanics was thought to reign supreme, the mechanics embodied in James Watt's steam engine and his partner Matthew Boulton's Soho Manufactory. Watt's engine is normally thought of as a product from the realm of mechanical systems engineering. Yet the centerpiece of his invention, the separate condenser, Watt himself regarded as a chemical invention, an expression of his interest in practical chemistry, namely the elasticity of steam.

The new understanding about the behavior of basic elements of matter and how molecules bond and interact would soon catch the attention of chemists with an eye toward social improvement and commercial opportunity. Industrialized chemistry was on its way, spurred by the association of theorists and experimentalists with entrepreneurs and by the new industrial organizations. One of the biggest breakthroughs occurred in 1828 when the German chemist Friedrich Wöhler synthesized urea, a small organic molecule found in urine, from the inorganic chemical ammonium cyanate. It is an accident actually, and Wöhler was delighted with it. His accident is a landmark in the history of science and was an important step along the long road to undermining the vital force theory. "Vitalism" holds that the processes of life do not follow the laws of physics and chemistry—that the molecules of life cannot be made in the laboratory.

Wöhler and his friend Justus von Liebig made organic chemistry a systematic science, a feat that "must count as a revolution equal to (and complementing) the insights of Lavoisier and his followers four decades earlier," says economic historian Joseph Mokyr. That four elements (oxygen, carbon, nitrogen, and hydrogen) "could combine together in almost infinitely many different ways" was a powerful combinatorial chemistry with practical implications. Urea became an important raw material for the emerging chemical industry. The manufacture of nitrogen fertilizers including urea through the Haber-Bosch process flowed directly from von Liebig's discovery of nitrogen as an essential plant nutrient. Today low-cost urea is widely used in India and China and other developing countries.

The revolution in chemistry that Wöhler and von Liebig set in motion commenced in academic laboratories, spread through the education of graduate students, and eventually took root in commercial enterprises. Wöhler synthesized urea in his laboratory at the Polytechnic School in Berlin. In 1836 he moved to the University of Göttingen, where he taught for nearly half a century and which became an international mecca for chemistry graduate students. Germany and neighboring Switzerland, in turn, became the dominant players in the rising synthetic chemistry industry during the second half of the nineteenth century. They were among the leaders in pharmaceutical manufacturing throughout the twentieth century. By the early twenty-first century, Germany and Switzerland still accounted for 15 percent of global production of branded drugs despite the rise of the American (34 percent), Japanese (11 percent), French (10 percent), British (6 percent), Irish (6 percent), and Italian (4 percent) industries. Indeed, the hundreds of British and America students who made their way to German universities to pursue doctoral studies in organic and inorganic chemistry until the outbreak of World War I helped lay the foundation for the chemical and pharmaceutical industries in their native lands.

Innovation Put to the Test

Pharmaceutical manufacturing initially concerned itself with making drugs from natural products such as quinine from cinchona bark and salicylic acid, the precursor of aspirin, from willow bark. With the rise of the germ theory of disease, antitoxins and therapeutic serums made their debut, followed by the antibiotics Salvarsan and sulfa drugs, which were products of synthetic organic chemistry, and the natural antibiotic penicillin. When bacteria began to show resistance to antibiotics, medicinal chemists created analogs that proved to be effective against resistant strains.

The discovery of the double helical structure of DNA in 1953 was soon followed by the availability of powerful spectrometers and separation techniques for studying molecules and advances in protein crystallography to study their precise structure. Innovations like recombinant DNA techniques, monoclonal antibodies, and polymerase chain reaction paved the way for employing molecular biology in drug design. These developments together with high-speed computation paved the way for computer-aided or "rational" drug design during the 1980s. By the 1990s, small molecule libraries, combinatorial chemistry, and high-throughput screening (HTS) were put to work in drug discovery thanks largely to advances in synthetic chemistry and robotics. HTS, in which batches of compounds are tested for binding activity or biological activity against target molecules, did not produce candidate molecules or drug approvals in the numbers many predicted. HTS typically does not reveal the oral bioavailability of candidate molecules, an important factor in drug failures. The emerging paradigm for drug discovery, still in an embryonic stage, will integrate data from various "omics" including genomics, transcriptomics and proteomics, and metabolomics to identify disease-related molecular biomarkers and cellular signaling pathways. The emerging model will also harness nanotechnology and the burgeoning power of pattern recognition and "deep learning" through artificial neural networks. Molecular medicine will need to be better integrated with diagnostics, devices, and services such as remote physiological monitoring of patients if its full power is to be realized.

At the time that Genentech and Amgen and other biopharmaceutical firms entered the drug-making arena in the 1970s and 1980s, the pharmaceutical industry was at the top of its game. The average profit margin of the Fortune 500 pharmaceutical companies was twice the median of other Fortune 500 industries. High barriers to entry from patent protections and the huge and rising costs of drug development secured the position of incumbents. Raw material suppliers and buyers groups had little power. Little competitive rivalry existed among major firms because they tended to concentrate on different market segments.

As more entrepreneurial biotech firms brought molecular biology and biomanufacturing to the fore, and as buyers groups, generic drug producers, and patients empowered by information from the Internet began to assert themselves, the traditional pharmaceutical industry faced a confluence of adversaries. Stock prices underperformed for a decade beginning in the late 1990s, part of what economist Michael Mandel termed an "innovation shortfall." By the end of the first decade of the twenty-first century, many blockbuster pharmaceuticals were going off patent (the "patent cliff"), a trend some analysts predicted would reduce revenues at some companies by 25 percent or more. Blockbuster drugs with a combined $170 billion in annual sales are slated to go off-patent by 2015. Three hundred billion dollars of drug sales are estimated to be at risk from patent expirations between 2012 and 2018. As a result, global spending on generic drugs is expected to nearly double by 2016 from $242 billion in 2011. That prospect led brand-name drug makers and generic drug makers into forging agreements in which the former pay the latter to stay out of their highly profitable market for a specified number of years, so-called pay-for-delay settlements. The US Supreme Court cleared the way in 2013 for antitrust lawsuits to proceed against brand name and generic drugmakers, and European antitrust regulators began to levy heavy fines for pay-for-delay practices.

As the pharmaceutical industry was facing these challenges and responding with innovation in marketing and deal making if not research and development, yet another troubling trend appeared. The United States accounts for some 40 percent of global prescription drug sales, but the growth rate of prescription drug spending began to slow in the 2000s. Harvard health economist David Cutler and his colleagues attributed the slowing rate of growth to the rise of less expensive generic drugs and the growing role of biologics and vaccine alternatives. They also cite the changing medication mix away from drugs prescribed principally by primary care physicians (20 percent of all drug spending in 2007 for cholesterol-lowering drugs, acid pump inhibitors such as omeprazole, respiratory drugs, antidepressants, and oral antidiabetics) toward those mostly prescribed by specialists (45 percent for anticancer drugs, antipsychotics, anti-epileptics, erythropoietins, and autoimmune treatments). Some blockbuster small-molecule drugs were relatively easy to imitate, but the rate of successful follow-on imitation has fallen since 2000 as the clinical and economic landscape shifted. Drugs prescribed by specialists, some of them large-molecule biopharmaceuticals rather than small-molecule synthetics, tend to be more complex and difficult to imitate. This has produced what some call an imitation deficit based on the decline of follow-on drugs rather than an innovation deficit based on US Food and Drug Administration (FDA) approvals for first-in-class molecules, which has ranged between seven and twelve approvals per year for three decades. Overall FDA drug approvals have remained fairly steady for six decades.

New molecular entity (NME) applications to the FDA and their approvals declined beginning in the late 1990s before mounting a rebound. The FDA approved thirty NMEs in 2011 and thirty-nine in 2012, the most since 2004 (thirty-six approvals; see Figure 2.1). Twenty of the thirty-nine approvals in 2012 (51 percent) and twelve of the thirty approvals in 2011 (40 percent) were for first-in-class drugs, meaning drugs that use a new and unique mechanism of action for treating a medical condition, a key barometer of pharmaceutical innovation and productivity. Six of the approved drugs in 2011 were biologics, as were eleven in 2012 (Figure 2.1). Total health care spending on drugs in the United States, according to IMS Health, which provides information for the health care industry, reached $326 billion in 2012, representing a 3.5 percent decline in per capita spending on drugs from the previous year. The use of branded drugs declined in favor of lower-cost generic drugs, which represented

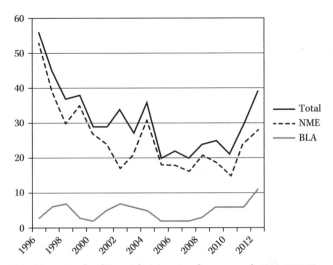

FIGURE 2.1 United States Food and Drug Administration drug approvals 1996–2012. New molecular entities (NMEs) and biologic license applications (BLAs) approved by the FDA's Center for Drug Evaluation and Research by year.

Source: The authors.

84 percent of dispensed prescriptions and saved $29 billion in spending by replacing brand-name drugs that went off-patent. The "patent dividend" for substitution of generic drugs for their brand-name counterparts saved an estimated $75 billion from 2008 through 2012. In May 2013, what consumers paid for health care in the United States fell for the first time in nearly four decades, largely as a result of a decline in prescription drug spending.

US spending on brand-name drugs totaled $235 billion or 71.5 percent of total spending in 2012. Small-molecule drugs spending totaled $245 billion and accounted for 75 percent of the market. Biological drugs spending reached $81 billion and accounted for 25 percent of the market. Spending growth was highest for biologics, injectable drugs, and specialty drugs, which together now account for about one-third of total spending. New drugs for cancer, asthma, glaucoma, diabetes, multiple sclerosis, cystic fibrosis, HIV, hepatitis C, and heart disease became available to treat the millions of patients with these diseases. New drug therapies for treating orphan diseases also were introduced to the market: one-third of the drugs approved in 2012 and at least one-third of the drugs approved in each of the past six years were orphan drugs. Despite their high cost, specialty drugs continued gaining market share against traditional pharmaceuticals (about 25 percent versus 75 percent, respectively). Among these drugs are treatments for cancer, autoimmune diseases, HIV, multiple sclerosis, and cystic fibrosis. But per capita utilization of drugs in the United States continued its decline. Usage rates of drugs to treat chronic disease fell. Also, patients age sixty-five and older took fewer prescription drugs.

The relatively high number of drug approvals in 2011 and 2012 was encouraging to some industry observers who saw the development as evidence that the drug pipeline is being replenished. The long and burdensome regulatory approval process is being successfully streamlined through the FDA's fast-track and breakthrough designation, accelerated approval, and priority review processes. The industry is responding to changes in the health-care enterprise, and patients are deriving the benefits. Other industry observers were not so sanguine that a corner has been turned and that the long decline in the industry's research and development productivity will be reversed anytime soon. The number of new drugs approved per billion US dollars spent on pharmaceutical R&D "has halved roughly every nine years since 1950, falling around 80-fold in inflation-adjusted terms," based on an analysis by pharmaceuticals analyst Jack Scannell and his colleagues in a widely discussed article about the decline in pharmaceutical R&D efficiency. They call it "Eroom's Law," which they describe as a kind of inverse of Moore's Law ("Moore" spelled backwards), the axiom that computing power per cost input doubles every two years. "Eroom's Law indicates that powerful forces have outweighed scientific, technical and managerial improvements over the past sixty years, and/or that some improvements have been less 'improving' than commonly thought." The disappointing number of drug approvals following the industry's adoption of automated systems such as combinatorial chemistry and HTS in the search for high-affinity targets, they argue, is a case in point.

No clear understanding or explanation exists for the phenomenon analysts like Scannell have described other than the fact that biology is extraordinarily complex and the additional fact that no two people (including monozygotic twins) are exactly alike, genetically or otherwise. The pharmaceutical industry has delved deep into biological complexity with the aid of automation, yet its current drug development model requires that a sufficient number of patients show a positive response to justify the massive investment in a given therapeutic strategy that is both safe and effective for these patients. The problem is that as safe and effective therapies are developed, approved, and marketed for a given indication, disease heterogeneity among substantial numbers of patients makes developing more effective therapies over existing treatments disproportionately more challenging. "Interpatient heterogeneity has always been one of the major obstacles to designing uniformly effective treatments for cancer," wrote Bert Vogelstein and his colleagues in a comprehensive review of cancer genome landscapes published in *Science*. Failure rates of late-stage drug trials are high and growing even as drug trial costs rise. As a result, drug companies are beginning to use genetic screening of tissues and tumors to select patients for clinical trials who are likely to benefit from experimental medicines. The FDA has approved cancer drugs with the requirement that physicians use companion diagnostic tests to determine whether the drug will be effective based on the patient's genetic profile, which should serve to improve outcomes and reduce costs. FDA critics are calling for the agency to embrace adaptive design clinical trial protocols that use powerful new statistical tools in trial design. Adaptive trials also take disease complexity and patient response into account during the trial and make adjustments, which is not typical of randomized controlled trials as they have been conducted over six decades. The genomic and statistical sciences are poised to drive disruptive innovation in clinical trials.

Some analysts predict that total pharmaceutical R&D spending—$60 billion in the United States in 2010, the largest share of any industry—will be less in the future than it is today. Why? Because "stock prices indicate that investors expect the financial returns on current and future R&D investments to be below the cost of capital...and would prefer less R&D and higher dividends," in the view of Scannell and colleagues. Shareholders will no longer brook multimillion- or multibillion-dollar drug development investment over a decade or more ending abruptly in late-phase trial failures. After all, shareholders through the board of directors ultimately appoint executives and control how resources are allocated, "so their perceptions matter."

With advances in biomedical research and genetics, smaller but genetically defined patient groups stand to see therapeutic benefits from drugs designed specifically for these smaller groups and subgroups that share certain genetic features. Can the pharmaceutical industry, long accustomed to blockbuster drugs and mass markets, adapt to such a dramatic shift in what constitutes the optimal drug therapy for a given patient or a small group of patients and still be a viable commercial enterprise? Would a shift to developing drugs against targets informed by human genetics rather than animal models raise the chances of clinical success, as Kari Stefansson and his Amgen colleagues suggest? Would structural reforms within the industry make an appreciable difference? Would new partnerships among industry players and between industry and nonprofit research institutes and universities serve to slow Eroom's Law and possibly arrest or even reverse it?

Cells, tissues, and "organs" embedded in silicon chips are taking the exploration of human biochemistry and physiology to unprecedented heights, but Moore's Law seems unusually alien to conventional drug development, at least to the industry that emerged in the late nineteenth century and is today's vast global enterprise. Yet Moore's Law *is* applicable to the automated systems the pharmaceutical industry has employed the past two decades to identify drug targets, signaling pathways and candidate molecules for development. Critics of Scannell and his colleagues counter that industrialized drug discovery methods such as HTS of chemical libraries have made numerous contributions. Industrial and academic scientists studying the impact of HTS in biomedical research found that in the 1990s HTS identified drug targets—typically cell-surface receptors (membrane and transmembrane receptors) or enzymes—involving tyrosine kinases, proteases, cytokine and G-protein-coupled receptors, and reverse transcriptase. Drugs targeting these molecules and their signaling pathways were approved in the 2000s to treat numerous diseases, including cancer, diabetes, HIV, pulmonary hypertension, and thrombocytopenia. Yet HTS as it is currently undertaken in industry, what Scannell and colleagues label "the 'basic research–brute force' bias" or "molecular reductionism" approach to drug discovery, has not led to increased R&D efficiency or slowed the growth of R&D inefficiency, at least not yet.

Drugs currently in the pipeline, particularly the surge in drug candidates for treating cancer, may well show the future value of HTS in drug discovery. Some 60 percent of the lead compounds in the Novartis drug pipeline in 2012 (139 drugs in clinical development, including 73 NMEs) were reportedly identified through HTS. Novartis led the industry in drug approvals with fifty-six in the United States, Europe, Japan, and China from 2007 through 2012.

Biology to the Rescue?

Protein-based biological drugs or biologics, fruits of the revolution in molecular biology, make up a large and growing share of experimental drugs, constituting a quarter of Novartis's large drug pipeline. Between 20 and 25 percent of drugs approved by the FDA and the European Medicines Agency have been biologics in recent years. Biologics—antibodies, antibody-drug conjugates, recombinant fusion proteins, enzymes, and hormones among them—are large-molecule drugs. They are delivered by injection or infusion, not taken orally like most small-molecule synthetic drugs that built the pharmaceutical industry. Biologics have had a higher success rate than small-molecule drugs in clinical trials. The FDA approved 32 percent of biologics applications compared to 13 percent for small-molecule drugs between 1993 and 2004, a period that saw a 19 percent overall rate of approval. In 2000 just one of the top eight selling drugs was a biologic; eight years later half were. Biologics achieved two milestones in 2010: Annual sales of recombinant therapeutic proteins around the world surpassed the $100 billion mark, an amount that is expected to double in the years ahead, and thirty biopharmaceutical products recorded sales of more than $1 billion each. Biologics are expensive. The average annual cost of a brand name biologic was estimated by one analyst to be $35,000 in 2012, the result of price increase rates far exceeding the rate of inflation based on the consumer price index.

The biologics field has its share of imitators compared to first-in-class molecules—eight of the nineteen products approved in 2010 and 2011 were imitation biologics. Generic-like biologics or biosimilars, for which the FDA is developing regulatory guidelines, are follow-on biological products that the agency describes as "interchangeable" with an FDA-licensed biological product. They are forecast to constitute a multibillion-dollar market. Yet precise interchangeability of biosimilars with branded biologics is certain to be less precise than is the case for generics with branded small-molecule drugs. Achieving bioequivalence is hampered by an evolving "drift" from the desired molecular structure and function that occurs in manufacturing large-molecule biologics compared to manufacturing small-molecule synthetic drugs. Such molecular drift could potentially produce dangerous immunogenic reactions in patients. In addition, the hurdles for approval are higher than for synthetic generics, and postmarket surveillance and tracking will be more demanding.

The Biologics Price Competition and Innovation Act, incorporated into the Patient Protection and Affordable Care Act of 2010, is described by the FDA as an abbreviated approval pathway. The law protects the original biologic drug from being copied, specifying a twelve-year period of clinical test data exclusivity, and provides a regulatory pathway for generic drug makers to make their own version without going through the extensive clinical testing process required for NMEs. But the pathway is not without obstacles. Lawyers of the firm that developed the original drug can scrutinize the biosimilar production process and may try to block it. For their part, some incumbents have challenged provisions of the new law, using a variety of legal claims and patent maneuvers to prevent or delay the entry of biosimilars into the market and thereby ensuring that patent attorneys are kept busy. Abbott Laboratories petitioned the FDA in 2012, claiming that its blockbuster anti-inflammatory antibody Humira (adalimumab) cannot be copied because regulators would need its manufacturing trade secrets to approve a biosimilar, violating its constitutional rights. Regulatory bodies around the world have approved Humira for the treatment of rheumatoid arthritis, psoriasis and psoriatic arthritis, Crohn's disease, ulcerative colitis, ankylosing spondylitis, and other autoimmune and arthritic conditions—in all for nine indications by the end of 2012.

Humira was the first fully human antibody drug to be granted FDA approval, which it received in 2002. It had $9 billion in sales in 2012, making it the world's top-selling drug. Humira works by blocking tumor necrosis factor—alpha (TNF–α), a key cytokine in systemic inflammation because TNF–α regulates immune cells. It is a prime example of Moore's Law at work in drug discovery through robotics-based HTS. In 1965, the year Intel's Gordon Moore described the trend in computing power productivity, the psychologist and Internet pioneer J. C. R. Licklider published *Libraries of the Future*, a report commissioned by the Council on Library Resources to explore how computers could be used in libraries around 2000. By the year 2000, the digital world Licklider imagined, enabled by Moore's Law, featured libraries of biological molecules made by automated library builder systems. Humira was isolated and optimized by selecting and screening recombinant antibody libraries. The adalimumab antibody (Humira) was generated using guided selection of human antibodies from phage display libraries, an automated laboratory technique that uses bacterial viruses to connect proteins with the genetic information that encodes them. Today DNA libraries are standard tools for next-generation semiconductor-based sequencing and the promise it holds for pharmacogenomics, biomarker discovery, and individualized therapies.

Monoclonal antibody-based drugs are the top-selling drugs in the fast-growing biologics market, well ahead of sales of hormones, growth factors, fusion products, and cytokines. Abbott's patent on Humira is slated to expire in 2016. Biopharmaceutical firms and agencies in Europe, Japan, and Brazil are partnering and pooling resources to copy the drug. Antibody-based brand-name TNF blockers, which include Pfizer and Amgen's drug Enbrel and Merck and J&J's Remicade, are estimated to constitute a global market in excess of $20 billion. As one of the most rapidly evolving fields of product development in the biopharmaceutical industry, biosimilars stand to erode the brand-name market, saving the health care systems in the United States and Europe (which approved its first biosimilar in 2006) billions of dollars. Chemical generic copies of small-molecule drugs reduce their price by up to 90 percent. Biosimilars are predicted to reduce prices by only 20 to 30 percent due to the greater resources required to copy the branded biologic, inviting brand-name manufacturers to compete with biosimilars by cutting prices while they pursue more innovative and profitable drug therapies.

Large-molecule biologics and their biosimilars are more expensive to manufacture than their small-molecule synthetic counterparts. Industrial biomanufacturing requires large-scale fermentation and cell culture facilities operated under current good manufacturing practices (cGMP) and strict product quality and consistency controls. Like other therapeutic antibodies, Humira (adalimumab) is made using Chinese hamster ovary cell culture, fermentation processes, and bioreactors. Chinese hamster ovary cell culture is the top protein-expression system for the top-selling biologics. Thanks to advances in gene synthesis, culture media, feeding strategies, protein purification, bioreactors, and automation, it has seen steady improvement in productivity since Amgen employed the technology to produce its blockbuster drug Epogen in the 1980s.

Although North America is the current leader, cell culture production capacity is growing faster in Europe and in Asia, most notably in China, India, Singapore, and other "pharmerging" countries in the Asia Pacific region. The expected surge in biosimilars will require expanding capacity well beyond its current threshold and will also elevate the role of contract manufacturing organizations in long-term corporate strategies. There is one place that stands out in projections of where biopharmaceutical manufacturing is most likely to occur in the future, a place where there are relatively few contract manufacturing organizations and thus where the opportunity for growth is perceived to be greatest. US biopharmaceutical companies view China as the number-one destination for their future manufacturing needs provided that the hurdles of quality and intellectual property protection can be crossed. They are already designing and installing cGMP facilities for contract biomanufacturing in China.

China has become the factory of the world, producing everything from basic consumer goods to Apple iPhones and iPads. It will likely remain so even as relentless advances in automation replace people with machines controlled by the technological consequences of Moore's Law. Is the Middle Kingdom destined to become the world's factory for biomolecular products? We have transitioned from Birmingham's Soho Manufactory at the dawn of the Industrial Age to Shenzhen's BGI at the dawn of the genomics era. The world's largest DNA sequencing center, BGI (formerly the Beijing Genomics Institute) is positioning itself to be out front when genome sequencing takes hold in the clinic. On the drug discovery front, BGI and the pharmaceutical giant Merck expanded their collaboration, launched in 2010, to focus on biomarkers and genomic technologies that can be brought to bear in tailoring drug therapies to patients most likely to respond. BGI formed a partnership with the Gates Foundation in 2012 in the fields of agricultural genomics and global health. It is a reminder that the scientific preeminence China once possessed, chronicled by the historian, biochemist, and embryologist Joseph Needham in his seven-volume *Science and Civilization in China*, has not been forgotten by China's scientific, business, and political elites. "China has been long one of the richest, that is, one of the most fertile, best cultivated, most industrious, and most populous countries in the world," wrote Adam Smith in *The Wealth of Nations*. "It seems, however, to have been long stationary." What was true in 1776 when Smith published his book and for two more centuries is true no longer.

Biological Technology Near and Far

The term *biotechnology* had its roots in the idea of large-scale processing of raw materials by skilled workers with the aid of living things. As the story goes, it was the Hungarian agricultural scientist and attorney Karl Ereky who coined the word *biotechnology*, just after the Great War. He introduced it in the title of a book published in German in 1919, which translates as *The Biotechnology of Meat, Fat and Milk Production in an Agricultural Large-Scale Farm*. He drew upon the disparate fields of labor and management studies, philosophy and economics, and the rapidly developing sciences of biology and chemistry. Ereky imagined an oncoming age of chemistry and biology that would be comparable to the Stone Age and the Iron Age—a technological "great leap forward."

In his book, Ereky called attention to the presence of nucleic acids in all living organisms and linked them to the production of proteins, the workhorses of the organism. "[W]e see that proteins, of either vegetal or animal in origin, contain the same amino-acids," he wrote. But his main argument centered around his firm conviction that prosperity would flow from the blending and application of the natural sciences, technology, and economics. By one account, his obsession was biomanufacturing in the form of a "large scale agricultural plant, which is led by up-to-date expertise and uses modern machines." In brief, he wanted to bring the mechanized, mass production system to the manufacture and assembly of products from bio-based raw materials perhaps not so unlike the way Henry Ford had successfully brought mass production to automobile manufacturing in Detroit. In Ereky's case, it began with fattening 50,000 pigs on an industrial scale with sugar beets as the key input. Ereky scholar Robert Bud notes that Paul Linder, editor of the *Zeitschrift für Technische Biologie* and a colleague of the German biophysicist Max Delbrück, adopted Ereky's term in his efforts to promote the use of microbes as fermentation factories.

Although Ereky envisioned the mass production of products by manipulating biologically based raw materials nearly a century ago, modern biomanufacturing commenced when technologies like recombinant DNA and monoclonal antibodies began to be employed to produce biopharmaceutical drugs. It soon became clear that the new large-molecule therapeutic products to treat cancer, infections, autoimmune and metabolic disorders, and other conditions required a new manufacturing infrastructure that could scale up the host cells and living systems that actually performed the molecular manufacturing. The new process was novel compared to that used to synthesize small molecules, the discrete chemicals that form the foundation of traditional drug manufacturing, though it is similar to antibiotic manufacturing.

Biomanufacturing is the production arm of bioscience industries, industries that are changing the way we grow crops, process foods, develop drugs and devices, produce materials, chemicals, and fuels, and protect the planet. They employ highly skilled knowledge workers in the bioproduction supply chain who are better trained, better paid, and often more satisfied with their work than their peers in traditional manufacturing industries. These workers know how to employ current good laboratory practices using cell culture and cloning, polymerase chain reaction, cDNA library construction, DNA sequencing, and Southern and Western blots. They also are versed in the great challenges of product purification and quality control/quality assurance. It is hard to overvalue the importance of cGMP given the potential consequences of a contaminant entering the bioproduction process. Expensive high-profile setbacks in the process of making drugs and vaccines such as the contamination of a Genzyme bioreactor in 2009 are stark reminders that quality manufacturing cannot be taken for granted. Mother Nature is keen to exploit the slightest mistake in a production protocol.

The broad field of pharmaceutical, crop, food, fuel, and industrial bioprocessing and biomanufacturing will be among the defining production technologies of the twenty-first century. Biomanufacturing employs highly skilled and technically trained people who work in gleaming laboratories, clean rooms, and high-tech production facilities. Scientists and technicians in these facilities, whether they are employed by biopharmaceutical or bioprocessing firms or by their contract manufacturers, bring to industrial scale the production of molecules of living systems created in the laboratory to make life-saving, life-extending, and life-enhancing agents. Production processes of chemically synthesized pharmaceuticals have remained largely constant for several decades. In contrast, production processes of biologics have advanced substantially.

The world of monoclonal antibody and related fusion protein bioprocessing "is becoming flatter," says Brian Kelley, a vice president of bioprocess development at Genentech. Biopharmaceutical companies anywhere in the world now have access to a consensus processing platform that features high cell densities and high levels of protein expression, combining to increase bioreactor production capacity and reduce costs. Monoclonal antibodies are becoming a class of therapeutic products unlike any other since the advent of genetic engineering in the 1970s. They have "unlimited production capacity and low production costs, whose pricing will have no direct link to drug substance production." Rather, prices of these drugs "reflect the innovator companies' clinical investment in addition to costs incurred from failed pipeline products," though such claims are coming under scrutiny given the exceptionally high cost of therapeutic proteins, especially those used to treat cancer. Kelley asks whether the biopharmaceutical industry is on the cusp of defining a processing platform that has matured sufficiently to last several decades, much as the cGMP blood plasma processing platform that has remained largely unchanged since the 1960s. Meanwhile, plant cells are poised to join mammalian cells as production factories. In 2012 the FDA approved Pfizer's recombinant enzyme taliglucerase alfa for treating Gaucher's disease, the first approved drug produced in a plant manufacturing system, specifically using carrot plant root cells.

The trend of pharmaceutical manufacturing and biomanufacturing migration from the United States to offshore locations prompts the question: Will biotech go the way of IT? Genentech's Kelley argues that the globalization of an efficient consensus platform and skills for antibody bioprocessing means that US companies are setting up in foreign jurisdictions not so much to cut labor costs but for other reasons, including for tax advantages. Moreover, tapping local talent in countries in which higher education is on the rise and that represent growth markets for pharmaceutical and biopharmaceutical products is a strategic move.

Not only is the industry expanding its biomanufacturing and bioprocessing activities in China, India, Singapore, Brazil and other rapidly emerging economies, it has been building research and development capacity in these countries as well, part of an accelerating trend. More than 85 percent of R&D employment growth among US multinational corporations between 2004 and 2009 occurred overseas while domestic R&D employment grew by less than 5 percent. Nearly a third of R&D employment in large US technology companies is now based overseas, according to National Science Foundation figures. Technology company executives, and especially biopharmaceutical company executives, complain that they cannot find the right talent in the United States. The question is whether their hiring and layoff practices owe more to the condition of domestic talent pools or to what they see as overseas market opportunities.

Nothing expresses a company's confidence in the ability of domestic brainpower to lead the next generation of drug discovery more than setting up R&D centers in these locations, places where a large and rising middle class will consume a growing proportion of the world's prescription drugs. The Novartis Research Campus in Shanghai, launched in 2009, is the third pillar of the company's R&D strategy, joining its R&D centers in Cambridge, Massachusetts, and Basel, Switzerland, its headquarters. Pfizer, Abbott, Merck, and other pharmaceutical giants have also expanded their R&D activities in China in recent years. In addition, they are forming more alliances with Chinese pharmaceutical companies as they combat shrinking profit margins. As China, India, and other developing nations become more wealthy and modernize their own R&D base and regulatory system and improve patent protection, they naturally become more attractive markets for drug and biotech companies. Free-market conservatives say big pharma investment in offshore R&D centers should serve as a warning to US policymakers and regulators. The R&D and manufacturing base of the biopharmaceutical industry doesn't have to stay in the United States. It can move to other markets that offer "a more attractive 'ecosystem' for life sciences innovation."

To emerge and flourish, innovation ecosystems depend on networks of local talent and entrepreneurship. In the era of globalization, India has supplied the offshore workforce for information technology and business processing services as China has for manufacturing. What country or countries will supply the workforce talent for future bioscience-based industries? Will it be India with its proven system of Institutes of Technology and a wellspring of highly educated and technically skilled young people? India's universities and agricultural institutes as well as its technology institutes have had biotechnology in their curriculum for many years. They are providing a feeder system for the country's biotech centers in Bangalore, Mumbai, Hyderabad, Pune, and Chennai. Will it be China with its rapidly developing postsecondary educational system; its biotech hubs in Beijing, Shanghai, and Shenzhen and emerging centers in Hangzhou and Chengdu; and its war chest of government funds to invest? Will it be Brazil with its booming bioenergy industries and expanding agbiotech sector? Or will it be less populous countries where innovation from life sciences research takes root and where native talent is finely honed to bring ideas and productivity to bear?

The Global Quest for Bioscience Brainpower and its Consequences

The universities of the United States, the world's leading economy, spent nearly $40 billion in life sciences research and development in 2011, $20 billion of it for medical research. The ten fields in the life sciences including the biological and agricultural sciences accounted for 57 percent of all R&D spending by American universities, a reflection of the federal research investment in field (Figure 2.2). An analysis of FDA-approved drugs from 1998 to 2007 showed that nearly a quarter were discovered through university research. In addition, FDA-approved drugs and vaccines discovered through research carried out at public research institutions from 1970 to 2009 tend to have a large clinical benefit in treating cancer and infectious disease. Pharmaceutical and biotechnology companies actively seek collaborations and partnerships with universities to get closer to cutting-edge science undertaken by faculty, postdoctoral research associates, and graduate students, many of them foreign born. Massachusetts Institute of Technology (MIT) and Harvard University's Broad Institute formed a five-year partnership with Bayer Healthcare in 2013. They will equally share the rights to any findings that come out of their joint research collaboration in oncogenomics, with Bayer having an option to an exclusive license to potential drug candidates arising from the research. These collaborations, often multimillion-dollar agreements, have grown in number in recent years and help to offset the shortfall in university funding from public sources.

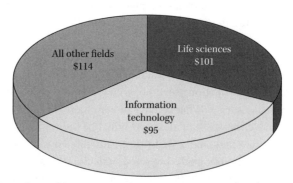

FIGURE 2.2 Estimated United States Domestic R&D Expenditures by Field, 2008 (billions of dollars). Includes domestic business, non-defense federal, and non-defense academic. Most but not all double-counting removed.

Source: Michael Mandel, South Mountain Economics LLC, from National Science Foundation data, with permission.

Countries around the world are taking a cue from the American model of higher education, and some, China in particular, are funding it at higher levels. They are viewing their universities as idea seedbeds for future innovation in the life sciences through the students they educate and graduate. The rising middle class in pharmerging countries will have access to higher education locally or abroad or through online services. The brainpower of young people in these countries is poised to change the innovation game. "When the workday ends at BGI's factory in Shenzhen, the headquarters of the largest genome mapping company in the world, it's like a bell has gone off at math camp," wrote Shanghai-based freelance writer Lauren Hilgers in describing how BGI's young Chinese scientists fresh out of college will map any genome. "The company's scientists and technicians spill out of the doorways of the building, baby-faced and wearing jeans and sneakers. Some still have braces." Technology-based industries are already taking this reality into account when they decide where to expand their businesses, where to set up their research operations, and what collaborations on distant vistas make sense.

In 2008 the National Research Council report *Innovation in Global Industries* observed that the globalization of innovation in biotechnology "is occurring in a much different way and for different reasons than the globalization of innovative activity in other manufacturing sectors, such as automobiles or IT." Bioinnovation tends to occur disproportionately in existing centers of excellence, in innovation clusters where proprietary ideas and processes flowing from research universities and institutes hook up with investment capital and managerial talent. But there are plenty of signs that bioindustries are not impervious to the sometimes disruptive forces of globalization. Whether it be R&D, manufacturing, clinical trials, or training programs, life sciences activity has been migrating from the shores of the United States, the country that gave birth to the biotechnology industry. "The shifting of employment to countries like India and China that has occurred in manufacturing, back-office work and computer programming is now spreading to a crown jewel of corporate America: the medical and drug sectors," *New York Times'* biotech watcher Andrew Pollack wrote in 2005. "The life sciences industry, with its largely white-collar work force and its heavy reliance on scientific innovation, was long thought to be less vulnerable to the outsourcing trend."

Will biotech indeed go the way of IT? Princeton University economist Alan Blinder drew public attention to the issue of offshoring with his *Foreign Affairs* article "Offshoring: The Next Industrial Revolution?" in 2006. As the United States continues its decades-long transition from a manufacturing to a service economy, assuming it does, the fraction of service jobs that can be moved offshore "is certain to rise as technology improves and as countries such as China and India continue to modernize, prosper, and educate their work forces." Over time, the number of service-sector jobs that will be vulnerable to competition from abroad will likely exceed the total number of manufacturing jobs, Blinder surmised. Economists J. Bradford Jensen and Lori Kletzer, analyzing tradable services, grouped life scientists with physical scientists and social scientists and attached an offshoring potential of 76 percent to the group compared to 93 percent for computer and mathematical occupations and 80 percent for engineering and architecture.

Yet based on current trends, only a small percentage of life science professional jobs (not including manufacturing) will be located overseas by 2015 compared to a quarter of jobs created by US firms for computer programmers and engineers. That may be in part because research undertaken in foreign countries by US-based biopharmaceutical companies tends to complement research done domestically rather than serving as a substitute for it and thus providing foreign employment, though that trend is changing with the availability of better-educated local talent. Policies that restrict visas or the ability of foreign graduate students to stay in the United States after completing formal education and training may well encourage growth in R&D activity in other parts of the world at the expense of domestic jobs. Be that as it may, offshoring in the biopharmaceutical industry appears to be growing, as one study described it, "across all components of the value chain," which translates into opportunity for rising nations.

It is in the rising nations of India and China where the opportunity is greatest. Their populations are expected to benefit from improved socioeconomic conditions and public health, which should translate into greater demand for bioscience innovations. Now they are themselves innovating and making the products used by their own populations. They are employing their own people to do R&D in addition to engineering and manufacturing products and offering services for populations in advanced industrial economies. "Indian and Chinese companies are performing the most advanced types of R&D for multinational corporations," said Indian-American tech entrepreneur and academic Vivek Wadhwa. "As a result, scientists from those corporations are rapidly developing the ability to innovate and create their own intellectual property." Wadhwa led a study by Duke University and Harvard University researchers that explored the globalization the pharmaceutical industry and the rise of China and India in the field. They found that the global technology landscape has changed dramatically over the past decade and that we are at the beginning of a new wave of globalization.

Two trends are dovetailing. One is the steady improvement in life sciences educational and certification standards worldwide. The other trend is the inroads that information technology and automation continue to make all along the value chain, from molecular design and drug discovery to scale-up to bioprocessing and biomanufacturing. These dovetailing trends mean that the offshoring of biotechnology will likely proceed, even if at a slower pace than IT. The headlines announcing job cuts at big pharma firms and biotech companies in the United States, Europe, and Canada represented by "Biopharma Job Outlook Bleak Through Mid-Decade" have counterparts represented by headlines like "Amid Widespread Belt Tightening, Pharma Hires in China." Hiring in China and other emerging economies does not substitute for the massive shedding of jobs in the West—pharmaceutical companies laid off an estimated 150,000 employees between 2009 and 2012. It does signal the industry's recognition of opportunities afforded by a growing middle class in countries like India and China along with government investments in health care and public education and, in China's case, a rapidly aging population.

Both professional service and manufacturing jobs related to life science industries are vulnerable to countries and regions offering multinational companies lower-cost labor as well as lower land, construction, health care, operating, and maintenance costs together with tax incentives. Asian microchip manufacturers have captured a major share of the global market. Looking to apply the tools powered by Moore's Law to biotechnology and genomics, Shanghai Biochip formed a subsidiary, called ShanghaiBio Corporation, which is a contract research organization possessing a large array of biological R&D outsourcing services, preclinical trial services as well as support for phase I through IV clinical trials. The company uses microarrays to analyze gene expression, performing data analysis for its big pharma clients, GlaxoSmithKline, Merck, Johnson & Johnson, and Lilly among them. It also is developing analytical and interpretive tools for genomics and biomarker identification. ShanghaiBio has one of the largest tissue banks in Asia housing more than 25,000 samples with detailed clinical information obtained through its access to hospitals and clinics in Shanghai and beyond, enabling it to procure specialized samples much faster than would be possible in the West.

China is already the second-leading nonhuman primate experimentation country and is bidding to become the world's top supplier of research primates. Plus it has access to large experimental animal testing facilities. Shanghai's Fudan University hosts one of the largest animal research operations in the world, a mouse facility that houses 45,000 cages with as many as five mice per cage. It is the Chinese arm of the Fudan-Yale biomedical research partnership directed by Yale University geneticist and Fudan alumnus Tian Xu. The partnership's mission is to map the mouse genome using a mass production gene knockout technology Xu developed, assigning each gene a specific role in the mouse's body. Massive gene-role assignments in mice would have major implications for human health. Xu spends three months a year in his Fudan University laboratory. He says his Fudan students are just as smart and capable as his Yale students (personal communication to Hoffman).

China is on a fairly steep and accelerating growth curve thanks to the infusion of government funding for scientific and technical education. It also has benefited from the proliferation of training programs stemming from partnerships and collaborations with global pharmaceutical and biotechnology corporations and contract research organizations based in the West. Critical masses of talent are also being formed by industry–university collaborations. Chinese "sea turtles" who have been educated and employed as staff scientists or scientific managers in the United States, Europe, or Australia are being lured back to their native country by generous government programs or entrepreneurial opportunity or both. Based on a survey, some 80,000 Western-trained PhDs in the life sciences have already returned to China, and two-thirds of Chinese life sciences professionals working in the United States are thinking about returning to China for good or moving back and forth between China and the United States in search of the best opportunities.

Increasingly those opportunities will be in China. "Chinese scientists have played a significant part in the success of biotech R&D in the West," said Grace Wong, founder, president, and chief scientific officer of Massachusetts-based ActoKine Therapeutics. "Now many are returning to China, taking their skills, experience and professional contacts with them." China is ready "to take its place as a leader in global biotechnology research and development," which is reflected in the large numbers of young people eager to learn and speak English, hungry to learn about biotechnology in an environment that is adapting its educational and research infrastructure to make careers in the field not only possible but highly rewarding.

How rapidly R&D outsourcing and manufacturing services as well as native industries will grow in the future—and thus how broadly the Indian and Chinese talent pool expands into industrial bioscience—depends in part on the evolution of the industry itself. Peter Singer and his colleagues at the University of Toronto conducted the first detailed, independent studies of health biotechnology in India, China, and several other developing countries. Although their bioscience research base and entrepreneurial activities are growing, both countries suffer from poorly developed technology commercialization processes. Their domestic capital markets are small and inefficient as a result of risk-averse cultural attitudes, the scarcity of risk capital from abroad, and cumbersome government regulations. Perhaps most challenging for their future growth prospects are their intellectual property systems. Despite reforms to comply with standardized global agreements set by the World Trade Organization, countries like India and China have yet to reassure outside investors that patents will be protected in domestic courts. Indeed, developing countries often choose to craft their intellectual property laws in ways that enable them to gain competitive advantage against developed countries until they themselves are more fully developed. In addition, programs that encourage academic scientists to become entrepreneurs have only recently been adopted and face embedded cultural obstacles.

India is strong in vaccine and generic drug development and production. Sales of Indian pharmaceuticals are predicted to grow to $50 billion by 2020 from $13 billion in 2009. Growth is driven by generic small-molecule and biogeneric drug production with an assist from India's Supreme Court, which denied patent protection for "ever-greening" or the incremental improvement of branded drugs. China is developing and manufacturing drugs based on traditional Chinese medicine as well as synthetic chemistry. Both countries are building critical mass capabilities for small-molecule generic drugs and biogenerics or biosimilars for drugs such as interleukins, interferons, insulin, human growth hormone, erythropoietin, tissue plasminogen activator, and others. That market is expected to surge with the imminent loss of patent protection for global blockbuster biophamaceuticals worth some $80 billion in global sales. Both countries are building or refurbishing manufacturing facilities to comply with the standards of international regulatory agencies—the FDA, the European Medicines Agency, the World Health Organization—"to facilitate access to international markets not only for biogenerics but also novel protein products currently in their pipelines," Singer's research group reported. Both countries use "hybrid" models of biotechnology development in which income from a firm's human resource-intensive R&D outsourcing and biomanufacturing services is used to fund expensive in-house drug development for products designed to serve both the domestic and international markets. Governments in developing countries are seeking to emulate developed countries in establishing university–industry technology transfer mechanisms, regulatory and intellectual property mechanisms, and science parks in urban regions.

Both population giants India and China have formed international academic and development alliances as well as corporate partnerships to spur activity in their bioscience sectors. The Indian government established a five-year $114 million jointly funded partnership with the London-based Wellcome Trust, Britain's leading biomedical philanthropy. The joint program provided funding for seventy-five fellowships each year from postdoctoral fellows through senior researchers to strengthen the research base of Indian biomedical science and its public health infrastructure. India has formed a host of other international collaborations, including a biodesign program with Stanford University; a plant biotech research program with Canada; a joint fund with the European Commission to support agricultural biotech; a collaboration with Japan in bioinformatics; a partnership with Norway in vaccine development; and a joint fund with Australia to support researchers in the fields of stem cells, vaccines, and transgenic crops. While relatively small in dollar amount, these relationships set the stage for substantially larger investments in the future.

The answer to the question we posed about whether biotechnology will be outsourced to international venues in the manner of IT services to India is no, at least not yet. Bioscience innovation does not move at the speed of electrons in data networks, though as the BGI–Merck collaboration suggests, bits, bytes, sequences, biomarkers, and drug development are no longer strange bedfellows. If broken down into its operating components, offshoring has made significant inroads into biomanufacturing and R&D services, and the competitive offshoring pressure continues to mount. Some of the pressure comes from the behavior of foreign governments toward US corporations. They entice foreign corporations with generous subsidies and require them to comply with local-content and technology-transfer policies in order to gain access to their markets. In many cases, these policies are inconsistent with their obligations to the World Trade Organization.

But others place the responsibility squarely at the feet of corporate America. Harvard University's Gary Pisano and Willy Shih contend that decades of outsourcing manufacturing, particularly in high-technology industries like computers and more recently biotechnology, have jeopardized the US innovation capacity. This is happening just when it is needed most to help the economy climb out of a deep economic hole and on to a pathway of steady growth. Countries can leverage what they learn in local manufacturing into design and development. Offshoring has been particularly deleterious to what they call the "industrial commons," which characterize clusters of knowledge and innovation. Once an industrial commons has taken root in a region, "a powerful virtuous cycle feeds its growth." Smart people flock there because that's where the jobs and knowledge networks are. Firms launch or migrate there to tap the talent pool, stay abreast of advances, and be near suppliers and potential partners. Offshoring seriously disrupts the industrial commons because the innovation process, of which production process engineering and manufacturing are critical components, is so interlinked and interdependent. With outsourcing and offshoring, the commons loses a critical mass of work, skills, and scientific knowledge and can no longer support providers of upstream and downstream activities, which are, in their turn, forced to devolve or move away as well.

Successful innovation normally involves feedback loops. Knowledge is transferred from R&D into production, but knowledge is also transferred from production back to R&D. In the long run, Pisano and Shih assert, "an economy that lacks an infrastructure for advanced process engineering and manufacturing will lose its ability to innovate." And along with the infrastructure go engineering and manufacturing jobs. Education and training programs in biotechnology and biomanufacturing have become standard curricular offerings at state universities, community colleges, technical schools, and private centers throughout the United States. Studies of future trends in biotech knowledge work describe the industry's desire to upgrade and broaden worker expertise. Paradoxically, even as work becomes more complex in scientific laboratories, it becomes more routine downstream at the production end involving less creative input, thanks in part to automation. That may broaden the number of job opportunities by making a bachelor's degree or an associate's in science degree or biotechnology certificate rather than a graduate degree the preferred entry requirement, but it also makes outsourcing and offshoring production activities more attractive. Yet as laboratory science grows even more complex and interdisciplinary, such as by using nanotechnology, synthetic biology, and neural networks to design and evaluate biopharmaceuticals, aspects of that complexity are likely to wend their way downstream to production, placing greater emphasis on both education and technical training and thus upgrading the essential nature of at least some production work.

To the extent US-based companies do their research and product development overseas and their production there as well, US workers lose out, says economist Michael Mandel, a leading thinker on innovation policy. Take the issue of production scale-up. The United States may be the global leader in entrepreneurial start-ups, but scale-up is where the rubber meets the road when it comes to jobs, in the view of retired Intel chairman and CEO Andy Grove. "Startups are a wonderful thing, but they cannot by themselves increase tech employment." Just as important is the phase where companies scale up. Scale-up is where the serious and large-scale hiring occurs, and it is no longer happening in the United States the way it used to. Just plowing capital into young companies that eventually build their factories elsewhere does nothing to create large numbers of jobs in the United States. What one analyst calls the "innovate-here-produce-there" model may work for IT and commodity products, but innovation is tightly linked to production processes in advanced manufacturing of complex products like biomedical devices and biopharmaceutical drugs. As a result, future innovation ecosystems are being seeded afar.

What should be done? The first task is "to rebuild our industrial commons," Grove asserts, echoing the view of the academics Pisano and Shih that innovation systems are regionally rooted. That has been the case since the dawn of the Industrial Age in Britain (Matthew Boulton's Soho Manufactory in Birmingham) and in America (Alexander Hamilton and Tench Coxe's industrial development on the Great Falls of the Passaic River near Paterson, New Jersey). Grove thinks a system of financial incentives and taxes should be created and employed to discourage offshore production and encourage the domestic scaling of operations. "Such a system would be a daily reminder that while pursuing our company goals, all of us in business have a responsibility to maintain the industrial base on which we depend and the society whose adaptability—and stability—we may have taken for granted." If what Grove is suggesting sounds protectionist or idealistic to some, "so be it."

Before there is scale-up there is startup, and before there is startup there is basic research. US research universities are being emulated around the world for their role in regional innovation. The President's Council on Science and Technology described them as "hubs of the American innovation ecosystem." But the shifting currents of innovation are bound to upset comfortable assumptions. The Council issued a cautionary note: "If U.S. willingness to support basic scientific research is undermined by policies that fail to optimally use the fruits of that research to build the U.S. economy, the United States will in effect cede leadership to other countries." Overall employment in US bioscience-related industries has rebounded since the economic downturn beginning in 2008 (Figure 2.3). Yet employment in public

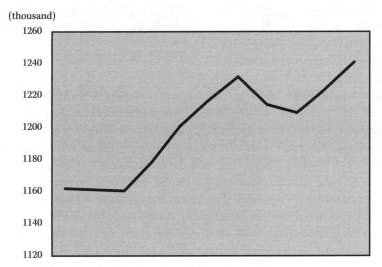

FIGURE 2.3 Employment in United States bioscience-related industries 2002–2012. Cumulative data drawn from all or parts of the U.S. Bureau of Labor Statistics NAICS codes 3253, 3254 334510, 3391, 541380, 541711, 541712, and 621511.

Source: The authors.

biotech companies involved in drug development, half of them US companies, shrunk by nearly 30 percent over the years 2008 to 2012. Scientific workforce downsizing and offshoring may be market imperatives, but they are not cost-free. To the extent the benefits of federal investment in research and education pass through the private sector to offshore operations and employment rather than to domestic innovation and economic growth, the federally funded science legacy of Vannevar Bush is put in jeopardy and the voices of the Michael Mandels and Andy Groves will grow louder.

Subjects of Interest: The Globalization of Sponsored Drug Trials

American pharmaceutical and biotechnology companies spent $60 billion in 2010 on drug discovery and development. In recent years, around 30 percent of the total has been spent researching candidate molecules, with 95 percent of research taking place in the United States, Europe, and Japan, and the rest in emerging markets. As much as half of total R&D expenditures by pharma companies per annum goes to the cost of conducting clinical trials, the most critical part of the development arm of R&D. That arm is reaching across the world even faster than the business development arm of the pharmaceutical and biotechnology industries because conducting clinical trials in lower-cost venues is one way companies have been able to manage trial costs. Clinical trials constitute nearly 60 percent of total drug development costs compared to 30 percent in 1980 by one estimate. The trend has alarmed some industry critics concerned about issues such as informed consent and shared benefits, but it has also opened up opportunities for drug companies to collaborate, innovate, and participate in developing the health-care infrastructure of both poor and pharmerging countries.

ClinicalTrials.gov is a registry of federally and privately supported clinical trials conducted in the United States and around the world. The online registry was launched in April 2000 following Congressional passage of the FDA Modernization Act of 1997 that required "the US Department of Health and Human Services, through the National Institutes of Health, to establish a registry of clinical trials for both federally and privately funded trials of experimental treatments for serious or life-threatening diseases." In 2004 the International Committee of Medical Journal Editors (ICMJE) mandated that clinical trial studies be registered in a trial registry database such as ClinicalTrials.gov as a condition of publication. The FDA Amendments Act of 2007 requires registration of summaries of trial protocols for "applicable clinical trials" involving trials (other than phase I) of drugs, biologics, and devices that are approved, licensed, or cleared by the FDA. Though ensuring compliance with these reporting requirements continues to be a challenge, these developments have led to an explosion of publicly available clinical trials data, making it possible to follow global trends in clinical trials comprehensively for the first time.

ClinicalTrials.gov shows that more than 5,000 industry-sponsored phase II through IV human clinical trials were conducted at 127,000 study sites around the world in the years 2005 to 2007. North America accounted for half of all clinical trial sites. Data from all clinical trials registered at ClinicalTrials.gov for the years 2007 to 2010 show that industry sponsored 15,000 of the 41,000 trials conducted (37 percent). North America accounted for 57.5 percent of all trials registered, down from 62 percent for the 2004 to 2007 period. For decades the United States, Western Europe, and Canada dominated clinical trials sites for both pharmaceutical and biopharmaceutical drugs. But since the turn of the century, sites have proliferated in Eastern Europe, Latin America, and Asia at the expense of Western Europe and North America. Clinical trials are undergoing globalization. The off-shoring of clinical trials to countries that represent emerging economies and future drug markets is considered a form of foreign direct investment in those countries— in their human and social as well as physical capacity to support an activity previously found only in advanced scientific and technological countries.

In one of the earliest studies of clinical trials and globalization, Fabio Thiers and his MIT colleagues revealed a notable trend: a shift away from wealthy or costly locations in the West and Australia to Eastern Europe, Latin America, and Asian countries. They analyzed 36,000 recruiting and completed studies sponsored by the public and private sectors in more than 140 countries based on the ClinicalTrials.gov data registry. In their analysis of clinical trials, sponsored by the biopharmaceutical industry, they looked at the clinical-trial capacity; trial density per population; the size of the trial, the global span of the trial (regional, national or multinational, etc.); and the type of the trial (from early stage trials to postmarketing trials). They found that while the traditional countries have more trial sites and greater trial capacity, such capacity is growing in developing nations, particularly India and China, which enjoyed rapid growth beginning in the 1990s. India and China possess an expanding nexus of contract research organizations, clinical research organizations, and health-care institutions collaborating with local and international pharmaceutical companies. The rapid outsourcing of clinical trials to Chinese-based clinical research organizations "is helping Chinese researchers gain deeper experience in working with Western models of drug development and raising their credibility," observed consultants who studied biotechnology in China. The idea that Western pharmaceutical companies can outsource standard clinical trials to Chinese clinical research organizations without consequences to upstream innovation is being tested.

Thiers and his colleagues found that nearly all of the fastest-growing countries for clinical trials are developing, while most of those countries growing slowest are developed. In a response to their study, Johan Karlberg of the University of Hong Kong used a larger dataset of 188,000 study sites for 80,000 industry-sponsored phase I through IV trials. Karlberg found that, although the United States, Germany, France, Canada, and the United Kingdom remain on top, emerging economies are increasingly the preferred clinical trial sites for pharmaceutical and biopharmaceutical companies with the potential to become major players. Industry participation in clinical trials registration jumped exponentially after the ICMJE registration mandate as a condition of journal publication of trial results was implemented.

The globalization of clinical trials carries with it both health benefits and potential hazards to research subjects and a country's general population. Countries may benefit from the diffusion of medical knowledge and effective practice, but inadequate regulatory oversight of research activities in emerging regions is a concern. One of the challenges in which genomics researchers have an interest is what Thiers et al. described as "the difficulty in drawing valid scientific conclusions with pooled data from ethnically and culturally diverse populations." In addition, the informed consent process is not consistent from one country to the next, nor is it always dependable, particularly when human subjects live in impoverished circumstances and are desperate for any kind of assistance to escape, even temporarily, from dire conditions. Sometime they are referred to as "drug naïve" patients.

Consistent with the globalization of clinical trials, a growing percentage of investigators conducting FDA-regulated studies are no longer based in North America and Western Europe. The trend reflects "the rising, and costly, logistical complexity of finding, building, managing, and monitoring relationships with investigative sites dispersed around the world," says Kenneth Getz, who monitors clinical trials for the Tufts University Center for the Study of Drug Development. Getz and his colleagues identified more than 26,000 unique investigators who conducted FDA-regulated clinical trials worldwide in 2007, with growth surging in the late 1990s. They also found that the proportion of investigators based in North America has declined steadily—from 96 percent of the total global pool of FDA-regulated investigators in 1990 to 54 percent in 2007. Steady growth in the proportion of investigators in emerging regions, in particular Latin America, Eastern Europe, and Asia, is due to many factors such as lower relative study conduct costs, harmonization of good clinical practice guidelines, the presence of clinical research organizations, and "the availability of large numbers of well-trained professionals and treatment-naïve patients."

As global clinical trials increasingly become the norm for the clinical development of new drugs, experts ask what effect this will have on biomedical innovation within advanced economies and their dominant biotechnology urban regions. Some dismiss the phenomenon as having anything but a marginal effect on productivity because clinical trials are viewed as less technology-intensive and more mundane, albeit critically important, activities in the drug discovery and development value chain.

One factor that is central to the continuing trend of biotechnology and pharmaceutical companies offshoring their clinical trials is the host country's intellectual property protection laws. A country's intellectual property regime is typically tied to its openness to foreign direct investment and imported technologies, its ability to integrate and absorb external technology flows, its domestic R&D efforts, and its productivity and economic growth. Boston researchers conducted a multifactorial analysis to determine the role of intellectual property in the globalization of human clinical trials. Such trials are being accelerated by stronger intellectual property regimes and the growth of medical and economic infrastructure in new markets, their considerable cost advantages (often saving half the cost or more), and the saturation and cost of clinical trials in traditional locations. For these reasons, global trials are becoming the norm. A country's capacity in terms of the human capital and "e-readiness" needed to conduct the trials matter more than its scientific capability as measured by the number of native first authors of articles resulting from randomized clinical trials.

The pace of offshoring clinical trials is dependent to a large extent on the pace of globalization and how economic downturns such as occurred following the financial crisis of 2008 slow its pace. Tufts' Getz, writing in early 2009, anticipated a deceleration of the trend. He also identified a "more alarming trend," that of sponsors engaging more investigators to recruit fewer patients per trial. Whatever the reason—the nature of the disease, new study designs, costs—the trend "does not bode well for investigative sites that cannot justify participating in future clinical trials because the compensation based on the volume of patients per study does not cover their operating costs." Thus drug trials are in a state of flux. The rise of genomic or personalized medicine is likely to complicate them further and exacerbate the inefficiencies.

Genomics, Pharmacogenomics, and the "Rise of the Rest"

Innovation in the biosciences has "gone global." It has migrated around the world from its US origins. That migration has been brought about by the increasing internationalization of science; the easier flow of capital, ideas, and talent from one country to another; and support from national, regional, and local governments for research and development. However much they may fall short, efforts are underway to harmonize intellectual property and regulatory systems around the world. Then there are the relentless and revolutionary effects of digital technologies and networks on the process of innovation everywhere.

In his book *The Post-American World*, journalist and TV host Fareed Zakaria made it central to his thesis that the "post-American" order he envisions is not about America's decline but about what he called the "Rise of the Rest," mainly India and China. The United States remains by far the world's supreme military power, yet the spread of capital, labor, innovation, ideas, and information is reducing the level of influence the United States once had in world affairs. Again, the shifting currents of innovation are bound to upset comfortable assumptions. East Asia as a whole is expected to surpass the United States in terms of aggregate gross domestic product sometime around the middle of the current decade. US investment in research and development as a fraction of its gross domestic product used to be first in the world. Today it is eighth, fourth among large economies. Some analysts project Asia to surpass the United States' 31 percent share of global R&D spending by several percentage points. Economically, by 2030 Asia will be well on its way to returning to being the world's powerhouse, "just as it was before 1500," according to the US National Intelligence Council's megatrends report "Global Trends 2030."

The globalization of genomic medicine as well as clinical trials will bring added impetus to the changing geography of bioscience innovation. Genomic medicine arising from the Human Genome Project, the International HapMap Project, the Cancer Genome Atlas Project, the Human Epigenome Project, the ENCODE Project, the Collaborative Oncological Gene-Environment Study, and other genomics collaborations, accelerated by advances in high-throughput technologies and the plummeting costs of genetic testing, is just beginning to be translated into practice. As genomics-based personalized medicine unfolds, developing countries see many opportunities in adopting it. Foremost will be its potential application to diagnosing and treating common health problems of their populations. These countries are forming strategic R&D alliances with multinational pharmaceutical and biotechnology companies, crafting their own research and training programs and translating their discoveries in genomic medicine into products and services for local and national as well as global markets.

The Pharmacogenetics for Every Nation Initiative (PGENI; pgeni.org) was launched with support from the National Institutes of Health as a first step to making pharmacogenetics applicable on a global level. PGENI's stated goals include enhancing the understanding of pharmacogenetics in the developing world, helping to build local infrastructure for studies in the field, and promoting the integration of genetic information into public health. PGENI expects to assist public health officials in more than 100 countries by highlighting clinically important polymorphisms in their populations, information that can guide specific public health intervention initiatives.

Even emerging economies in the developing world are investing in large-scale human genomic variation studies. Peter Singer and his research team looked at Mexico, India, and South Africa. Mexico's program INMEGEN (National Institute for Genomic Medicine) is an example of how pharmacogenetic approaches can be used to serve public health and future economic growth. Mexican president Felipe Calderón reinforced that message while inaugurating an INMEGEN laboratory in 2012 shortly before he left office. Working with the [Carlos] Slim Initiative for Genomic Medicine involving the Broad Institute, INMEGEN uses identification of unique single nucleotide polymorphisms to trace ancestral genetic variations between populations from different regions of Mexico. Such studies can reveal important differences in drug metabolism and disease risk within Latin American populations and between Latin Americans and other peoples. Singer et al. reported that similar studies by India's Genome Variation consortium showed high levels of genetic divergence between groups of Indian populations that cluster largely on the basis of ethnicity and language. This data together with epidemiological and phenotypic data would allow for the construction of countrywide ethnically related drug response and disease predisposition maps. The South African National Bioinformatics Institute is a collaborator in Human Heredity and Health in Africa, an initiative trying to pin down genes and genetic risk factors associated with tuberculosis, heart disease, HIV, sleeping sickness, and other diseases that afflict the most genetically diverse continental population on Earth.

In the world of clinical trials, "bridging" studies are sometimes undertaken in developing countries after a drug or vaccine is developed for Western populations and approved by regulatory authorities there to see if it is safe and effective given the different ethnic populations of these countries. Because ethnic (genetic) factors can determine whether a new agent is safe and effective and at what dose, regulators in those countries are sometimes reluctant to accept the results of trials conducted in other parts of the world. Sponsors, for their part, are reluctant to repeat the trials in their entirety. To resolve the dilemma, the International Conference on Harmonization (ICH.org) published guidelines to provide a framework for ethnic factors on the safety and efficacy of drugs at particular dosages.

As globalization becomes more important to the viability of pharmaceutical companies, the design and analysis of bridging studies with respect to "geotherapeutics" pose new challenges to statisticians, says Taiwanese biostatistician Jen-pei Liu. Pharmacogenomics makes it difficult to replicate single-population results in ethnically mixed and disproportionate patient populations. Trial design needs to take that heterogeneity into account. Bridging studies are likely to become more complex as functional genetic variations of drug metabolism, such as those found in ethnic subpopulations of Mexico and India, move into the public health arena.

Even so, differences among nations reign. Japan, despite its homogeneous population, insists that no clinical trials be waived for regulatory approval unless ethnic equivalency is shown. Singapore, despite its ethnic diversity, decided to ignore the issue in order to access drugs approved in the West as rapidly as possible and thereby elevate its image as an Asian biotech hub. Taiwan with its political and geographic vulnerabilities opted for the middle ground, considering the problem a regional rather than national one. "Among the genomic/globalized world states, race has gained new life," says physician scientist Wen-Hua Kuo after exploring how Japan, Taiwan, and Singapore are taking different approaches to addressing the issue of race and ethnicity in the genomic world. Kuo contends we are entering the age of the "techno-politics of genomic nationalism" reflected in two technical dynamics: the use of genomics for understanding ethnic factors in clinical trials and the use of genomics to serve a nation's cultural and international competitiveness interests. Both dynamics were at play in the way Japan and Taiwan established genomics biobanks.

However nation-states decide to address issues arising from drug trials, genomics, and race, studies that compare the impact of drugs on different ethnic populations are here to stay. They will become fundamental to gauging the safety and efficacy of experimental drugs and vaccines and the distribution patterns of approved medicines worldwide. DNA genotyping is making personal genomic studies possible with their promise of helping patients get the right dose of the right drug at the right time and in affordable manner in developing countries. In some countries, particularly large and populous countries like India, China, and Brazil, isolated ethnic groups are yielding rare genetic variants that serve as clues of disease etiology, or risk, or as biomarkers for drug and vaccine safety and efficacy. A global alliance plans to expand the responsible sharing of clinical genomic information in the developing as well as the developed world. Launched in 2013, the alliance involves some seventy institutions in thirteen countries, including the United States, the United Kingdom, and China.

Innovation, Regulation, and Sovereignty: Can They All Get Along?

US FDA global product regulation has grown four-fold over the past decade. Food and drug products regulated by the agency originate from more than 150 countries, 130,000 importers, and 300,000 foreign facilities, according to information on its website. Forty percent of finished drugs come to the United States from overseas. Eighty percent of active ingredient manufacturers for pharma are located outside the United States. "Over the next decade, FDA will transform from a predominantly domestically-focused agency operating in a globalized economy to a modern public health regulatory agency fully prepared for a complex globalized regulatory environment." That FDA statement is backed up by the fact that the agency has established permanent offices in Mexico, Costa Rica, Chile, China (three offices), India (two offices), Jordan, Italy, Belgium, the United Kingdom, and South Africa. Even as the FDA is laying down innovation pathways and guidance for breakthrough therapies designation in its regulatory science, it is seeking to create cooperative pathways with the regulatory authorities of sovereign nations around the world.

Is the globalization of clinical research and drug trials destined to continue? There are countervailing forces at work. The growing focus on biologically based medicines together with the shift toward more specialized treatments targeted at smaller subgroups of patients with a particular genetic make-up could alter the clinical trial landscape. Tuft University's Ken Getz thinks these trends might even reverse the globalization of clinical trials over time. But the current trend is clear. More than half of clinical trial subjects and sites for drugs and biologics were located outside the United States in fiscal 2008, based on a report by Daniel Levinson, the US Department of Health and Human Services Inspector General. Some 80 percent of the drug and biologics marketing applications approved by the FDA contained foreign clinical trial data. Of patient enrollees in drug trials, 57 percent were foreign; of enrollees in biologics trials, 87 percent were foreign. Foreign clinical trials for FDA-regulated drugs and biologics appear likely to grow, in Levinson's view.

The FDA inspects clinical investigators at less than 1 percent of foreign sites, only about one-third the rate of inspections at domestic clinical trial sites, in part because of the larger number of patient enrollees at US sites. The challenges in conducting foreign inspections and data limitations inhibit FDA's ability to monitor foreign clinical trials the same way trials would be monitored in the United States. Moreover, sponsors are conducting more early-phase clinical trials outside the United States without an FDA investigational new drug (IND) application. Levinson's report revealed serious problems in the FDA's ability to monitor clinical trials. It saw the likelihood that clinical trials at foreign sites will grow in number. The report recommended that the FDA monitor trends in foreign clinical trials not conducted under INDs and, if necessary, take steps to encourage sponsors to file INDs. The latter could be accomplished by working more closely with foreign regulatory bodies, performing more inspections at foreign trial sites, and exploring new models of collective or collaborative oversight.

Another of the report's recommendations was that the FDA require drug trial sponsors to submit standardized electronic clinical trial data. Two years after the report's release, ten big pharma companies—Abbott, AstraZeneca, Boehringer-Ingelheim, Bristol-Myers Squibb, Eli Lilly, GlaxoSmithKline, Johnson & Johnson, Pfizer, Genentech (Roche), and Sanofi—announced the formation of Transcelerate BioPharma. The nonprofit will create clinical data standards for the industry and standardize other aspects of the clinical trial process, including patient enrollment, patient risk measurement, and physician training. Open collaboration models such as TransCelerate BioPharma "may hold the key to transforming pharmaceutical R&D through sharing knowledge, experiences, best practices, intellectual property and information," as Tuft's Getz sees it.

In response to the Health and Human Services Office of Inspector General report published in an appendix, Joshua Sharfstein, FDA's principal deputy commissioner, said that the agency has ongoing efforts that will address the recommendations or has initiated development of new procedures that will incorporate the recommendations. Sharfstein noted, however, that under the FDA's existing statutory authority, the agency cannot require sponsors to file an IND for studies conducted outside the United States. In its oversight activities of clinical trials, the agency must be "respectful of the sovereignty of individual countries and consider the role of national regulatory authorities."

This means that globalization of clinical trials further complicates the already very complicated commercial activity. Participating drug companies are obligated to deal with domestic and foreign regulatory oversight, foreign health care systems, and differing ethical norms and human subjects protections. They also must take into account ethnic subpopulations, patients differing in their response to drugs based on variations in their genes (pharmacogenomics), and even "genomic nationalism." Regulators in the United States, Europe, and Japan have issued guidelines for pharmacogenomics given that genetic variants can affect a drug's absorption, distribution, metabolism, and excretion. The FDA advised that, due to known differences in the prevalence of these gene variants among racial or ethnically distinct groups, "plans should be considered at the outset" to obtain consent and collect DNA for drug development and clinical trials.

The famous NASA composite picture from space called "Earth at Night" shows patterns of illumination across the globe with the brightest regions in developed areas of the earth's surface, including the seaboards of Europe, the East and West coasts of the United States, and Japan. The map was first published in 2000 and then updated in 2012. During the interim, half of the total economic growth worldwide took place in developing countries. Urban centers of developing countries, particularly in China and India, radiated more light in 2012 than they did in 2000. So it is with global innovation patterns in the biosciences. Places like Singapore, Bangalore, Hyderabad, Hong Kong, Shanghai, Seoul, Taipei, Dubai, São Paulo, Belo Horizonte, West Havana, and other emerging venues in the field are expressing themselves brightly like highly expressed genes on an innovation microarray. It is a safe bet that in another dozen years others will join them.

The rise of the global middle class, the expansion of educational opportunity, and borderless communication mean that humanity's capacity to innovate and the incentives to innovate are greater today than at any other time in history. That was the way then US Federal Reserve Bank Chairman Ben Bernanke concluded a commencement address to college graduates in 2013. The currents of bioscience innovation are shifting as the human capital of smart, talented, and creative people in developing countries gain access to educational and research opportunities where they live. Of those being educated in the universities of advanced technological societies, more will return home to work in university laboratories and companies or launch their own bioscience startups. Bioscience brainpower is building critical mass in urban centers around the world, as we will see in the next chapter. All the while, the DNA molecule is being induced to disclose its creative and variable role in health and disease. Our expanding knowledge of the economy of nature in the human genome is poised to reconstitute the highly stressed economy of pharmaceutical innovation. How it happens, and where, will come to light as learning, discovery, and enterprise spread over the land.

References

Listed sequentially based on their order in the chapter.

"The Global Use of Medicines: Outlook Through 2016," IMS Institute for Healthcare Informatics, July 2012, accessed November 16, 2012, http://www.imshealth.com.

David Maris, "Who Is Popping All Those Pills?" *Forbes,* July 24, 2012.

Robert W. Fogel, *The Escape from Hunger and Premature Death, 1700–2100* (Cambridge, UK: Cambridge University Press, 2004).

Roderick Floud, Robert W. Fogel, Bernard Harris, and Sok Chui Hong, *The Changing Body: Health, Nutrition, and Human Development in the Western World since 1700* (Cambridge, UK: Cambridge University Press, 2011).

Patricia M. Danzon and Sean Nicholson, "Introduction," in *The Oxford Handbook of the Economics of the Biopharmaceutical Industry,"* Patricia M. Danzon and Sean Nicholson, eds. (New York: Oxford University Press, 2012).

Joel Mokyr, "The Contribution of Economic History to the Study of Innovation and Technical Change: 1750–1914," in *Handbook of the Economics of Innovation*, Vol. 1, Bronwyn H. Hall and Nathan Rosenberg, eds. (Boston: Elsevier/North-Holland, 2010).

David Philip Miller, " Seeing the Chemical Steam Through the Historical Fog: Watt's Steam Engine as Chemistry." *Annals of Science* 65.1 (2007):47–72.

Paul R. Jones, "The 1989 OESPER Lecture: The Strong German Influence on Chemistry in Britain and America," *Bulletin of the History of Chemistry* 4 (1989):3–7.

"The Pharmaceutical Industry in German," Germany Trade & Invest, Berlin, 2011, accessed November 15, 2012, http://www.gtai.com.

Innovation Put to the Test

Eugene C. Butcher, Ellen L. Berg, and Eric J. Kunkel, "Systems Biology in Drug Discovery," *Nature Biotechnology* 22 (2004):1253–1259.

John Markoff, "Scientists See Promise in Deep Learning Programs," *The New York Times,* November 24, 2012.

Vaibhav A. Narayan et al., "Beyond Magic Bullets: Better Innovation in Health Care," *Nature Reviews Drug Discovery* 12 (2013):85.

John W. Mullins, *The New Business Road Test: What Entrepreneurs and Executives Should Do Before Writing a Business Plan* (Upper Saddle River, NJ: Prentice Hall/Pearson, 2008).

Michael Mandel, "The Failed Promise of Innovation in the U.S.," *Bloomberg BusinessWeek,* June 3, 2009, accessed November 18, 2013, http://www.businessweek.com.

"Cliffhanger: Big Pharma Struggles to Protect its Blockbusters as They Lose Patent Protection," *The Economist,* December 3, 2011, accessed November 26, 2012, http://www.economist.com/node/21541018.

Charlotte Harrison, "Dangling from the Patent Cliff," *Nature Reviews Drug Discovery* 12 (2013):14–15.

Lyle Denniston, "Opinion Recap: "Pay to Delay" in Deep Trouble," SCOTUSblog, June 17, 2013, accessed June 17, 2013, http://www.scotusblog.com/2013/06/opinion-recap-pay-to-delay-in-deep-trouble/.

Foo Yun Chee, "EU Fines Lundbeck and Others 146 Million Euros for Blocking Rival Drugs," Reuters, June 19, 2013, accessed June 21, 2013, http://www.reuters.com/article/2013/06/19/eu-lundbeck-idUSL5N0EV1J620130619.

Murray Aitken, Ernst R. Berndt, and David M. Cutler, "Prescription Drug Spending Trends in the United States: Looking Beyond the Turning Point," *Health Affairs* 28.1 (2009):151–160.

Asher Mullard, "2012 FDA Drug Approvals," *Nature Reviews Drug Discovery* 12 (2013):87.

Steven G. Morgan, Colleen M Cunningham, and Michael R Law, "Drug Development: Innovation or Imitation Deficit?" *BMJ* 345 (2012):e5880, accessed November 16, 2012, http://dx.doi.org/10.1136/bmj.e5880.

Bernard Munos, "Lessons from 60 Years of Pharmaceutical Innovation," *Nature Reviews Drug Discovery* 8 (2009):959–968.

"The Use of Medicines in the United States: Review of 2011," IMS Institute for Healthcare Informatics, April 2012, accessed November 19, 2012, http://www.imshealth.com.

"Declining Medicine Use and Costs: For Better or Worse?" IMS Institute for Healthcare Informatics, May 2013, accessed May 9, 2012, http://www.imshealth.com.

"Consumer Price Index—May 2013," US Bureau of Labor Statistics, June 18, 2013, accessed June 18, 2013, http://www.bls.gov/cpi/.

"Research on Savings from Generic Drug Use," US Government Accounting Office, GAO-12-371R, January 31, 2012, accessed February 2, 2013, http://www.gao.gov/products/GAO-12-371R.

Jack W. Scannell et al., "Diagnosing the Decline in Pharmaceutical R&D Efficiency," *Nature Reviews Drug Discovery* 11 (2012):191–200.

Bert Vogelstein et al., "Cancer Genome Landscapes," *Science* 339 (2013):1546–1558.

Peter Huber, "The Digital Future of Molecular Medicine: Rethinking FDA Regulation," Manhattan Institute, May 2013, accessed June 15, 2013, www.manhattan-institute.org/html/fda_06.htm.

Celia Henry Arnaud, "Diagnostics-Drugs Pairings Advance Personalized Medicine," *Chemical & Engineering News* 90.30 (2012): 10–13.

Alexander Kamb, Sean Harper, and Kari Stefansson, "Human Genetics as a Foundation for Innovative Drug Development," *Nature Biotechnology* 31 (2013):975–978.

John LaMattina, "Analysts Get It All Wrong…Again," *Forbes,* April 4, 2012.

Ricardo Macarron et al., "Impact of High-Throughput Screening in Biomedical Research," *Nature Reviews Drug Discovery* 10 (2011):188–195.

Laura DeFrancesco, "Drug Pipeline: Q412," *Nature Biotechnology* 31 (2013):190.

"High Throughput Screening: Innovation in Novartis R&D," Norvartis White Paper, accessed November 20, 2012, http://nibr.com/newsroom/media/index.shtml.

Biology to the Rescue?

"Novartis Has a Substantial Drug Pipeline for All Diseases," *BioSpectrum,* November 9, 2012, accessed November 20, 2012, http://www.biospectrumasia.com/.

"Biologics at Novartis: Innovation in Novartis R&D," Norvartis White Paper, accessed November 20, 2012, http://nibr.com/newsroom/media/index.shtml.

Gary Walsh, "New Biopharmaceuticals: A Review of New Biologic Drug Approvals Over the Years, Featuring Highlights from 2010 and 2011," *BioPharm International* 25.6 (2012):34–38.

J. A. DiMasi et al., "Trends in Risks Associated with New Drug Development: Success Rates for Investigational Drugs," *Clinical Pharmacology & Therapeutics* 87.3 (2010):272–277.

Erwin A. Blackstone and Joseph P. Fuhr, "Innovation and Competition: Will Biosimilars Succeed?" *Biotechnology Healthcare* (Spring 2012):24–27.

Jonathan D. Rockoff, "Abbott is Challenging Biotech Drug Companies," *Wall Street Journal* June 21, 2012.

Hennie R. Hoogenboom and Patrick Chames, "Natural and Designer Binding Sites Made by Phage Display Technology," *Immunology Today* 21.8 (2000):371–378.

Jane Osbourn, Maria Groves, and Tristan Vaughan, "From Rodent Reagents to Human Therapeutics Using Antibody Guided Selection," *Methods* 36 (2005):61–68.

Bharat B. Aggarwal, Subash C. Gupta, and Ji Hye Kim, "Historical Perspectives on Tumor Necrosis Factor and its Superfamily: 25 Years Later, a Golden Journey." *Blood* 119.3 (2012):651–665.

Blackstone and Fuhr, "Innovation and Competition: Will Biosimilars Succeed?" 27.

Patricia Seymour, "Biomanufacturing Capacity for Biosimilars: Is There Enough?" Cambridge Healthtech Institute's Third Annual BIOAnalytical Summit 2012, March 19–22, Baltimore, accessed November 25, 2012, http://www.bptc.com.

"BioPlan's Top 1000 Global Biopharmaceutical Facilities Index," accessed November 25, 2012, http://www.top1000bio.com/index.asp

"Strategies for Improving Antibody Production in CHO Cells," *The Cell Culture Dish,* January 20, 2012, accessed November 25, 2012, http://thecellculturedish.com/2012/01/strategies-for-improving-antibody-production-in-cho-cells/.

Eric S. Langer, "China Ranked Top Biopharma Outsourcing Destination," *BioProcess International* 9.S6 (2011):36–40, accessed November 25, 2012, http://www.bioprocessintl.com.

"BGI, Merck Announce Biomarker Collaboration," *Bio-IT World,* September 13, 2011, accessed November 25, 2012, http://www.bio-itworld.com/2011/09/13/bgi-merck-announce-biomarker-collaboration.html.

"BGI and Gates Foundation Collaborate," *Bio-IT World,* September 25, 2012, accessed September 19, 2013, http://www.bio-itworld.com/2012/09/25/bgi-gates-foundation-collaborate.html.

Adam Smith, *An Inquiry into the Nature and Causes of the Wealth of Nations* (London: W. Strahan and T. Cadell, 1776).

Biological Technology Near and Far

M. G. Fári and U. P. Kralovánszky, "The Founding Father of Biotechnology: Károly (Karl) Ereky," *International Journal of Horticultural Science* 12.1 (2006):9–12.

Robert Bud, *The Uses of Life: A History of Biotechnology* (Cambridge UK: Cambridge University Press, 1993), 35.

Jonathan D. Rockoff, "Drug Manufacturing Mending After Questions of Quality," *Wall Street Journal*, June 11, 2010.

Andrew Gonce and Ulf Schrader, "Plantopia? A Mandate for Innovation in Pharma Manufacturing," *McKinsey Quarterly*, August 2012.

Brian Kelley, "Industrialization of mAb Production Technology," *mAbs* 1.5 (2009):443–452.

Asher Mullard, "2012 FDA Drug Approvals," *Nature Reviews Drug Discovery* 12 (2013):87.

Science and Engineering Indicators 2012, National Science Board, National Science Foundation, accessed February 27, 2013, http://www.nsf.gov/statistics/seind12/.

Kathy Chu, "Pfizer Seeks Alliances in China," *Wall Street Journal*, February 25, 2013.

Paul Howard, "Biotech R&D: The Next U.S. Industry to Outsource?" Medical Progress Today.com, December 7, 2011, accessed November 27, 2012, http://www.medicalprogresstoday.com/2011/12/biotech-rd-the-next-us-industry-to-outsource.php.

The Global Quest for Bioscience Brainpower and its Consequences

"Universities Report Highest-Ever R&D Spending of $65 Billion in FY 2011," National Science Foundation, NSF 13-305, November 2012, accessed November 28, 2012, http://www.nsf.gov/statistics/infbrief/nsf13305/.

Robert Kneller, "The Importance of New Companies for Drug Discovery: Origins of a Decade of New Drugs," *Nature Reviews Drug Discovery* 9 (2010):867–882.

Ashley J. Stevens et al., "The Role of Public-Sector Research in the Discovery of Drugs and Vaccines," *New England Journal of Medicine* 364.6 (2011):535–541.

"New Chemistry: Getting the Biopharmaceutical Talent Formula Right," PriceWaterhouseCoopers, LLP, February 4, 2013, accessed March 3, 2013, http://www.pwc.com/us/newchemistry.

Ben Fidler, "Broad Institute Joins up with Bayer to Find Targeted Cancer Drugs," Xconomy.com, September 10, 2013, accessed September 11, 2013, http://www.xconomy.com/boston/2013/09/10/broad-institute-joins-bayer-find-targeted-cancer-drugs.

Jim King, "Biotechs Follow Big Pharma Lead Back Into Academia," *Nature Biotechnology* 29.7 (2011): 555–556.

Lauren Hilgers, "BGI's Young Chinese Scientists Will Map Any Genome," *Bloomberg BusinessWeek,* February 7, 2013.

Raine Hermans, Alicia Löffler, and Scott Stern, "Biotechnology," in *Innovation in Global Industries: U.S. Firms Competing in a New World* (Washington, DC: National Research Council, National Academies Press), 231–272.

Andrew Pollack, "Biotech Shifts Jobs Offshore. Life Science Research is Outsourced to Asia," *The New York Times,* February 24, 2005.

Alan Blinder, "Offshoring: The Next Industrial Revolution?" *Foreign Affairs* 85.2 (2006): 113–128.

J. Bradford Jensen and Lori G. Kletzer, "Measuring Tradable Services and the Task Content of Offshorable Services Jobs," in *Labor in the New Economy*, Katharine G. Abraham, James R. Spletzer, and Michael Harper, eds. (Chicago: University of Chicago Press, 2010), 309–335.

"Global Outsourcing of Engineering Jobs: Recent Trends and Possible Implications," Testimony of Ronil Hira, Institute of Electrical and Electronics Engineers, to the Committee on Small Business, US House of Representatives, June 18, 2003.

Bernadette Tansey, "Are Biotech Jobs Next to Go?" *San Francisco Chronicle,* April 18, 2004.

Iain M. Cockburn and Matthew J. Slaughter, "The Global Location of Biopharmaceutical Knowledge Activity: New Findings, New Questions," in *Innovation Policy and the Economy*, Vol. 10, Josh Lerner and Scott Stern, eds. (Chicago: University of Chicago Press, 2010).

Frank L. Douglas and Gigi Hirsch, "Global Outsourcing: Defining China's Leading Edge," MIT Center for Biomedical Innovation, Working Papers, 2006, accessed February 8, 2010, http://web.mit.edu/cbi.

Vivek Wadhwa, "Losing Our Lead in Innovative R&D," *BusinessWeek,* June 10, 2008.

Vivek Wadhwa et al., "The Globalization of Innovation: Pharmaceuticals: Can India and China Cure the Global Pharmaceutical Market?" Social Science Research Network, June 11, 2008, accessed February 8, 2010, http://ssrn.com.

Alex Philippidis, "Biopharma Job Outlook Bleak through Mid-Decade," *Genetic Engineering & Biotechnology News,* July 10, 2012.

Jeremy Spivey, "Amid Widespread Belt Tightening, Pharma Hires in China," *Cutting Edge Info,* August 2, 2012, accessed November 27, 2012, http://www.cuttingedgeinfo.com/2012/pharma-hires-in-china/.

Ed Silverman, "Where, Oh Where, Has All the Pharma Gone?" *Pharmalot,* February 5, 2013, accessed March 3, 2013, http://www.pharmalot.com/2013/02/where-oh-where-has-all-the-pharma-talent-gone.

Wadhwa et al., "The Globalization of Innovation," 46–47.

"Biotech in China: Special Feature on China's Rising Biotech Industry." *Nature Biotechnology* (advertising supplement) February 2010:S11, accessed November 25, 2013, http://www. nature.com/nbt/advertorial/pdf/China.pdf.

Apoorva Mandavilli, "Monkey Business," *Nature Medicine* 12 (2006):266–267.

Michael Wines, "A U.S.–China Odyssey: Building a Better Mouse Map," *The New York Times,* January 28, 2011.

George Baeder and Michael Zielenziger, "China, the Life Sciences Leader of 2020," The Monitor Group, November 17, 2010, accessed November 30, 2010, http://www.monitor. com.

Grace Wong, "Developing China's Homegrown Biotechnology Workforce," *Nature Biotechnology* 26 (2008):353–354.

Halla Thorsteinsdóttir et al., "Introduction: Promoting Global Health through Biotechnology," *Nature Biotechnology* 22 (Suppl.) (2004):DC3–DC7.

Li Zhenzhen et al., "Health Biotechnology in China—Reawakening of a Giant," *Nature Biotechnology* 22 (Suppl.) (2004):DC13–DC18.

Halla Thorsteinsdóttir et al., "Conclusions: Promoting Biotechnology Innovation in Developing Countries," *Nature Biotechnology* 22 (Suppl.) (2004):DC48–DC52.

"India Pharma Inc.: Enhancing Value Through Alliances and Partnerships," PriceWaterhouseCoopers, LLC, October 2011, accessed April 1, 2013, http://www.pwc. com/gx/en/pharma-life-sciences/index.jhtml.

R. Jai Krishna and Ashutosh Joshi, "Novartis Loses Glivec Patent Battle in India," *Wall Street Journal,* April 1, 2013.

Karen P. Virk, "India's Biologics Manufacturing," Language Connections White Paper, Boston, August 2012, accessed April 21, 2013, http://www.languageconnections.com.

"Wellcome Trust and Indian Government Announce £80 Million Partnership to Boost Biomedical Research," Wellcome Trust, September 10, 2008, accessed February 9, 2010, http://www.wellcomedbt.org.

Killugudi Jayaraman, "India Partners to Fast Track Biotech," *Nature Biotechnology* 26.11 (2008):1202.

Marcella Bombardieri, "Yale in Forefront of Colleges' Fight for Overseas Clout," *Boston Globe,* May 14, 2006.

Aaron D. Levine, "Trends in the Movement of Scientists between China and the United States and Implications for Future Collaborations," Princeton University, October 2006.

Allison Proffitt, "Sequencing the Human Secret," Bio-ITWorld.com, September 28, 2010, accessed December 5, 2010, http://www.bio-itworld.com.

Laura D'Andrea Tyson, "Washington Should Get Tough with Countries that Force U.S. Companies to Outsource," *Harvard Business Review* blog, November 12, 2009, accessed February 9, 2010, http://blogs.hbr.org.

Gary P. Pisano and Willy C. Shih, "Restoring American Competitiveness," *Harvard Business Review* (July/August 2009): 114–125.

Gary P. Pisano, "The U.S. Is Outsourcing Away Its Competitive Edge," *Harvard Business Review* blog, October 1, 2009, accessed February 9, 2010, http://blogs.hbr.org.

Fiona Murray and Helen His, "Knowledge Workers in Biotechnology: Occupational Structures, Career & Skills Demands," Paper presented at the National Academies Workshop on Research Evidence Related to Future Skill Demands, May 3–June 1, 2007, in *Research on Future Skill Demands: A Workshop Summary* (Washington, DC: National

Academies Press, 2008), accessed November 29, 2012, http://www.nap.edu/catalog.
php?record_id=12066.

Michael Mandel, "The GDP Mirage," *Business Week,* October 29, 2009.

Andy Grove, "How to Make an American Job Before It's Too Late," Bloomberg Opinion,
July 1, 2010.

William B. Bonvillian, "Advanced Manufacturing Policies and Paradigms for Innovation,"
Science 342 (2013):1173–1175.

"Report to the President: Transformation and Opportunity: The Future of the U.S.
Research Enterprise," President's Council of Advisors on Science and Technology, The
White House, November 2012, accessed December 5, 2012, http://www.whitehouse.gov/
administration/eop/ostp/pcast/docsreports.

Anusuya Chatterjee and Ross DeVol, "Estimating Long-term Economic Returns of NIH
Funding on Output in the Biosciences," The Milken Institute, August 31, 2012, accessed
December 8, 2013, http://www.milkeninstitute.org/pdf/RossandAnuNIHpaper.pdf.

Subjects of Interest: The Globalization of Sponsored Drug Trials

Brady Huggett, "Public Biotech 2012: The Numbers," *Nature Biotechnology* 31 (2013):697–703.

"The Global Use of Medicines."

Center for Information and Study on Clinical Research Participation, accessed November 29,
2012, http://www.ciscrp.org/professional/facts_pat.html#3.

Rest-of-World Clinical Trial Outsourcing Conference, Princeton, New Jersey, March 23–24,
2009, accessed February 9, 2010, http://www.iibig.com/conferences/P0901/overview.html

Food and Drug Administration Modernization Act of 1997, accessed February 9, 2010,
http://www.fda.gov.

International Committee of Medical Journal Editors, Uniform Requirements for Manuscripts
Submitted to Biomedical Journals: Writing and Editing for Biomedical Publication,
accessed February 9, 2010, http://www.icmje.org.

Food and Drug Administration, Amendments Act of 2007, Public Law No 110-85, 2007,
accessed February 20, 2013, http://www.fda.gov.

Andrew P. Prayle, "Compliance with Mandatory Reporting of Clinical Trial Results on
ClinicalTrials.gov: Cross Sectional Study," *BMJ* 344 (2012):d7373, accessed February 20,
2013, http://dx.doi.org/10.1136/bmj.d7373.

Fabio A. Thiers, Anthony J. Sinskey, and Ernst R. Berndt, "Trends in the Globalization of
Clinical Trials," *Nature Reviews Drug Discovery* 7 (2008):13–14.

Robert M. Califf et al., "Characteristics of Clinical Trials Registered in ClinicalTrials.gov,
2007–2010," *JAMA* 307.17 (2012): 1838–1847, accessed November 29, 2012, http://jama.
jamanetwork.com.

Baeder and Zielenziger, "China, the Life Sciences Leader of 2020."

Johan P. E. Karlberg, "Globalization of Sponsored Clinical Trials," *Nature Reviews Drug
Discovery* 7 (2008):639–640.

Kenneth A. Getz, "The Elusive Sponsor–Site Relationship," Applied Clinical Trials Online,
February 1, 2009, accessed February 9, 2010, http://www.appliedclinicaltrials.com.

Ernst Berndt, Iain Cockburn, and Fabio Thiers, "The Globalization of Clinical Trials for New
Medicines into Emerging Economies: Where Are They Going and Why?" Conference
Paper, UNU-MERIT Conference on Micro Evidence on Innovation in Developing
Countries, Maastricht, May 31, 2007.

Genomics, Pharmacogenomics, and the "Rise of the Rest"

"Report to the President."

"Global Trends 2030: Alternative Worlds," US National Intelligence Council, Office of the Director of National Intelligence, December 2012, http://www.dni.gov/index.php/about/organization/national-intelligence-council-global-trends, accessed December 10, 2012.

Billie-Jo Hardy et al., "The Next Steps for Genomic Medicine: Challenges and Opportunities for the Developing World," *Nature Reviews Genetics* 9 (2008):S23–S27. For a review of this subject, see Dhavendra Kumar, ed., *Genomics and Health in the Developing World*, Oxford Monographs on Medical Genetics (New York: Oxford University Press, 2012).

Irma Silva-Zolezzi et al., "Analysis of Genomic Diversity in Mexican Mestizo Populations to Develop Genomic Medicine in Mexico," *Proceedings of the National Academy of Sciences USA* 106.21 (2009):8611–8616.

Indian Genome Variation Consortium, "Genetic Landscape of the People of India: A Canvas for Disease Gene Exploration," *Journal of Genetics* 87.1 (2008):3–19.

International Conference on Harmonisation of Technical Requirements for Registration of Pharmaceuticals for Human Use, accessed February 9, 2010, http://www.ich.org.

Jen-pei Liu, "Bridging Studies," in *Encyclopedia of Biopharmaceutical Statistics*, 2nd edition, Shein-Chung Chow, ed. (New York: Marcel Dekker, 2003), 134–138.

Sue-Jane Wang, "Application for Genomic Technologies for Bridging Strategy Involving Different Race and Ethnicity in Pharmacogenomics Clinical Trials," in *Design and Analysis of Bridging Studies*, Jen-pei Liu, Shein-Chung Chow, and Chin-Fu Hsiao, eds. (Boca Raton, FL: CRC Press, 2012), 151–196.

Wen-Hua Kuo, "Bring the State Back in the Global/Genomic World: Racial Difference and the Transforming States of Japan, Taiwan and Singapore," accessed February 9, 2010, http://ocw.mit.edu.

Wen-Hua Kuo, "Understanding Race at the Frontier of Pharmaceutical Regulation: An Analysis of the Racial Difference Debate at the ICH," *Journal of Law, Medicine, and Ethics* 36 (2008):498–505.

Wen-Hua Kuo, "Techno-Politics of Genomic Nationalism: Tracing Genomics and its Use in Drug Regulation in Japan and Taiwan," *Social Science & Medicine* 73 (2011):1200–1207.

Erika Check Hayden, "Geneticists Push for Global Data-Sharing," *Nature* 498 (2013):16–17.

Innovation, Regulation, and Sovereignty: Can They All Get Along?

Andrew Jack, "New Lease on Life? The Ethics of Offshoring Clinical Trials," *Financial Times,* January 28, 2008.

"Challenges to FDA's Ability to Monitor and Inspect Foreign Clinical Trials," Office of Inspector General, US Department of Health and Human Services, June 2010, accessed August 3, 2010, http://oig.hhs.gov.

Matthew Herper, "Ten Pharmas Aim To Cut Red Tape to Speed Drug Development," *Forbes,* September 20, 2012.

Man Tsuey Tse and Peter Kirkpatrick, "2012 in Reflection," *Nature Reviews Drug Discovery* 12 (2013):8–10.

Marc Maliepaard et al., "Pharmacogenetics in the Evaluation of New Drugs: A Multiregional Regulatory Perspective," *Nature Reviews Drug Discovery* 12 (2013):103–115.

Guidance for Industry: Clinical Pharmacogenomics: Premarket Evaluation in Early-Phase Clinical Studies and Recommendations for Labeling, US Food and Drug Administration, January 2013, accessed February 2, 2013, http://www.fda.gov.

Earth at Night 2012, NASA, December 5, 2012, accessed January 7, 2013, http://earthobservatory.nasa.gov/Features/NightLights/

Astronomy Picture of the Day: Earth at Night, NASA, November 27, 2000, accessed January 7, 2013, http://apod.nasa.gov/apod/ap001127.html

Ben S. Bernanke, "Economic Prospects for the Long Run," Chairman Ben S. Bernanke at Bard College at Simon's Rock, Great Barrington, Massachusetts, May 18, 2013, Board of Governors of the Federal Reserve System, accessed May 19, 2013, http://www.federalreserve.gov.

3

Regional Bioinnovation: Reaping the Harvest of the Local *and* the Global

It is thus that through the greater part of Europe the commerce
and manufactures of cities, instead of being the effect, have
been the cause and occasion of the improvement and cultivation of the country.

> —Adam Smith, *An Inquiry into the Nature and Causes of the
> Wealth of Nations*

The sparks of innovation illuminate the global map more than ever before from cities and their surrounding regions. Not only do people concentrate in these places but those that do tend to form networks of like-minded individuals and generate the ideas that lead to innovation. Indeed, the density of an urban population is sometimes looked at as an innovation barometer, the bustling human equivalent of ants in a colony or bees in a hive. Communities harboring curious and nimble minds and hard-driven entrepreneurs, the economic "adventurers" described by the French economist Jean-Baptiste Say two centuries ago, are the places where innovation is most likely to happen.

When we think of the earliest communities, we are drawn to the first permanent settlements formed in ancient Mesopotamia, the Fertile Crescent of what now constitutes modern Iraq and parts of eastern Syria, southeastern Turkey, and the western fringe of Iran. These sedentary communities came into being with the practice of agriculture beginning some 10,000 years ago and with the development of farm animal domestication, ceramics, and grain storage. They marked a turning point in the history of human beings as social and technological creatures. For the first time, population size was not limited to the ability of nature to provide food on its own. Agricultural settlements where inhabitants practiced genetic modification through crop plant and animal selection and breeding together with fermentation arose independently in different parts of the world.

Farming, settlement building, and tool making allowed populations to expand. Conquest allowed them to expand further. With the rise of ancient civilizations like the Sumerians, the Babylonians, the Egyptians, and the sea-faring Phoenicians, the practice of trade signaled the emergence of human beings as economic creatures. "Without trade, innovation just does not happen," says scientist, journalist, and author Matt Ridley. The exchange of goods and ideas is as central to future innovation as genetic recombination is to the future of sexual species. The trading of stone tools and seashell jewelry had already been going on for tens of thousands of years when agricultural production facilitated by implements—by *techné*—enabled trade on a vastly greater scale. Food could be stored, including aboard ship, to be used in trade and to feed the crew. Regional trading networks based on specialization, cooperation, and mutual benefit were built link by link, with seaports, cities, and towns serving as the nodes. Brain activity responsible for building trust in fellow traders at great distances gradually overcame the evolutionary impulse to fear the unfamiliar, the unknown. Innovation took a great leap forward.

During the Middle Ages in the West, regional trading networks were given new meaning with the rise of the Hanseatic League. *Hanse* was a medieval German word for guild or association. Northern European port cities, towns, and merchant communities founded the League in the twelfth century to protect their trading interests in the Baltic Sea, in part by eliminating tariffs. *Homo economicus*, that *über*-rational dimension of the human character looking to maximize return on investment, began to see how regional organization and cooperation can lead to competitive advantage. The European "Age of Discovery" from the fifteenth through the seventeenth centuries gave trade a truly global reach for the first time. Ships left European ports and traveled around the world in search of new trading routes, trading partners, and opportunities for outright conquest.

The Columbian Exchange, the massive global exchange of plants including crop plants, animals, insects, invertebrates, allergens, and infectious microbes between the Old World and the New World altered the biosphere, as atmospheric greenhouse gas accumulation is altering it today. The Columbian Exchange linked continental ecosystems together and facilitated the global dispersion of plant, animal, microbial, and human genes. It launched what some biologists consider the beginning of a new biological era: the *Homogenocene* arising from the homogenizing of ecosystems and loss of biodiversity.

The rapid rise of global trade spurred by the spice trade, species exchange, and the introduction of novel food crops and biological materials and fibers was a boon to both urban ecosystems and capitalism in Europe. The introduction of the potato, which was extolled by Adam Smith in *The Wealth of Nations*, may have accounted for a quarter of the growth in European population and urbanization in the 1700s just as capitalism was building its bones and sinews after two centuries of incubation. The initial setting for that incubation was Italy. As Venice yielded its trading monopoly with the East, its invention of a system of patent rights plus the beginning of banking in Florence and the rise of the bond market combined to make Renaissance Italy the seedbed of modern finance and innovation. Soon others got into the game. Britain and Holland chartered joint stock companies at the beginning of the seventeenth century. New limited liability companies gave investors legal protection against losing all their wealth if the venture failed, which vastly expanded the pool of individuals willing to take a risk. Deals were arranged between buyers and sellers of shares in return for a cut on each transaction. They were arranged by brokers and stock jobbers. In England, they often met at Jonathan's Coffee House in London. By the end of the eighteenth century the brokers were calling themselves the Stock Exchange, and Jonathan's was the site of the first systematic exchange of securities in London.

It is difficult to separate the birth of modern capitalism and the evolution of legal contracts, private property, chartered joint-stock companies, and accounting systems from a plant product and the business and social institution it created, the coffee house. Coffee beans flooded into Europe and England in the seventeenth and eighteenth centuries thanks largely to the British East India Company and the Dutch East India Company. Like Starbucks and the boutique coffee shops in Silicon Valley and other clusters of innovation today, not to mention the bars, the availability of newspapers and exchange of news in coffee and tea houses marked them as key establishments for negotiation and deal-making. Coffee and teahouses and taverns, the "penny universities" of the day, also figured in the conversations of the Lunar Society of Birmingham, which had strong links to London's coffee house society. The potter Josiah Wedgwood who made coffee pots and cups and who created a forerunner of the industry cluster wrote letters from the Baptist Head Coffee House. That house was a favorite haunt for scientists and entrepreneurs curious about chemistry and steam, including James Watt. With the university system tradition-bound and corporate R&D labs in the distant future, the coffee house was the innovation wheelhouse of British society, the hub of caffeine-stimulated neural and social networks.

Adam Smith, who worked on revisions of *The Wealth of Nations* at the British Coffee House on Cockspur Street in London, wrote in his masterpiece that he considered the discovery of America and the passage to the East Indies around the Cape of Good Hope as "the two greatest and most important events recorded in the history of mankind." Smith deduced that the global expansion of markets meant greater specialization whereby countries would attain a comparative advantage in what they do best. That would lead to ever-higher rates of exchange and barter, which would, correspondingly, raise the living standards of countries participating in the trading network. Indeed, as he wrote, sugar, coffee, and tea were accounting for an estimated 15 percent of the overall English consumption basket, which by his free-trade theory was a very good thing for both the Old World and the New.

As the modern nation-state took shape, incorporating previously sovereign cities and imposing a system of mercantilism to gain competitive advantage, the Hanseatic League went into decline. The expansion of trade embodied in the Columbian Exchange with its diffusion of species and economically productive genes in potatoes, maize, tobacco, and sugarcane hastened its demise in the sixteenth and seventeenth centuries. Yet the idea to which the Hanseatic League gave birth—the integrated regional cooperative exchange and trading network—is manifested today wherever countries, states, provinces, and districts agree to downplay jurisdictional boundaries in the interests of cooperation for mutual advantage. The European Union is the most obvious example. Others of note are South America's Mercosur trading bloc; the Association of Southeast Asian Nations, the southeastern Asian nations free trade bloc; and the North American Free Trade Agreement. On a smaller scale, clusters of bioscience innovation and bioregions owe something to what the merchants, traders, and free-city officials of the Hanseatic League had in mind—an open-exchange network among participants; so it is with ScanBalt BioRegion, a twelve-country Nordic and Baltic initiative in health and life sciences development.

ScanBalt BioRegion advertises itself as "Borderless Biotech." But is biotechnology actually "borderless?" *Nature Biotechnology* editorialized that biotech clusters of the future will be "virtual" because "the need for geographic proximity is diminishing." The biological data deluge from genomic studies would seem to uphold that perspective, a perspective that joins history and geography with genomics. The spice trade launched the "Age of Discovery" in the early fifteenth century; today, spices like ginger, nutmeg, and tumeric are being intensively studied, particularly in India, for genomic markers to assist plant-breeding programs, and that information is widely shared. Then the Columbian Exchange opened up gene transfer channels between two worlds along sea lanes, with profound and inescapable economic and social consequences. Today, the DNA of the plants that bear those genes—apple, banana, barley, bell pepper, cacao (chocolate), carrot, cassava (manioc), chili pepper, cotton, grape, maize, orange, papaya, peanut, pineapple, potato, pumpkin, rice, rubber, sorghum, soybean, squash, sugar beet, sugarcane, tomato, wheat—has been fully (or nearly) sequenced (Table 3.1). So has the DNA of many domesticated animals in the exchange, including cat, chicken, cow, dog, goat, guinea pig, horse, pig, sheep, and turkey. The genomes of pathogens responsible for cholera, malaria, measles, smallpox, typhus, yellow fever, and other infectious diseases that devastated New World populations in the immediate post-Columbian period have also been sequenced. The genome of Atlantic cod, which bridges the Old and New Worlds from the Baltic Sea to the eastern coast of North America (which it helped to build), was sequenced in 2011. Meanwhile, thousands of human beings of various ethnic stripes, infants included, have been decoded over the past decade, with the number expected to increase exponentially as sequencing technologies grow in productivity and decline in price.

In the Genomic Exchange era we have entered, plant, animal, microbial, and human genetic and regulatory sequences still travel around the world in the cells of organisms aboard ships as they did in Columbus's day, but they also travel over high-speed data networks. Genomic sequence information about crop plants, livestock, natural materials and fibers, and pathogens is of great value for human health, agricultural productivity, food security, and disease treatment and

TABLE 3.1 } Genes in transit from Columbus to global data networks: Key species of the Columbian Exchange and their genome

Types of organism	Old World to New World	New World to Old World	Sequence reported	Base pairs (BILLION) *	Genes (protein coding plus others)*
Human	*Homo sapiens*	*Homo sapiens*	2001	3.210	41,500
Domesticated animals		Alpaca	2010	1.90	25,000
	Cat		2007	2.460	20,300
	Chicken		2004	1.070	17,500
	Cow		2009	2.70	27,100
	Dog (large)	Dog	2005	2.530	24,400
		Guinea pig	2012	2.720	24,300
	Horse		2009	2.470	22,900
	Pig		2012	2.810	26,200
	Sheep		2001	2.860	24,000
		Turkey	2010	1.050	16,000
Domesticated plants	Apple		2010	1.870	—
	Barley		2012	5.10	30,400
		Cassava/manioc	2012	0.420	30,700
		Chile pepper	2013	3.50	37,000
		Cotton (New World)	2012	0.770	41,000
	Grape		2007	0.490	25,600
		Maize/corn	2009	2.050	39,000
	Oil palm		2013	1.550	34,800
		Pineapple	2012	—	25,000
		Potato	2011	0.840	39,000
	Rice		2011	0.380	30,500
		Rubber	2010	2.10	69,000
	Sorghum		2009	0.740	33,500
	Soybean		2010	1.10	50,000
	Wheat		2012	17.0	94–96,000

(Continued)

TABLE 3.1 } (CONTINUED)

Types of organism	Old World to New World	New World to Old World	Sequence reported	Base pairs (BILLION) *	Genes (protein coding plus others)*
Infectious disease agents	Bubonic plague (*Y. pestis*)		2001	0.005	4,400
		Chagas disease (*T. cruzi*)	2005	0.090	25,200
	Chicken pox (*V. zoster*)		1986	0.001	70
	Cholera (*V. cholerae*)		2000	0.004	4,000
	Diphtheria (*C. diphtheriae*)		2003	0.003	2,400
	Leprosy (*M. leprae*)		2009	0.003	2,800
	Malaria (*P. falciparum*)		2002	0.002	5,500
	Measles (measles virus)		2000	0.002	6
	Smallpox (variola virus)		1996	0.0002	200
		Syphilis (*T. pallidum*)	2010	0.001	1,100
	Typhus (*R. prowazekii*)		1998	0.001	900
	Whooping cough (*B. pertussis*)		2011	0.004	3,900
	Yellow fever (*Aedes flavivirus*)		2009	0.001	2

*Numbers rounded off in most cases.

Note: Genomic sequencing ranges from low to high coverage and from initial drafts to well-characterized genomes based on multiple studies.

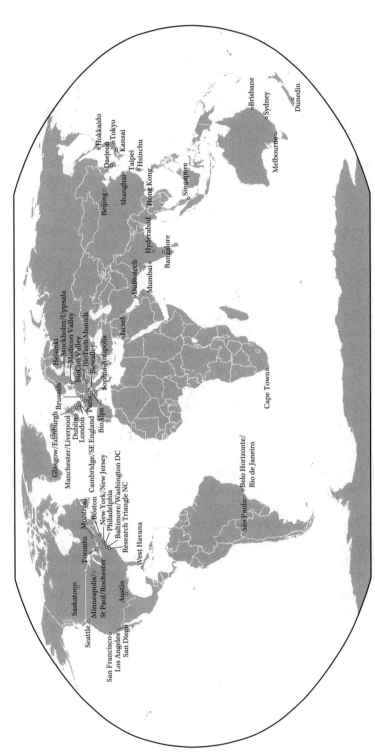

FIGURE 3.1 A sampling of global bioscience hubs and clusters where much of research, investment, and production, particularly in the biomedical field, is concentrated. Clusters are fertile ground for high productivity and employment, increasing returns on investment, and entrepreneurial startup activity.

Source: The authors. Map is modified from figure 1 of Andrea Rinaldi, "More Than the Sum of Their Parts." *EMBO Reports* 7 (2006):135.

prevention. Innovators can access such information from public sequence repositories like the National Institutes of Health's (NIH) GenBank, which holds DNA and RNA sequences from more than 250,000 species. The Genomic Exchange era has the potential for creating new bioindustries based on the knowledge of life code and how the code builds and maintains proteins, cells, and organisms.

We have entered a world of "life at the speed of light," as genomics pioneer J. Craig Venter describes it, a world in which the DNA code freed from its cellular housing blankets the earth through wires, radio waves, and data clouds. With biological data readily available to innovators having an Internet connection, is bioscience innovation in the *Homogenocene* as likely to occur in one place as another? Could a biotech startup succeed on Blueseed, a proposed marine platform for startups to be moored in international waters off the coast of San Francisco? Does the globalization of natural ecosystems mean that regional ecosystems for bioscience innovation are a thing of the past?

"Every entrepreneur in every startup is competing everywhere, accessing science everywhere, from day one," says biotech investor and guru Steven Burrill. "Capital is global. And the Web provides for real time, global collaboration never before imagined." What Burrill describes is true, but innovation is not getting easier just because capital and biological data can be sourced from anywhere. The internationalization of scientific R&D, enabled by high-speed networks, vast databases and data mining tools, and open-source initiatives does not mean bioscience industries are no longer so concerned about "place." The globalization of scientific culture has not "leveled the playing field" for startups, spin-offs, or expansions of existing firms. Biotech boosters, wanting to see everyone get into the act, have a tendency to downplay the clustering dimension of technology development in the field.

Yet place still matters, especially for drug and biomedical device development (Figure 3.1). The top ten innovative biotech startups in 2012 based on the Series A round of venture capital funding were US companies, most located in traditional biotech hubs and all of them located near research universities and hospitals. "Geography and place-specific interactions shape industries," says Maryann Feldman who has studied the geography of innovation for more than two decades. The presence of a strong university–industry nexus and what economists call "path dependence"—the self-reinforcing nature of existing technological activities in a given locale—can limit the ability of communities to build from scratch economic strength based on new technologies. Evolutionary biologist and geographer Jared Diamond describes the tendency of technology to "catalyze itself" because advances depend on previous mastery of simpler problems. Such mastery tends to be embedded locally and regionally, especially in medical technology and biotechnology, and cannot be easily accessed except through personal interaction. Electrons producing information and images over fiber optic networks or via satellite transmission, valuable as they are, cannot replace hands-on mastery, at least not yet.

Local Knowledge and the Location Paradox

Nations succeed not in isolated industries, however, but in *clusters* of industries connected through vertical and horizontal relationships.

—Michael Porter, *The Competitive Advantage of Nations*, 1990

Nearly two decades after Netscape went public, signaling a landmark in the development of the World Wide Web, location still matters. "I call it the location paradox," says Harvard University's clusters and competitiveness authority Michael Porter. "If you think of globalization, your first reaction is to think that location doesn't matter any more. There are no barriers to investment. But the paradox is that location still matters." The United States is still the most important space for innovation in the world, and economic regions are highly specialized. Anything that can be easily accessed from a distance no longer is a competitive advantage, Porter argues. The more there are no barriers, the more things are mobile, the more decisive location becomes because proximity has fueled innovation ever since ancient Uruk, the first city. The location paradox, he says, "has tripped up a lot of really smart people."

Porter expressed his view in 2006 when globalization was arguably at its apex, "the world is flat" was the catchphrase, and the Great Moderation era of assumed permanent macroeconomic stability was in full stride. Two years later the tide turned. The global financial crisis and severe global recession commenced a period of slow-growth or no-growth economies in developed countries. The contours of the post-crash global economy are discernible. Low consumer demand, high energy costs, and political instability are shaking up global supply chains, as are natural disasters and environmental stresses like earthquakes, hurricanes, tsunamis, floods, droughts, and fires. Robots are doing more of the jobs that used to pay wages. Risk-taking is not as robust as it needs to be.

When the world was purportedly flattening and Porter was pushing back on the notion that place is no longer important for innovation, urban economist Richard Florida and colleague Tim Gulden mapped global innovation and wealth production and found that the world is actually "spiky." Patenting activity, scientific citations, and light emissions are concentrated in urban regions where technology, talent, money, and imagination tend to converge. Urban regions provide most of the horsepower and brainpower that drive the global economy. Urbanization accelerated after the financial crisis of 2008 and the global recession that followed. Giant multinational firms like Caterpillar and Siemens began to focus their global operations on regional economies and "localnomics" where markets are stable and where the research and development infrastructure is strong.

Technology tends to catalyze itself more readily in places where it has already established a foothold. Regions with their clustering of industries, suppliers, services, institutions, and demanding buyers and customers demonstrate that more comes from knowing more, and knowing more is easier in proximity. "Knowledge is our most powerful engine of production," wrote the nineteenth-century British economist Alfred Marshall. Tacit knowledge arising from local practice allowed the level of enterprise in a place like Manchester to supersede that of other aspiring industrial districts in England during the Industrial Revolution.

Tacit knowledge, in contrast to the explicit, codified knowledge found in books, journals, databases, and so on, resides in people and their practices, in their interactions, particularly their direct face-to-face interactions. It is based on the idea that "we can know more than we can tell," wrote the scientist and philosopher Michael Polanyi. What innovation futurist John Seely Brown describes as the "plain know-how" of tacit knowledge is not only central to the success of learning organizations but also a fundamental characteristic of geographic regions that excel at a particular kind of economic activity, regions like Hollywood and Silicon Valley in California, the Emilia-Romagna region of northern Italy, Bangalore in India, and Hsinchu in Taiwan. Because tacit knowledge is deeply embedded in people, in organizations, and among organizations and players in a regional cluster, it is not easily shared outside that venue. Tacit knowledge is tied to specific ways of doing things in particular places. Face-to-face interactions and handshakes facilitate the exchange of tacit knowledge and the "learning by doing" or direct observation that it implies. Investors, for example, tend to like to invest in things they know well, which means their investment may well occur locally. New telecommunications systems can help with explicit codifiable knowledge, but the tacit dimension of knowledge is intangible and highly local. It permeates the culture of clusters. The relationship between explicit and tacit knowledge may take the form of an equation: The more easily transferable codified knowledge is, the more valuable is the tacit form of knowledge. That is the location paradox and what Michael Porter means when he says that the more things are mobile, the more decisive location becomes.

Polanyi saw tacit knowledge as fundamental to the creative role played by the imagination—the innate ability to connect seemingly disparate elements or fields of knowledge. Such exercise of tacit knowledge can reveal new ways of looking at the world and, from the business standpoint, achieve a competitive advantage. Some scholars describe "architectural knowledge," a form of knowledge with many tacit features that can be shared among organizations in an industry cluster. Such know-how constitutes a set of nonproprietary norms, habits, customs, understandings, and practices that evolve organically. Together these norms confer a competitive advantage on a region. The tendency of Silicon Valley's high-tech firms to share technology and patents, collaborate on establishing joint standards, and facilitate the mobility of managers and skilled workers from one firm to the next illustrates the sharing ethos of architectural knowledge. That sharing ethos is present in all successful industrial clusters. It is a safe bet that such an ethos will also be a critical component for the growth of regionally based green industries, as bioregions scholar Philip Cooke notes in his book *Growth Cultures*.

The history of American economic innovation in the twentieth century "is really a history of clusters—think of movies in Hollywood, cars in Detroit, and technology in Silicon Valley," wrote *New Yorker* financial columnist James Surowiecki in describing why people in the culture industry continue to flock to New York City. Aspiring talent gravitates to New York "because it's where previous generations of artists and designers, now powerful, gravitated to. The result is a classic case of what economists call network effects: success in the past creates success in the future." Knowledge network and clustering effects in biotechnology were intensively studied using a variety of methodologies in the 1990s and the 2000s before the financial crisis of 2008 and the ensuing economic turmoil. Those studies yielded valuable insights:

- Biotechnology and venture capital firms are strongly concentrated, with local venture capitalists tending to finance local companies.
- The essential ingredient of scientific expertise has meant that from the beginning academics have wielded more power in biotechnology than in traditional industries. Firm proximity to research is often influenced not only by the scientist's role in the firm but also by his or her status in the scientific community.
- Scientific stars played a strong role in determining where the biotechnology industry developed during the 1980s, particularly in the process of the geographic concentration of the nascent industry and spillover effects.
- Biotechnology firms are embedded in multiple networks of strategic alliances. They gain competitive advantage from continuous scientific and technical innovation. Access to new knowledge and capabilities in biotechnology occurs through information spillovers typically within clusters and also through international strategic alliance networks.
- Biotechnology is seen as a pioneer of entrepreneurial innovation in which knowledge clusters and bioregions rather than multinational pharmaceutical firms drive development in the field.

The rise of digital media that allows many firms to conduct operations wherever they happen to be located prompted urban theorist Joel Kotkin to find "declustering" in the economy a decade ago. The idea that high-wage economies must be rooted in localized and specialized networks, exemplified by Wall Street, Hollywood, and Silicon Valley, is "increasingly tenuous" in the digital age. In brief, with the new telecommunications technology, it is increasingly easy for firms to operate in a dispersed manner. "To attract the best workers, managers in a declustered economy need to let their skilled workers live where they like," he advised. City officials and economic-development professionals should focus on improving the local quality of life, public services, cultural opportunities, and tax structures and forget about clusters.

If Kotkin's counsel makes sense for some industries, say, design and engineering firms, financial and business services, media, and consulting, it is not for others. There is little evidence of declustering in biotechnology and the life sciences, particularly with respect to drug discovery and development. Tacit and architectural knowledge are likely to be a more pronounced feature of knowledge-intensive and highly regulated industries like biotechnology than in conventional manufacturing or craft industries or in service-based industries in which documents and reports and other codified knowledge play a large role. Innovation in medicine and biology tends to happen in places where the ecology for such innovation is already established or emerging. The cliché "location, location, location" is usually associated with the real estate industry. Will the cliché also determine the winners and losers as cities, regions, and countries strive to convert the new knowledge of the life sciences to their economic advantage? To what extent can industrial policy, urban-regional planning, and government subsidies make a difference?

Edinburgh: From the "Science of Man" to the Science of Life

Communities of creative people with diverse talents, technical know-how, and drive are key to innovation, if you read economic history the way Joel Mokyr does. Historians of technology used to see individual inventors as the main actors that brought about the Industrial Revolution, the James Watts and Robert Fultons of the world. That theory was discarded in favor of larger economic and institutional factors such as incentives, taxes, and demand. Mokyr thinks something else was at play. The crucial elements "were neither brilliant individuals nor the impersonal forces governing the masses, but a small group of at most a few thousand people who formed a creative community based on the exchange of knowledge." Engineers, mechanics, chemists, physicians, natural philosophers, businessmen, and skilled craftsmen "formed circles in which access to knowledge was the primary objective." Technological innovation spurs economic activity, growth economist Paul Romer argues, and more smart people involved in research, creative enterprise, and the workforce, including immigrants and guest workers, raises the prospect for innovation and growth.

Early in the Industrial Revolution, as we have seen, creative scientific communities like the Lunar Society of Birmingham in England and Benjamin Franklin's American Philosophical Society in Philadelphia saw the promotion of invention, manufacturing, and trade to be part of their raison d'être. Their members—the engineers, mechanics, chemists, physicians, philosophers, and inventors noted by Mokyr—convened at meeting halls, lodges, parlors, coffee houses, taverns, and private residences. Their discussions coincided with the rise of science and its application through new technologies, the development of skilled labor, a growing scientific and technological literacy in the general population, and the spreading of trade links that expanded the range of raw materials and consumable goods available in local markets.

Slowly but surely, over many decades beginning in the late eighteenth century and with transformational momentum by the second half of the nineteenth century, these forces combined to release populations from the Malthusian trap in which rising income stimulated population growth, which in turn put added pressure on natural resources, namely the food and fuel needed to sustain the larger population. The "trap" tended to hold economic progress in check. Only a revolutionary expansion of scientific knowledge and technological know-how flowing from that expansion with attendant changes in social norms would give society a chance to break free from the Malthusian trap. It was the systematic process of innovation in thinking and doing that made all the difference. Thomas Malthus's predictive failure wasn't accidental, says economist Paul Krugman. "Technological takeoff was the product of a newly inquisitive, empirically minded, *scientific* culture—the kind of culture that could produce people like Malthus."

The process of innovation in thinking had its roots in the old city of Edinburgh, today a vibrant center of Scotland's global ambitions in the life sciences revolution. Edinburgh was the birthplace of Adam Smith and home to the eighteenth-century Scottish Enlightenment, with its penchant for experiment and trade. For half a century until the outbreak of revolution in France in 1789, Edinburgh "ruled the Western intellect" wrote James Buchan in his acclaimed book *Crowded with Genius*, his account of Edinburgh and the Scottish Enlightment. Edinburgh was a place where the old medieval order with its static "Great Chain of Being" was put to rest in favor of a "new theory of progress, based on good laws, international commerce and the companionship of men and women." The luminaries included the philosophers and historians Francis Hutcheson, David Hume, Adam Smith, William Robertson, and Adam Ferguson; the political economist James Steuart; and the geologist James Hutton. They constituted the core of an intellectual wellspring that championed innovation in thought as well as action and had a profound influence on the founding of the United States and the writing of its Constitution, a document that constitutes the most profound innovation in government the world has ever produced. James Watt was a typical product of the intellectual ferment, pioneering as he did the modern process of machine innovation as his fellow Scots of a more philosophical bent advanced notions of moral sentiment, gender equality, natural science, and civil society. In brief, the Scots gave new life to innovation—in human affairs, agricultural science, mechanics, and trade.

Adam Smith heartily approved of exploration, trade, and advances in food production thanks to the Columbian Exchange, writing, "Such are potatoes and maize, or what is called Indian corn, the two most important improvements which the agriculture of Europe, perhaps, which Europe itself, has received from the great extension of its commerce and navigation." What intrigued Smith was how much national wealth differed from country to country, how commercial activities in these countries, driven by self-interest, produced such different levels of commercial prosperity. His answer to the riddle of why a worker in Britain or Holland enjoyed more commercial luxury than an Indian prince or African king was, of course, "the division of labour." Specialization was the critical component to the wealth of nations, as Buchan put it, to "a wonderful machine whose parts are unconscious of their mutual connection, or of any larger purpose to their activity." The father of free-market economics himself might be surprised by how far some take the idea of specialization today in a world more specialized than it has ever been. Looking back over the past 100,000 years, specialization is "the central story of humanity" because it stimulates novelty, as Matt Ridley sees it. Exchange and specialization make it possible for successive generations to live better than their ancestors and pay off the debts left to them.

In his Adam Smith Award address in 1997, Michael Porter observed that Smith "laid the foundations of economics around the notions of specialization within enterprises, specialization across countries, and the power of unencumbered competition." Smith's pin factory "legitimized the place of business and profitmaking in society." Today, economics has moved into the realm of fiscal, monetary and trade policy, and the arcane workings of capital markets. In short, the field of economics is heavily influenced by macroeconomics, the study of the sum total of national economic activity—growth, inflation, unemployment, and policies that affect these measures.

In his talk, Porter focused on microeconomics, namely the study of companies, industries, and competitiveness in the context of economic geography, the spatial pattern of production, distribution, and consumption of goods and services at local, regional, national, and international levels. Economic geography looks at the economies of places and the connections between and among those places. Adam Smith himself observed the rising productivity of places like Birmingham, Manchester, Sheffield, Leeds, and London. Indeed, Smith wrote about economic conditions in cities and countries all over the world, including the self-administered "free-towns" of northern Europe that formed the Hanseatic League.

By the "new" microeconomics of competition in the title of his talk Porter said he meant to stress both elements of competition that are truly new, such as globalization, and those that have been around but have been underappreciated by most economists as they continue to focus on aggregate inputs and outputs and employment and largely ignore the role of competition and place in the economy. "Competition is dynamic and rests on innovation and the search for strategic differences," he said. "Close linkages with buyers, suppliers, and other institutions are important not only to efficiency but to the rate of progress." Porter noted, as did Smith and Alfred Marshall before him, that the business environment and idiosyncrasies of place are often tied to productivity, prosperity, and competitive advantage in industries as diverse as shoemaking, semiconductor manufacturing, and bioscience.

Economists favoring limited government are fond of quoting Adam Smith from a lecture he gave early in his career while teaching at the University of Glasgow: "Little else is requisite to carry a state to the highest degree of opulence from the lowest barbarism but peace, easy taxes, and a tolerable administration of justice: all the rest being brought about by the natural course of things." But Porter insisted that the old distinctions between Smith's celebrated laissez-faire approach and government intervention, say, through industrial policy in which desirable industries are subsidized—these distinctions are simplistic. Government has a host of proper activities in a modern economy. It can help create a good business environment through sensible tax, trade, and intellectual property policies; set standards; regulate product quality, safety, and environmental impact; eliminate barriers to entry; and provide robust education and vocational training. Local, county, state, and national policies need not be mutually exclusive but can be synergistic, producing a compounding effect.

Porter described a cluster orientation as "very different from industrial policy." The idea is to help clusters improve their productivity by removing obstacles and constraints to productivity growth. But should local, county, and state governments support clusters directly, even build them? Speaking as he was during the high-tech boom of the late 1990s, Porter thought so. For example, when certain suppliers and assets are located in places far away from firms in a given field, "the challenge is to build a cluster," Porter said. "This involves wooing suppliers, encouraging local institutions to make supporting investments, [and] finding ways to build the local stock of specialized inputs." One of the larger national experiments designed to build regional strength in biotechnology was the BioRegio competition Germany launched in 1996, with the winning regions receiving tens of millions of dollars in direct subsidies. Other European countries followed in Germany's wake. Though such initiatives may be beneficial at the margins, so far empirical evidence is lacking that government subsidies for cooperation in R&D projects between organizations located in the same region constitute a clear pathway for successful regional innovation.

The Royal Society of Edinburgh named Porter an Honorary Fellow in 2005 in recognition of his contributions to the field of competitive strategy. Scotland's national academy of arts and letters, the Society was founded in 1783 by royal charter for the "advancement of learning and useful knowledge." Adam Smith was one of its founding members. By the time of his recognition by the Society, Porter was already well known among Scottish economic development and planning officials. In 1993 Scottish Enterprise, the lead economic development agency for Scotland, commissioned him to study and provide recommendations for where and how cluster strategies might be introduced in Scotland. Like many such reports produced by commissions, foundations, and economic development authorities in the 1990s and early 2000s, his report was put on the shelf. Unlike other reports, however, Porter's was lifted from the shelf, dusted off, and actually put to work. His recommendations for cluster initiatives involving semiconductors, food and drink, oil and gas, and biotechnology were adopted.

Scotland had cast a covetous eye on biotechnology for more than a decade. But unlike other aspiring countries and regions, it has backed up its dreams with action. "A Fertile Environment for Culture to Develop" headlined a 2007 story in the

Financial Times about how Scotland is stacking up in the life sciences. "Scotland is home to one of the biggest and fastest expanding life science 'clusters' in Europe," helped along by a strategic alliance between industry, government, universities, and the country's national health service. More than 80 percent of Scotland's 600 biotechnology organizations and 30,000 employees are situated within fifty miles of the cities of Edinburgh, Dundee, and Glasgow, most of them within these metropolitan districts. The multibillion-dollar industry enjoys a growth rate well above the average growth rate of the Scottish economy.

In Adam Smith's day, the old medieval city of Edinburgh, perched on a hill overlooking the Firth of Forth, expanded to house its growing population. The expansion came to be known as New Town, today considered a gem of neoclassical urban planning. Though Edinburgh has experienced many expansions since, perhaps none is more telling of the hopes it has for its economic future than the commercial park located on its outskirts, the Center for Biomedical Research rebranded as Edinburgh BioQuarter. The goal is to build one of the world's top ten hubs for the life sciences. If all goes as planned, the 100-acre site is expected to house 1.4 million square feet of academic, institutional, and commercial life science space on its campus, a scale that has never been seen before in the UK life science sector, according to the chief executive of Scottish Enterprise.

One of the key tenants is Edinburgh University's $85 million MRC Centre for Regenerative Medicine. Nearly half the funds are from the Scottish government, which has cast its lot with stem cell research in a big way. When it opened 2012, the Centre could accommodate 250 stem cell scientists as well as provide cellular development and therapeutic scale-up capabilities for the commercial sector. Among its resident stem cell scientists is Ian Wilmut, one of the creators of Dolly the cloned sheep at the Roslin Institute not far from Edinburgh.

Edinburgh BioQuarter was a player in the five-year $100 million Scottish collaboration with Wyeth, the US pharmaceutical group Pfizer acquired in 2009. Signed in 2006 with Scottish Enterprise and the universities of Dundee, Edinburgh, Aberdeen, and Glasgow and the National Health Service, the Scottish-Wyeth collaboration set out to accelerate the development of drugs for patients with cancer, heart disease, depression, osteoporosis, and other diseases, with the Scottish academic participants testing them in clinical studies. Wyeth chose Scotland for its translational medicine needs because of its strong medical research tradition, the ability to work with its universities, and the scope to work with a high proportion of Scotland's millions of patients, most of whom have records in electronic databases. Pfizer subsequently formed a partnership with NHS Scotland in clinical research.

To build Edinburgh BioQuarter and put the country's biotechnology and pharmaceutical industries on the fast track, Scottish Enterprise and Scottish Development International turned to Alexandria Real Estate Equities, perhaps the leading life science lab space and office development company in the world. Based in Pasadena, California, the multibillion-dollar company lists among its clients universities and independent not-for-profit institutions; pharmaceutical, biotechnology, medical device, and life science product and service companies; bio-defense and translational medicine firms and consortia; and governmental agencies. Alexandria notes that its international operating platform "is based on the principle of 'clustering,' with assets and operations strategically located in key life science hub markets." It claims to own and operate more than 11 million square feet of office/laboratory properties, including properties under development in Cambridge, Massachusetts, San Francisco, and New York City.

Harry Potter author J. K. Rowling, an Edinburgh native, buried a time capsule at Edinburgh University in the BioQuarter in 2011 to mark the groundbreaking for a research clinic for patients with neurodegenerative diseases. Rowling gave $16 million to the clinic project to "unravel the mysteries of MS." Her mother died from multiple sclerosis at age forty-five. The time capsule contains written accounts from patients living with neurodegenerative diseases and contributions from clinicians about what the future may hold for treatments. What the future holds for Edinburgh BioQuarter, an ambitious attempt to build a bioscience cluster from a science park development rather than hope one would emerge naturally from anchor firms, as most do, remains a mystery without Harry Potter-like foresight. After putting its commitment to the project on hold following the financial crisis of 2008, Alexandria Real Estate Equities cut it by a third four years later. At the time, except for several small companies located in or associated with the BioQuarter and a new bioincubator that housed a few private tenants, most of the developments, the Anne Rowling Regenerative Neurology Clinic among them, have been in the public rather than the private sphere. The field of dreams "build it and they will come" development mantra is being put to the test.

At the dawn of the Industrial Age, Adam Smith wrote there is "scarce a common trade which does not afford some opportunities of applying to it the principles of geometry and mechanics, and which would not therefore gradually exercise and improve the common people in those principles." His fellow countryman James Watt applied the principles of math, mechanics and chemistry to his steam engine design and lent his name to a principle of electricity, the watt. Today researchers at Edinburgh's Heriot-Watt University use mechanics, electricity, and biology to print pluripotent stem cells from a modified inkjet printer. Microvalve-based 3D printing of cells and biological molecules through layer-by-layer deposition opens up a new arena for biofabrication of tissues and organs like bone, cartilage, kidneys, and livers. In the near term, the technology can be used to create organoids for drug development and toxicity testing.

Taking into account genomic medicine and an aging population, growing energy demand, the earth's resources under stress, disappearing species, and pandemic flu threats, one can be confident that Smith would revise his list of useful sciences today. He would include biology along with math, mechanics, chemistry and electricity among the principles forming "the necessary introduction to the most sublime as well as to the most useful sciences." Edinburgh and Scotland are betting on it. So are many other nations, states, and cities around the world.

Geography, Deep Craft, and Increasing Returns

The financial crisis of 2008 and ensuing global recession hit the biosciences hard. Investment dropped precipitously, initial public offerings all but dried up, and established clusters of bioscience innovation retracted as firms like Alexandria Real Estate Equities downsized their development projects. A global life sciences cluster report released in 2011 by Jones Lang LaSalle, a real estate services firm based in Chicago, concluded that established clusters in the United States, the United Kingdom, France, Switzerland, and Germany experienced "a notable decline in speculative construction of laboratory facilities, demonstrating market awareness of constricted demand following the closing of many start-up operations during the recession." The Jones Lang LaSalle analysts also found that investments focused on core R&D aspects of the value chain continue to fuel activity in such clusters. As companies evaluate the financial equation surrounding innovation, "they often times are able to offset the higher real estate costs of established clusters with the advantages of deep intellectual capacity in such locations, and thus the enhanced odds of drug discovery." The global economic downtown and prolonged growth slump have only reinforced location as key to accessing superior research and talent pools.

In his classic metaphor of the pin factory, Adam Smith described how the division of labor led to exponentially higher levels of productivity and increasing returns to scale. "The greatest improvement in the productive powers of labour, and the greater part of the skill, dexterity, and judgment with which it is any where directed, or applied, seem to have been the effects of the division of labour," he wrote, proceeding to describe how the division of labor works in the manufacture of pins. The subdivision and specialization of labor produces in total many times the output compared to the total output of laborers exercising each step in the manufacturing process on their own.

As economist Paul Krugman notes in his 1991 study *Geography and Trade*, it was Cambridge economist Alfred Marshall who presented the classic economic analysis of industry localization. First, localization concentrated a number of firms in the same place, allowing a pooled market for workers with specialized skills. Second, industry centers allowed for greater availability and lower cost of nontraded inputs specific to an industry. Third, industry centers generated technological spillovers through local information networks. Marshall described how industrial districts with their interaction of specialization, variation, and knowledge sharing shifted innovation's center of gravity from the individual firm to the district itself. Industry, economics, and geography have been linked ever since.

Think of a successful technology cluster today as a learning center, a constellation of pin factories and their suppliers that both compete and collaborate. Through the division of labor, knowledge architecture, communications channels, and increasing returns to scale through higher productivity, firms within the cluster enjoy a competitive advantage they would not enjoy without the proximity of buyers, competitors, suppliers, and research institutions in the neighborhood. Buyers know what is needed in the market and push innovation; research scientists and engineers develop the enabling technologies; entrepreneurs and manufacturers adopt the technologies; and suppliers provide the essential materials and equipment. A technology cluster is also home to venture capitalists and angel investors who provide funding and guidance for startups, patent and licensing professionals well versed in intellectual property law relating to the technologies, regulatory and clinical affairs consultants (if needed), and marketing communications experts who know how to introduce innovative products into the marketplace. A well-developed cluster can or can come close to providing a holistic innovation game.

Two academic papers published in 1990 provided a theoretical framework for the rise of innovation clusters, including in the biosciences. Economist Paul Romer published his seminal paper "Endogenous Technological Change" highlighting the role of knowledge and ideas in economic growth. Economist and mathematician Brian Arthur published a study titled " 'Silicon Valley' Locational Clusters: When Do Increasing Returns Imply Monopoly?" Arthur is a pioneer of the science of complexity and self-organization. He studies the economics of high technology and how businesses evolve, including within high-tech regions. In his study of Silicon Valley, Arthur found the cluster is characterized by enough diversity and is permeable enough for new entrants to offset any tendency toward monopoly of individuals firms within the cluster. The picture that emerges from the Valley's increasing returns, Arthur wrote, is a Darwinian one, "with geographical attractiveness bestowing 'selectional advantage.'" Once such an advantage is present, the process becomes a virtuous self-reinforcing circle: The local system of economic, social, and institutional interactions generates new rounds of innovation, investment, and development. In his later study of the nature of technology, what it is, and how it evolves, Arthur calls it the locational advantage of "deep craft," a kind of architectural knowledge or know-how based on practice, shared beliefs, and a vibrant subculture of hands-on experience. It is reflected in the fact that the initial buildout of new bodies of technology tends to occur in one country or region, or a few at most.

Path dependence, which is a kind of positive feedback in which the past exerts a strong influence on the present and future, is a characteristic feature of bioscience industries, as it is of economic and social institutions. It is perhaps a larger factor in cluster formation, equilibrium, and growth in the biosciences than it is for clusters based on the technologies because of the innate complexity of biology. The nine established life sciences clusters in the United States listed by Jones Lang LaSalle in its analysis—the San Francisco Bay Area, the Boston–Cambridge area, Los Angeles, New York/New Jersey, Philadelphia, Raleigh-Durham, San Diego, Seattle, and Washington, DC/suburban Washington—are all "coastal hubs" that "will forever play an important role as the headquarters cities for many of the industry's largest players." Emerging life sciences clusters in the United States and Canada share many features with establish clusters, including technologies, facilities, end products, workforce, investments, and incentives, but on a smaller scale.

Entrepreneurship provides the fuel for the emergence and ongoing vitality of technology clusters. Maryann Feldman and her colleagues studied the micro-dynamics of mostly small biotechnology clusters in Canada to see if entrepreneurial start-up companies tend to locate where the action is, that is, in clusters, or prefer to distance themselves from competitors or potential competitors located in clusters. They looked at increasing returns, social cohesion, and the location of new entrants in geographic and technological space. They found that firms in the same technological specialization have a strong tendency to concentrate geographically—in biotech clusters located from Vancouver to Newfoundland. In addition to benefiting from the increasing returns present in the cluster, entrepreneurial start-ups also benefit from the social cohesion of cluster and knowledge spillovers that occur when research and development is concentrated. Such factors tend to offset any concerns startups may have about expropriation of their knowledge by nearby competitors. The creation of a local knowledge pool around an area of technological specialization, which is what incumbent firms do, is key. The more innovative a firm is in that specialized space, the less likely it is to be lured away from an environment that helps to equip it with a competitive advantage. Specialization is an industry cluster's centripetal force, the pull to the center that keeps firms from ranging too far afield and thus losing the momentum that a timely technological niche can provide. The complexity of knowledge is also a factor. Complex knowledge systems, as in the case of biotechnology, require face-to-face interactions to assure the essential flow of information to innovate.

Indeed, a kind of centripetal force has been at play in human migration and settlement patterns since the Neolithic age. The Industrial Revolution accelerated the pull of people from rural areas into cities and the industrial districts that were typically located nearby, where natural resources and energy supplies could be readily accessed. Innovation and the growth of cities have been closely related ever since. Today, in the post-Industrial Age with more than half the world's population now living in urban regions, and with ongoing migration from rural to urban areas, especially in China, it is the city and its surroundings that constitute the crucibles of technological innovation. Urban density increases the returns to innovation at the same time that information technology increases its scope. "Humanity is a social species and our greatest gift is our ability to learn from one another," says Harvard economist Edward Glaeser who has been studying cities and innovation for more than two decades. Citing the growth economics work of Paul Romer and Nobel Laureate Robert Lucas, Glaeser and his colleagues wrote in "Growth in Cities" that if geographic proximity facilitates transmission of ideas, "then we should expect knowledge spillovers to be particularly important in cities. After all, intellectual breakthroughs must cross hallways and streets more easily than oceans and continents."

In the paper, published in 1992 just as urban cluster studies were taking off in the United States and Europe, Glaeser's research group challenged the idea that specialization was key to innovation in urban regions. Looking at the growth of large industries in 170 US cities between 1956 and 1987, Glaeser et al. found that local competition and urban variety, not regional specialization, encouraged employment growth in industries, a view consistent with that of urban theorist Jane Jacobs who saw industrial diversification as the wellspring of growth. Philip Cooke argues that it is neither *specialization* nor *diversification* alone that accounts for the emergence and growth of global bioregions but rather that opportunities for the swiftest innovation and growth occur in conditions of *proximate and related variety*, that is, evolutionary or iterative trial-and-error experimentation involving specialized but different technology platforms. Such platforms may be based on molecular biology, genetic engineering, genomics and bioinformatics, regenerative medicine and tissue engineering, synthetic biology, or others.

Cities and regional hubs thrive by enabling learning. They have become only more important as knowledge, including knowledge produced by the study of biology, has become more valuable in the marketplace. If cities are "giant Petri dishes, where creative types and entrepreneurs rub up against each other, combining and recombining to spark new ideas, new inventions, new businesses and new industries," as Richard Florida describes them, that is especially true for the biosciences. Cities with dense employment patterns and vigorous local competitive activity generate more patents per capita. In brief, the density of knowledge workers is correlated with increasing returns to scale in the innovative process. Economies of scale improve with population density just as they do in social insect communities like anthills and beehives. They become more than the sum of their parts, based on the mathematical properties of the multiple networks that become linked together.

Such factors of local and regional knowledge production are likely to become even more important for endogenous economic growth, particularly as economies absorb ever-more complex and complementary technologies into their operations. As Brian Arthur frames it, the economy is an expression of its technologies. They form its skeletal structure, supporting the various and sundry activities—the interaction of cells and tissues and bodily fluids—that together make up an organic social system, a knowledge-sharing ecosystem.

Science Parks Launched to Seed Clusters

New York City is one of the world's great financial, entertainment, and media centers. But it wants to be more. The city's new Alexandria Center for Life Science, built by Alexandria Real Estate Equities, is a state-of-the-art science park on Manhattan's East Side. The 310,000-square-foot, fifteen-floor facility is the first of three buildings to be developed as part of Alexandria's "life science cluster campus" in the most densely populated large city in the United States. The campus is designed to foster innovative collaborations and connections among the city's academic and medical institutions and scientific talent and provide access to investment funding for the life science industry.

A year after the grand opening of the center in 2010, Mayor Michael Bloomberg announced that a partnership between Cornell University in Ithaca, New York, and the Technion Institute in Israel had won the contest to build an engineering and applied science campus on a grant of land on Roosevelt Island in the city's East River together with $100 million for infrastructure improvements. The vision for the NYCTech Campus is for it to bolster job creation and generate 600 spinoff companies and $23 billion in economic activity over the next three decades. Within months, Bloomberg announced the establishment of an urban science and progress center in a vacant office building in downtown Brooklyn. The Polytechnic Institute of New York University, the City University of New York, Carnegie Mellon University in Pittsburgh, the University of Warwick in the United Kingdom, the University of Toronto, and the Indian Institute of Technology in Bombay are actively involved in the project, which will tackle city life design problems ranging from energy to transportation to public health. At the end of his tenure in 2013, Bloomberg presided at the opening of the New York Genome Center in downtown Manhattan.

A decade after the Nobel Laureate and then Memorial Sloan-Kettering Cancer Center president Harold Varmus published "The DNA of a New Industry" in *The New York Times* in which he lamented the city's poor ranking in biotechnology, the Big Apple was pouring money into the field and into research and innovation more broadly. Like the planners and developers of Edinburgh's BioQuarter and Singapore's Biopolis, the city's public officials, economic development authorities, and academic administrators were not about to stand back and let nature take its course. Because economically productive knowledge clusters are so coveted, they are not left to Adam Smith's "invisible hand" with the evolutionary dynamics of economic agents operating freely. Much to the dismay of libertarians and pure free-market devotees, they are supported and even planned and funded by state, regional, and local governmental bodies looking to tap into their scientific research base for future industries and jobs. After all, that is how Research Triangle Park (RTP) in North Carolina became an American industrial planning success story, a highly productive cluster of innovation. How did it happen, and what was the thinking of the leaders that made it happen?

In 1957, North Carolina Governor Luther Hodges appointed a committee to explore the possibility of a park to make North Carolina a magnet for research in space and technology. Economic development recruiters fanned out to gauge the interest of research-oriented companies of moving their operations to a prospective park that would be located in a triangle connecting the University of North Carolina at Chapel Hill, Duke University in Durham and what was then known as North Carolina State College in Raleigh, today North Carolina State University. As they traveled, they left behind brochures highlighting the research being done at these institutions. With the Soviet's successful launch of Sputnik in October 1957 and the federal government's response, the tide of history favored the project, and in 1959 it broke ground. By the time of its fiftieth anniversary in 2009, RTP was the largest and probably the best-known research park in the United States. The public–private park encompasses more than 7,000 acres of North Carolina pine forest within which some 150 organizations employ some 40,000 people with combined annual salaries amounting to nearly $3 billion. At least 80 percent of its organizations engage in R&D, and more than 93 percent of its employees work at those R&D organizations.

RTP's first-mover advantage may not serve regions looking at RTP as a model for development. Not that other regions are deterred. The Association of University Research Parks based in Reston, Virginia, listed more than 350 members for is twentieth anniversary in 2006 and estimates there are more than 700 science, research, and tech parks around the world in various stages of development. The International Association of Science Parks, based in Malaga, Spain, has more than 350 members representing science and technology parks and business incubators from seventy countries. These parks and incubators house 55,000 tenants. "Throughout the world, the biosciences rank prominently in governments' economic development agendas," wrote Andrea Rinaldi in a survey of science parks, incubators, and bioscience clusters undertaken before the global recession trimmed such investments, in some cases dramatically.

The fact that the number of private biotech companies in the United States and Europe has been basically flat (3,361 in 2007 and 3,294 in 2011) has led some to suggest that venture capital is being more evenly spread around the world. After an analysis of available data from China, Brazil, India, and South Africa from 2000 to 2012, Justin Chakma from the life sciences venture firm McNerney & Partners and two colleagues from the Wharton School found just modest private venture capital and private equity funding for innovative startups in the health-care sector ($1.7 billion in total during that period). That compares to $14 billion their governments and multinational and domestic pharmaceutical companies invest in R&D annually. They estimated total private life sciences venture capital and private equity investment in 2011 to be $572 million for China and $340 million for India compared to total capital deployed that year of $27.4 billion by China (eighty-nine funds) and $9.6 billion by India (thirty-five funds). Science parks in China (eg. Tsinghua) were more successful in attracting venture capital investment than those in India (Mumbai, Gujarat). The low level of venture capital and private equity investment compared to total capital deployed suggests that life sciences commercialization is a nascent activity in these emerging markets. Ernst & Young reported in an analysis published in 2012 that the global venture capital universe is shrinking in the number of firms and in biotech investments as a portion of overall dollar investments (12.2 percent in 2010 and 10.7 percent in 2011 in the United States and Europe). Yet venture capital industries in China and India are growing. The question is whether, and if so how soon, venture capital firms in these countries will turn to investing substantial sums in life sciences start-up companies. A start-up and scale-up culture is the foundation on which a cluster can emerge and flourish.

Most research parks and biotechnology incubators are established through direct or indirect public subsidies with help from investment funds. In his analysis (2006), Rinaldi found $241 million being spent for a park in Beijing, China; $175 million for a park in Klong Luang, Thailand; $140 million for a park in Hinjewadi, India; $50 million for a center of excellence in Astana, Kazakhstan; and untold millions for DuBiotech in Dubai, United Arab Emirates. As noted above, in their 2013 analysis Chakma et al. found that incubator science parks in China have fared better than those in India in terms of venture capital investment, perhaps owing to poor linkages in India between incubators and leading academic research centers and undercapitalization of biotech startups.

South Africa has a small biotechnology industry and no biotechnology parks. Domestic analysts advised its government to build them so that emerging clusters in the Western Cape and Gauteng regions "could become biotechnology hubs for the African continent." The Russian government got into the act in 2010 when then-president Dmitri Medvedev announced that a large technopark would be built from scratch on 150 acres of donated land near Moscow. The Skolkovo Innovation Center, pitched as Russia's version of Silicon Valley and funded largely through the government's modernization and innovation budget, will focus on energy, IT, telecommunications, biotechnology, and nuclear technology. General Electric, Siemens, IBM, and Microsoft have agreed to establish research centers at Skolkovo with Microsoft announcing it would invest in Skolkovo startups. Whether Skolkovo bears the fruits of innovation or ends up, as one Russian economist put it, "just like one of those science expositions we had in the Soviet Union era—designed to sit there and look pretty," time will tell.

Some of the research parks created over the past several decades are obvious success stories. Taiwan's Hsinchu Science Park, which opened in 1980, today is home to some 400 companies. But there have been some fairly high-profile failures as well. Technology policy conservatives are critical of biotech parks and particularly of public dollars to build them. They consider the global drive to build them a fad, and chasing fads is bad policy. Policymakers would be better off focusing on more mundane issues, beginning with the elimination of barriers to competition and making it easier for entrepreneurs to start businesses. In short, market forces should decide what innovation should occur and where it should occur. Let nature take its course through the creative juices of capitalism and the animal spirits of the entrepreneur. That way, the economic ecosystems that evolve have a better shot at long-term sustainability than if they are planned and kept running with government subsidies.

What does the future hold for such investment activities in which reliable return-on-investment metrics and standards have not been developed, much less applied? Many of the megaparks being built in developing nations such as China, India, and some Persian Gulf states appear to be based on old models, says Anthony

Townsend of the Institute for the Future, a Silicon Valley-based think tank. Many of them "are just big real estate deals" rather than investments in people and innovation. The Institute for the Future and the Research Triangle Foundation released a scenario planning report in conjunction with an International Association of Science Parks conference celebrating the fiftieth anniversary of RTP in 2009. The model of self-contained research parks and incubators that dominated the last fifty years of technology-based economic development is being challenged by "deep shifts in the global economy, science and technology, and models of innovation," wrote the authors of "Future Knowledge Ecosystems," one of which was Townsend. The rise of what they call "research clouds" and "biology by design" may eventually put place-anchored innovation embodied in research parks at risk.

Such views are still in the minority. New science parks have been sprouting across the United States and around the world. "Every city and state with a university wants to jump on this bandwagon," said a leading developer. The moment of truth for conventional science parks and expensive high-tech meccas came with the disappearance of easy capital following the collapse of the global financial system, the recession that followed, and the protracted slow-growth period the recession ushered in. Questions about the return on investment of what one Massachusetts Institute of Technology urban planner dubbed "new century cities" like Tsukuba Science City near Tokyo, Spain's 22@Barcelona project, Seoul, Korea's Digital Media City, Snowpolis in Vuokatti, Finland, and Singapore's Biopolis were suddenly cast in a new light, a light of tighter money. "When legislators are choosing between cutting off funding for health care for low-income children as opposed to cutting funding for science parks, in most cases the children are going to win—and they probably should," says the Milken Institute's Ross DeVol. By 2011 the International Association of Science Parks found in a poll of members that science park funding had stabilized or increased and that most science parks around the world were growing again.

The planners of RTP wanted to address that generation's challenge of a mobile workforce—the "brain drain" of educated workers from the American South to other parts of the country. Today states and regions are faced with the mobility of talent but also the mobility of knowledge, the digitally transmissible forms of it. The first step in cultivating a regional knowledge ecosystem is to find out what one's "know-how" assets are, what tacit knowledge and core competencies, for example, are locked up in local manufacturing and technology firms. The Institute for the Future's Townsend sees coming what he calls the "therapeutic city," urban space shaped by technology, health services, and aging populations in advanced economies. "Therapeutic cities will form the core of new innovation clusters, where transdisciplinary teaching, research, and treatment come together" because biomedical innovation requires close and frequent contact with patients, the "'early adopters'" of biopharmaceutical drugs and biomedical technologies.

Whether carefully planned strategies for building regional advantage through science or research parks can achieve unqualified success, the experience of RTP suggests that they cannot achieve it overnight, nor can they achieve it purely as real-estate ventures or marketing campaigns with catchy names beginning with "Silicon" or "Bio." Successful science parks take wise investment, lots of time, and plenty of luck. They require the foresight to imagine where research, technology, markets, and human capital will be decades down the road. When such strategies work well, they can contribute a great deal to innovation, including the luring of firms as anchor tenants to help spawn startups, seed clusters, and establish networks for global interactions.

International partnerships matter as well. Matching up local firms or researchers with international counterparts performing complementary work can benefit geographically distant institutions and regions. The fact that China has declared bioscience an area of strategic importance is a positive development for bioregions in North America and Europe because it invites them to look beyond local markets for scientific exchange, joint ventures, talent recruitment, and other opportunities, thereby expanding their own knowledge pool. In brief, successful regional innovation means reaping the harvest of the local *and* the global.

Megacenters, Megacities, Money, and Cross-Border Collaboration

How many bioscience parks end up as success stories like RTP and how many succeed in seeding bioscience clusters won't be known for decades. It is hard to imagine that the United States will dominate the field the way it has since the biotechnology industry was born in California in the 1970s. Asia and Latin America as well as Europe and Australia are investing in the field and collaborating through cross-border and transoceanic consortia. Yet US leadership remains intact because of its leading universities and its deeply rooted knowledge centers that attract highly educated foreign-born scientists, engineers, and others to the country. Biological knowledge with economic potential is increasingly concentrated in "megacenters" of research and industry, says Philip Cooke. These megacenters go beyond competitive business clusters championed by Michael Porter. The megacenters envisioned by Cooke capture the knowledge value chain from exploration through examination to the exploitation of new knowledge. Bioscience megacenters, with the Boston–Cambridge region the prime example, possess leading international universities and associated research institutions, large hospitals, and numerous clinics and laboratories at their core, plus dense linkages among institutions and both established and start-up firms. Basel-based Novartis moved its R&D headquarters to the Boston–Cambridge hub in 2002. Paris-based Sanofi announced plans to expand its R&D presence there in 2013, joining Novartis, Merck, Pfizer, AstraZeneca, and Takeda. Whether expanding in such megacenters constitutes an innovation imperative for big pharma or merely reflects a "me too" response to their R&D productivity problems will not be known anytime soon. Whether the innovation these megacenters foster comes at the price of continued health care cost inflation or whether it will spur the development and diffusion of relatively low-cost, high-impact technologies also will not be known anytime soon.

Meanwhile, life sciences industries in the food, materials, energy, green chemistry, and environmental remediation sectors are evolving in their own right. Their interaction with each other and with health care could lead to the emergence of what Cooke calls "platform regions" that draw on the strength of converging technologies in their industrial base. The strength of the St. Louis region in the plant sciences and genomics research coupled with local research programs in biofuels and funding by the US Department of Energy is a case in point. The agency awarded $44 million in 2010 to a national biofuels and bioproducts consortium led by the Donald Danforth Plant Science Center, which bills itself as the world's largest independent plant research center.

Where bioscience industries grow is not determined solely by government research funding and capital investment, though they figure heavily in the viability of established and emerging clusters. No industry is as dependent on capital markets as biotechnology, at least based on the model for funding biotech startups in the biopharmaceutical field for the past three decades. Investment sources, which are increasingly global, vary from the familiar ones—venture capital, angel investment, public appropriations and subsidies—to large and powerful pools of capital such as pension funds, hedge funds, private equity funds, and petrodollars (e.g., Dubai's DuBiotech, Kazakhstan's National Center for Biotechnology). Government bonds entered the investment picture in a big way in the United States with California's $3 billion Proposition 71 bonding initiative passed by voters in 2004 to support stem cell research. Beneath the current of cross-border science and bioindustry collaboration lies a global financial system that is certain to have a lot to say about where the industry will reside in the future and who will own what parts of it.

Cross-border investment has taken on new features in recent years with sovereign wealth funds, which reflect the accumulation of capital from domestic production and export, mainly of oil. Estimated to be $5.4 trillion in 2013, sovereign wealth funds can serve established and emerging domestic industries, including industries associated with expanding health services for populations in need of maternal and infant care, infectious and chronic disease treatment, and geriatric care. China, Russia, and other state capitalist countries with sovereign wealth funds at their disposal are currently using them to invest in physical infrastructure projects like roads, railways, and dams, with considerable success. Their investments directly in innovation through science and technology parks like Moscow's Skolkovo Innovation Center and Beijing's Zhongguancun Science Park are far more speculative. "Governments are good at providing the seed corn for innovation," opined the *Economist*, "but they are bad at turning seed corn into bread." As an example it cited Malaysia's $150 million BioValley, which opened in 2005 and has since earned the moniker "Valley of the Bio Ghosts."

Many factors are at play in Asia's growing prominence as an R&D location for pharmaceutical and bioscience firms headquartered in the West. The large supply of skilled professionals at relatively low costs in India and China is a major attraction, particularly in the context of their public investments in science and technology. New and improved research laboratories are part of the Chinese government's fifteen-year plan to build basic education in biology and invest in strategic areas like stem cell and regenerative medicine. They are designed to lure home the "sea turtles," repatriated Chinese scientists who have pursued education and worked in Australia, North America, and Europe. Given its success in building the Biopolis science park into a gleaming global center for life scientific research and bioindustry, a $3.5 billion investment (Figure 3.2), Singapore can be seen as a fulcrum for India–China cooperation in the fields of pharmaceuticals and biotechnology. It constitutes the heart of a ScanBalt BioRegion-like metacluster, a regional hub where vast pools of talent from India and China interact, taking advantage of the large markets of Southeast Asia, South Asia, and China.

It is not the "death of distance" so much as the strategic connection of knowledge workers and knowledge centers with each other that is altering the competitive landscape around the world. More than that, it is a network or universe of like-minded clusters to form a matrix or global metacluster. Such connections are expanded and accelerated by new communications technologies. Communications devices and networks help to weave regional learning webs as well as worldwide webs. In a special issue on the internationalization of science and regional collaboration, *Nature* reported that China's rapid growth since 2000 "is leading to closer research collaboration with Japan (up fourfold since 1999), Taiwan (up eightfold), South Korea (up tenfold), Australia (more than tenfold) and with every other research-active country in the Asia-Pacific region." The rapid expansion of each nation's research base and regional links, driven by relatively strong economies investing in innovation, can be expected to produce a highly skilled regional labor force by 2020.

The growth of regional collaboration has many implications, among them the fact that it amplifies the development of emergent research economies. Researchers in Asia will no longer need recognition from European and US authors if their research is being cited and used by partners within the region. That translates into Asian students seeking attractive opportunities closer to home rather than traveling to Europe or the United States. Singapore, for example, is an attractive venue for students from all over Southeast Asia. Students from China, India, and other countries in the Association of Southeast Asian Nations now number around 11,000 full-time students enrolled in Singapore's colleges and universities, some 20 percent of the total, with the prospect of finding a job in the city-state upon graduation quite good.

Such concepts as collaborative networks and metaclusters among developing countries are themselves at an embryonic stage of development, but emerging markets are growing, if unevenly, and their governments have turned increasingly to higher education and research as the foundation for future prosperity. Moreover, there is plenty of evidence that Bangalore, Singapore, Hong Kong, Shanghai,

FIGURE 3.2 Singapore's Biopolis, Southeast Asia's gleaming multi-billion dollar biomedical sciences hub, what one observer called a "life-sciences blockbuster cluster."
Source: Isaac Ow, 2009, with permission.

Seoul, and other East Asian megacities are setting the pace when it comes to integrating the process of discovery and innovation with the exploding digital world. It is perhaps not an exaggeration to assert, as *Nature* does, that new regional networks of emerging research economies are changing the global balance of research activity.

Since 2000, Singapore has spent nearly $20 billion to shore up its research infrastructure and create a talent pool to attract private investment. The country's biomedical manufacturing output more than tripled from about $5 billion in 2000 to more than $16.2 billion in 2009. The biomedical sector now represents 10 percent of the country's manufacturing output, up from 4 percent in 2000. Most big pharma companies have a presence in Singapore. With Syngenta opening an R&D facility in 2010, agricultural biotechnology is also building strength in the city-state, which experienced 5 percent average annual growth from 2007 to 2012.

Three-fourths of Singapore's resident population is ethnic Chinese. Sino-Singaporean trade, cross-border investment, and student exchange are on the upswing. The China-Singapore Joint Council for Bilateral Cooperation, responsible for promoting Sino-Singaporean cooperation, has set up councils to support science and technology innovation and regional development around China's numerous megacities. Contract workers from China helped to build Singapore's biomedical infrastructure. Now it is China's turn. With its "Thousand Talents" program designed to lure home its sea turtles with financial incentives and its major expenditures in science, technology, and higher education, China is seeding its future. Observers say it may take a decade or more for the seeding to bear fruit but also recommend that US life science industries create their own programs to keep foreign-born talent in the country. China's cooperation with Singapore, a place of exceptional dynamism whose government is dominated by China's ethnic cousins, makes the prospect of a competitive domestic bioindustry in China realistic despite state centralization and the country's cultural and educational impediments to innovation in the Western style.

Bioscience Innovation Confronts the Uncertainty Principle

The regional concentration and clustering of ideas, technology, and talent is part and parcel of what Brian Arthur calls the future "generative economy." Nothing else in the world generates and regenerates the way living things do, whether they be microbes, plants, animals, or ecosystems. Intricate knowledge about the processes of life holds enormous potential for energy, food, and biomaterials production as well as in health care and sustainable environmental policies and practices. Thus our understanding of how such knowledge is generated and applied is fundamental to our future economic well-being.

The rise of innovation hubs, research megacenters, and cross-border collaborations will shape the competitive economic geography of the century of biology just as the factory system shaped the emerging industrial landscape two centuries ago. But exactly how no one knows because economists still do not understand everything at work at the intersection of entrepreneurship, innovation, and geography. When we move beyond broad policies for promoting entrepreneurship that span taxes, education, immigration, technology and standards, financial markets, and patent laws toward specific strategies like clustering, "our ignorance becomes obvious." So concluded Edward Glaeser and colleagues Aaron Chatterji and William Kerr in their study of clusters of entrepreneurship and innovation. To understand clustering will require "experimentation and evaluation" borrowing from the design and rigor of drug trials.

Contemporary technology clusters and regional innovation ecosystems, like the industrial districts spawned by the Industrial Revolution, have variable lifespans. Bioscience clusters, particularly those that have emerged organically, have tended to be more stable than those based on other technologies. They possess a developed infrastructure; an essential proximity to universities, research centers, and hospitals; and premium human capital that includes PhD scientists and engineers, physician scientists, clinicians, and skilled laboratory technicians. Biology is highly complex, making efforts to apply it to solve problems invariably a collaborative affair requiring direct interaction of people all along the value chain.

Location quotients, which measure the job concentration and strength of an industry in a region versus the nation as a whole, show that biotechnology R&D is highly concentrated in a handful of states in the United States (Figure 3.3).

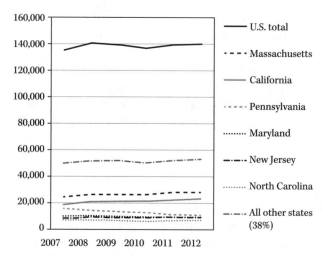

FIGURE 3.3 State-based localization of biotech R&D in the United States, 2007–2012. Data from the Location Quotient Calculator, U.S. Bureau of Labor Statistics, NAICS code 541711: Biotechnology research and development. More than 60 percent of biotech R&D employment is concentrated in Massachusetts, California, Pennsylvania, Maryland, New Jersey, and North Carolina, especially within specific counties in these states.

Source: The authors.

Among them are Massachusetts, California, Pennsylvania, Maryland, New Jersey, and North Carolina. Typically biotechnology R&D employment is concentrated in certain counties: Middlesex and Suffolk counties in Massachusetts, Montgomery counties in Pennsylvania and Maryland, Durham County in North Carolina, Somerset and Mercer counties in New Jersey, and San Diego and San Mateo counties in California. Cluster studies over the past two decades have tried to define the urban–regional locus of innovation in the field using a variety of metrics, with decidedly mixed success. "Will somebody please come out with a deeply researched and credible report that ranks U.S. regional biotech clusters on the criteria that matter the most?" implored Luke Timmerman, the BioBeat columnist for Xconomy. com. Those criteria in his view should include the number of public and private life sciences companies, NIH funding, R&D spending at public life sciences companies, the number of patents issued per capita, total life sciences employment, total venture capital dollars invested, and venture capital allocated for early stage/seed investments. Cluster criteria should also measure the number of life science startups formed each year, the scientific citation rates for local research institutions, and the number of FDA-approved drugs, devices, and diagnostics discovered in the region.

What happens to a cluster when its anchor tenant is purchased by big pharma? What happens when a generative force for startups is suddenly managed from afar? Washington, DC's MedImmune was acquired by AstraZeneca (2007), the Bay Area's Genentech by Roche (2009), Boston's Genzyme by Sanofi-Aventis (2011),

Rockville, Maryland's Human Genome Sciences by GlaxoSmithKline (2012), and San Diego's Amylin Pharmaceuticals by Bristol-Myers Squibb (2012). The answer is that it is too soon to know, but it cannot be a good thing if the training ground for would-be entrepreneurs looking to launch a local company is lost.

Another dimension to the uncertainty in the field is how anticipated future cuts in appropriations for federal funding agencies like the NIH and National Science Foundation and public research universities would affect regional centers of bioscience innovation. Again, it is too soon to know but not too soon to imagine what would happen. Severe cuts in federal research funding would jeopardize the productivity of innovation clusters and spell trouble for some state and regional initiatives intended to translate the fruits of basic research into effective therapies, valuable products and services, and well-paying jobs for a twenty-first century workforce. Studies by the Milken Institute show that NIH funding has been a critical component in the emergence and long-term prosperity of bioscience clusters in states and metropolitan regions. It does so through the direct effect of an increase in economic output and through indirect effects such as an increase in employment and private R&D. Given federal budgetary constraints well into the future, it may not be possible to entertain the idea of government funding for creation of a genomics cluster—a Silicon Valley for the emerging life sciences—which some call for. Yet federal support for basic research and targeted technology development has no substitute. Concerning the NIH budget "we're looking at pretty tight times ahead," said Arthur Levinson, chairman of Apple, former Genentech chief executive and chairman of the life sciences Breakthrough Prize Foundation he set up along with Facebook founder Mark Zuckerberg and other Silicon Valley icons. The foundation awarded eleven biomedical scientists $3 million each in 2013, its inaugural year. Nearly all of them work at universities located in established or emerging clusters of biomedical innovation.

To illustrate the division of labor, specialization, and the manufacturing process in *The Wealth of Nations*, Adam Smith told the story of a pin factory. His metaphor served well through the rise of steam power, railroads, electricity, oil, automobiles, aviation, mass production, jet engines, and into the age of information technology. Today, as biological engineering and biomanufacturing enter the economic lexicon, the commercial DNA foundry helps us to imagine the future of bioautomation. In a DNA foundry, machines take orders over the Internet, synthesize nucleic acids from hundreds to thousands of letters long in precise order, perform quality control, and package the order for shipment, with minimal human involvement.

Synthesizing DNA is a largely automated process. Reengineering cellular production pathways is not, at least not yet. The Defense Advanced Research Projects Agency, the Department of Defense's research arm that brought us the Internet, Stealth aircraft, and GPS technologies, launched a $30 million research program to jump-start the creation of cellular factories in 2011. The agency's Living Foundries program envisions using cells as factories for producing useful products such as enzymes, lubricants, coatings, polymers, fibers, and biosensors through the manipulation of their component parts. Funded investigators will use the tools of synthetic biology to turn living cells into manufacturing platforms, the economic endgame following decades of federal support for basic research on the mechanism of the cell. One of the projects funded in the first round was Craig Venter's "biological teleporter," a machine that converts digital-biological information transmitted in electromagnetic waves into proteins, viruses, and single microbial cells.

We are now at a juncture in the human story where the DNA that constitutes the genes that program the cells that build the brain circuits we employed to create farming tools in Neolithic settlements is just another manufactured commodity. For their part, cells and their components are slated to become production centers for goods and services we want: medicines, vaccines, regenerated tissues, micronutrients, biofuels, biosensors, bioluminescence, data storage, thin films, carbon capture, anti-aging cosmetics, advanced forensic technologies, and many others. In less than three centuries, we have journeyed from pin factories to industrial districts to dynamic technology clusters to DNA and cellular foundries. Fast-moving biological technologies will continue to shape post-industrial innovation and production in unforeseen ways. Where that innovation and production takes place is very likely to be wherever talent, knowledge, organizations, resources, and nurturing public policies come together to create, as Alfred Marshall famously put it, a certain something "in the air."

References

Listed sequentially based on their order in the chapter.

Matt Ridley, *The Rational Optimist: How Prosperity Evolves* (New York: HarperCollins, 2010).

Alfred W. Crosby, *The Columbian Exchange: Biological and Cultural Consequences of 1492* (Westport, CT: Greenwood Press, 1973).

Charles C. Mann, *1493: Uncovering the New World Columbus Created* (New York: Knopf, 2011).

Michael Samways, "Translocating Fauna to Foreign Lands: Here Comes the Homogenocene," *Journal of Insect Conservation* 3.2 (1999):65–66, accessed August 31, 2012, doi:10.1023/A:1017267807870.

Adam Smith, *An Inquiry into the Nature and Causes of the Wealth of Nations* (London: W. Strahan and T. Cadell, 1776).

Nathan Nunn and Nancy Qian, "The Potato's Contribution to Population and Urbanization: Evidence from a Historical Experiment," *The Quarterly Journal of Economics* 126 (2011):593–650, accessed August 15, 2012, doi: 10.1093/qje/qjr009.

Niall Ferguson, *The Ascent of Money: A Financial History of the World* (New York: Penguin Press, 2008).

Aytoun Ellis, *Penny Universities: A History of the Coffee-Houses* (London: Secker &Warburg, 1956).

Mark Pendergrast, *Uncommon Grounds: The History of Coffee and How It Transformed Our World* (New York: Basic Books, 1999).

Eliakim Little and Robert S. Little, *Littell's Living Age*, 6th series, Vol. VII. (Boston: Littell & Co., 1895).

Larry Stewart, "Putting on Airs: Science, Medicine and Polity in the Late Eighteenth Century," in *Discussing Chemistry and Steam: The Minutes of a Coffee House Philosophical Society 1780-1787*, Trevor H. Levere and Gerald L. E. Turner, eds. (New York: Oxford University Press, 2002), 207–255.

Jonathan Hersh and Hans-Joachim Voth, "Sweet Diversity: Colonial Goods and the Rise of European Living Standards after 1492," July 17, 2009, accessed August 14, 2012, http://dx.doi.org/10.2139/ssrn.1402322.

ScanBalt Website, accessed August 14, 2012, http://www.scanbalt.org.

"BIO '06 Visible from Space?" Editorial, *Nature Biotechnology* 24.5 (2006):474.

Genome Resources Website. National Center for Biotechnology Information, National Institutes of Health, accessed August 19, 2012, http://www.ncbi.nlm.nih.gov/genome.

"Sequenced Plant Genomes," CoGePedia: Accelerating Comparative Genomics, accessed August 14, 2012, http://genomevolution.org/wiki/index.php/Sequenced_plant_genomes.

"List of Sequenced Animal Genomes," Wikipedia, accessed August 14, 2012, http://en.wikipedia.org/wiki/List_of_sequenced_animal_genomes.

Bastiaan Star et al., "The Genome Sequence of the Atlantic Cod Genome Reveals a Unique Immune System," *Nature* 477 (2011), accessed August 14, 2012, doi: 10.1038/nature10342.

Dennis A. Benson et al., "GenBank," *Nucleic Acids Research* (2013) 41(D1):D36–D42.

G. Steven Burrill, "The Global Transformation: The International Nature of Research and Development is Changing the Life Sciences in Radical Ways," *The Burrill Report,* May 16, 2007, accessed September 16, 2012, http://www.burrillreport.com/printer_article-198.html.

Brady Huggett, "Innovative Startups 2012," *Nature Biotechnology* 31 (2013):194.

Maryann Feldman, "Place Matters," ScienceProgress.org, January 21, 2009, accessed August 31, 2012, http://www.scienceprogress.org/2009/01/place-matters.

Jared Diamond, *Guns, Germs and Steel: The Fates of Human Societies* (New York: Norton, 1997), 259.

Local Knowledge and the Location Paradox

Michael E. Porter, *The Competitive Advantage of Nations* (New York: Free Press, 1990).

"Q&A with Michael Porter." *BusinessWeek,* August 21, 2006, accessed April 5, 2013, http://www.businessweek.com/stories/2006-08-20/online-extra-q-and-a-with-michael-porter.

Richard Florida, "The World Is Spiky," *Atlantic Monthly* (October 2005):48–51.

Rana Foroohar, "The Economy's New Rules: Go Glocal," *Time,* August 20, 2012.

Lori Montgomery, "Siemens Plant in Charlotte Offers Lessons as Obama, Romney Talk Job Creation," *Washington Post*, September 4, 2012.

Alfred Marshall, *Principles of Economics* (London: Macmillan and Co., 1890).

Michael Polanyi, *Personal Knowledge: Towards a Post-Critical Philosophy* (Chicago: University of Chicago Press, 1962).

Martha Lagace, "Leatherbee Lecture: Seely Brown—The Crafts of Super Innovation," *HBS Working Knowledge,* April 30, 2001, accessed September 16, 2012, http://hbswk.hbs.edu/archive/2212.html.

Stephen Pinch et al., "From 'Industrial Districts' to 'Knowledge Clusters': A Model of Knowledge Dissemination and Competitive Advantage in Industrial Agglomerations," *Journal of Economic Geography* 3.4 (2003):373–388.

Philip Cooke, *Growth Cultures: The Global Bioeconomy and its Bioregions* (London: Routledge, 2007).

James Surowiecki, "If You Can Make It Here," *New Yorker* (October 22, 2007):64.

Paula E. Stephan and David B. Audretsch, "Company–Scientist Locational Links: The Case of Biotechnology," *American Economic Review* 86.3 (1996):641–652.

Lynn G. Zucker, Michael R. Darby, and Marilyn B. Brewer, "Intellectual Human Capital and the Birth of U.S. Biotechnology Enterprises," *American Economic Review* 88.1 (1998):290–306.

Walter W. Powell et al., "The Spatial Clustering of Science and Capital: Accounting for Biotech Firm-Venture Capital Relationships," *Regional Studies* 36 (2002):291–305.

Jason Owen-Smith and Walter W. Powell, "Knowledge Networks as Channels and Conduits: The Effects of Spillovers in the Boston Biotechnology Community," *Organization Science* 15 (2004):5–21.

Walter W. Powell et al., "Network Dynamics and Field Evolution: The Growth of Interorganizational Collaboration in the Life Sciences," *American Journal of Sociology* 110 (2005):1132–1205.

Philip E. Cooke, "Global Bioregional Networks: A New Economic Geography of Bioscientific Knowledge," *European Planning Studies* 14 (2006):1265–1285.

Joel Kotkin, "The Declustering of America," *Wall Street Journal,* August 15, 2002.

Heidi Ledford, "Ahead of the Pack," *Nature* 459 (2009):286–287.

Edinburgh: From the "Science of Man" to the Science of Life

Joel Mokyr, *The Gifts of Athena: Historical Origins of the Knowledge Economy* (Princeton, NJ: Princeton University Press, 2002).

David Wessel, "Professor Romer Goes to Washington," *Wall Street Journal,* January 25, 2001.

Paul Krugman, "A Bit More on Malthus," Conscience of a Liberal blog, *The New York Times,* July 1, 2009, accessed May 15, 2013, http://krugman.blogs.nytimes.com/2009/07/01/a-bit-more-on-malthus.

James Buchan, *Crowded with Genius: The Scottish Enlightenment: Edinburgh's Moment of the Mind* (New York: HarperCollins, 2003).

Smith, *An Inquiry into the Nature and Causes of the Wealth of Nations.*

James Buchan, *The Authentic Adam Smith: His Life and Ideas* (New York: W.W. Norton, 2006).

Ridley, *The Rational Optimist.*

Michael Porter, "The Adam Smith Address: Location, Clusters and the 'New' Microeconomics of Competition," *Business Economics* 33.1 (1998):7–13.

John Hodgson, "Ten Years of Biotech Gaffes," *Nature Biotechnology* 24 (2006):270–273, accessed March 19, 2013, doi: 10.1038/nbt0306-270.

Dugald Stewart, *The Collected Works of Dugald Stewart: Biographical memoirs of Adam Smith* (Edinburgh: T&T Clark, 1877).

Andrew Bolger, "Life Sciences: A Fertile Environment for Culture to Develop," *Financial Times,* September 20, 2007.

Neil McInnes, "Leading US Life Science Property Specialist Appointed to Drive Scotland's Bioscience Industry," *Scottish Enterprise,* May 7, 2007.

Alexandria Real Estate Equities, Inc. Website, accessed February 2, 2010, http://www.lab-space.com.

Jonathan Brown, "In Memory of Her Mother, J. K. Rowling's £10m for MS," *The Independent,* September 1, 2012.

Stephen Vass, "Builder Cuts Back Bioquarter Option," *The Herald* (Scotland), August 12, 2012.

Smith, *An Inquiry into the Nature and Causes of the Wealth of Nations.*

Will W. Shu and Jason King, "Organ Printing from Stem Cells," *Genetic Engineering & Biotechnology News* 32.13 (July 1, 2012), accessed August 23, 2012, http://www.genengnews.com/gen-articles/organ-printing-from-stem-cells/4171/.

Geography, Deep Craft, and Increasing Returns

"Global Life Sciences Cluster Report, 2011," Jones Lang LaSalle, November 28, 2011, accessed August 24, 2012, http://www.joneslanglasalle.com.

Smith, *An Inquiry into the Nature and Causes of the Wealth of Nations.*

Paul Krugman, "What's New About the New Economic Geography?" *Oxford Review of Economic Policy* 14.2 (1998):7–17.

Paul Romer, "Endogenous Technological Change," *Journal of Political Economy* 98.5 (1990): S71–S102.

W. Brian Arthur, "'Silicon Valley' Locational Clusters: When Do Increasing Returns Imply Monopoly?" *Mathematical Social Sciences* 19.3 (1990):235–251.

W. Brian Arthur, *The Nature of Technology: What It Is and How It Evolves* (New York: Free Press, 2009).

Barak S. Aharonson, Joel A. C. Baum, and Maryann P. Feldman, "Desperately Seeking Spillovers? Increasing Returns, Industrial Organization and the Location of New Entrants in Geographic and Technological Space," *Industrial and Corporate Change* 16 (2007):89–130.

Edward L. Glaeser, Hedi Kallal, Jose A. Scheinkman, and Andrei Shleifer, "Growth in Cities," *Journal of Political Economy* 100.6 (1992):1126–1152.

Philip Cooke, *Growth Cultures: The Global Bioeconomy and its Bioregions* (London: Routledge, 2007).

Richard Florida, "The Joys of Urban Tech," *Wall Street Journal,* August 31, 2012.

Gerald A. Carlino, Satajit Chatterjee, and Robert M. Hunt, "Urban Density and the Rate of Invention," *Journal of Urban Economics* 61 (2007):389–419.

Luis Bettencourt and Geoffrey A. West, "A Unified Theory of Urban Living," *Nature* 467 (2010):912–913.

Arthur, *The Nature of Technology.*

Science Parks Launched to Seed Clusters

Oliver Staley and Henry Goldman, "Cornell, Technion Are Chosen by New York City to Create Engineering Campus," *Bloomberg News,* December 19, 2012.

"Urban Science Centre Born in the Big Apple," *Nature News* blog, April 25, 2012, accessed August 27, 2012, http://blogs.nature.com/news/2012/04/urban-science-centre-born-in-the-big-apple.html.

"NY Genome Center Marks Its Formal Launch," *Genetic Engineering & Biotechnology News,* September 19, 2013, accessed September 19, 2013, http://www.genengnews.com/gen-news-highlights/ny-genome-center-marks-its-formal-launch/81248869/.

Harold Varmus, "The DNA of a New Industry," *The New York Times,* September 24, 2002.

Research Triangle Park Website, accessed February 2, 2010, http://www.rtp.org.

John W. Hardin, "North Carolina's Research Triangle Park: Overview, History, Success Factors and Lessons Learned," in *Pathways to High-Tech Valleys and Research Triangles: Innovative Entrepreneurship, Knowledge Transfer and Cluster Formation in Europe and the United States,* Willem Hulsink and J. J. M. Dons, eds. Wageningen UR Frontis Series, Vol. 24 (Dordrecht, The Netherlands: Springer, 2008), 27–51.

Association of University Research Parks Website, accessed February 2, 2012, http://www.aurp.net.

International Association of Science Parks Website, accessed February 2, 2010, http://www.iasp.ws.

Andrea Rinaldi, "More Than the Sum of their Parts," *EMBO Reports* 7.2 (2006):133–136.

Huggett, "Innovative Startups 2012."

Justin Chakma, Stephen M. Sammul, and Ajay Agrawal, "Life Sciences Venture Capital in Emerging Markets," *Nature Biotechnology* 31.3 (2013):195–201, doi: 10.1038/nbt.2529.

"Globalizing Venture Capital: Global Venture Capital Insights and Trends Report 2011," Ernst & Young, April 17, 2012, accessed May 6, 2013, http://www.ey.com.

Nirvana S. Pillay and Ramazan Uctu, "A Snapshot of the Successful Bio-Clusters Around the World: Lessons for South African Biotechnology," *Journal of Commercial Biotechnology* 19.1 (2013):40–52, accessed May 14, 2013, doi: 10.5912/jcb.544.

Charles Clover, "Moscow Restarts the Machine," *Financial Times,* December 1, 2010.

Pete Engardio, "Research Parks for the Knowledge Economy," *Business Week,* June 15, 2009.

Heidi Ledford, "Research Parks Feel the Economic Pinch," *Nature* 459 (2009):896–897, accessed September 16, 2012, doi:10.1038/459896a.

Nitin Dahad, "Science Parks and Innovation Clusters Grow in 2011," *The Next Silicon Valley,* January 15, 2012, accessed August 28, 2012, http://thenextsiliconvalley.wordpress.com/category/publishers-corner.

Aida Nurutdinova, "Future Research of Science Parks and Incubators: Overall Analysis," *International Journal of Advanced Studies* 2.1 (2012), accessed August 28, 2012, http://www.ijournal-as.com.

Anthony Townsend, Alex Soojung-Kim Pang, and Rick Weddie, "Future Knowledge Ecosystems: The Next Twenty Years of Technology-Led Economic Development," Institute for the Future, accessed February 2, 2010, http://www.iftf.org.

Anthony Townsend, "The Rise of the Therapeutic City," in HC 20/20 Perspectives, February 2011, Institute for the Future, accessed February 2, 2010, http://www.iftf.org.

Andreas B. Eisingerich and Leslie Boehm, "Group Analysis: Why Some Regional Clusters Work Better than Others," *Wall Street Journal,* September 17, 2007.

Megacenters, Megacities, Money, and Cross-Border Collaboration

Philip Cooke, "Rational Drug Design, the Knowledge Value Chain and Bioscience Megacentres," *Cambridge Journal of Economics* 29 (2005):325–341.

Cooke, *Growth Cultures.*

"View from the Debate over Clustering," Interview with Philip Cooke, European Industrial Research Management Association. EIMRA's *Innovation Quarterly* 13 (Spring 2008), accessed August 30, 2012, http://www.eirma.org/eiq/013/pages/eiq-2008-013-0012.html.

Kelsey Volkmann, "Danforth-led Algae Biofuel Research Wins $44M in Stimulus," *St. Louis Business Journal,* January 13, 2010.

"Mixed Bag: SOEs Are Good at Infrastructure Projects, Not so Good at Innovation," *The Economist,* January 21, 2012, accessed January 2, 2013, http://www.economist.com/node/21542929.

Jayan Jose Thomas, "Knowledge Economies in India, China, and Singapore: Issues and Prospects: Case Studies of Pharmaceuticals and Biotechnology," Institute of South Asian Studies. ISAS Working Paper No. 18, January 2007.

Jonathan Adams, "Collaborations: The Rise of Research Networks," *Nature* 490 (2012):335–336.

Gunja Sinha, "Singapore Injects $12.5 Billion," *Nature Biotechnology* 28.12 (2010):1229.

Erik Lundh, "Assessing the Impact of China's Thousand Talents Program on Life Sciences Innovation," *Nature Biotechnology* 29 (2011):547–548.

Bioscience Innovation Confronts the Uncertainty Principle

Arthur, *The Nature of Technology.*

Aaron Chatterji, Edward Glaeser, and William Kerr, "Clusters of Entrepreneurship and Innovation," NBER Working Paper Series No. 19013, issued May 2013, accessed June 9, 2013, http://www.nber.org/papers/w19013.

Location Quotient Calculator, US Bureau of Labor Statistics. NAICS code 541711, Research and Development in Biotechnology, accessed September 2, 2012, http://data.bls.gov/location_quotient/ControllerServlet.

Luke Timmerman, "Someone Needs to Rank U.S. Biotech Hubs, For Real," Xconomy.com, March 4, 2013, accessed March 5, 2013, http://www.xconomy.com.

Luke Timmerman, "U.S. Biotech Clusters Are Losing their Anchor Tenants, and It Hurts," Xconomy.com, August 6, 2012, accessed August 23, 2012, http://www.xconomy.com.

Alex Philippidis, "Fearing Anchors Away After M&A," *Genetic Engineering & Biotechnology News,* August 23, 2012, accessed August 23, 2012, http://www.genengnews.com.

Anusuya Chatterjee and Ross DeVol, "Estimating Long-term Economic Returns of NIH Funding on Output in the Biosciences," The Milken Institute, August 31, 2012, accessed January 31, 2012, http://www.milkeninstitute.org/pdf/RossandAnuNIHpaper.pdf.

Fariborz Ghadar, John Sviokla, and Dietrich A. Stephan, "Why Life Science Needs its Own Silicon Valley," *Harvard Business Review* (July/August 2012):25–27.

Arthur Levinson and Mark Zuckerberg interviewed in "Breakthrough Prize Awards Research to Cure Disease," National Public Radio, February 20, 2013, accessed February 20, 2013, http://www.npr.org.

Dennis Overbye, "At $3 Million, New Award Gives Medical Researchers a Dose of Celebrity," *The New York Times,* February 20, 2013.

Elizabeth Pennissi, "DARPA to Offer $30 Million to Jump-Start Cellular Factories," *Science Insider,* June 29, 2011, accessed August 29, 2012, http://news.sciencemag.org/scienceinsider/2011/06/darpa-to-offer-30-million-to-jump.html.

J. Craig Venter, *Life at the Speed of Light: From the Double Helix to the Dawn of Digital Life* (New York: Penguin, 2013).

4

Mendel's Journey from Peas to Petabytes

In all, thirty-four more or less distinct varieties of Peas were
obtained from several seedsmen and subjected to a two
years' trial.

    —Gregor Mendel, *Experiments in Plant Hybridization*

Numbered in the massive transfer of domesticated animals, crop plants, infectious
microbes and their genes represented by the Columbian Exchange was *Pisum sativum*, the garden pea. The pea was introduced to America from Europe in the sixteenth century. It was one of the first crop plants adopted by native inhabitants of
the New World. The pea is relatively easy to grow and cultivate, but its genome has
proved challenging to sequence. The genome of *Pisum sativum* is large and repeti-
tive, possessing perhaps a quarter more genes than the human genome. The effort to
create a reference pea genome is international, involving investigators from Europe,
the United States, the United Kingdom, Canada, and Australia. Scientists from
the Czech Republic are among the pea sequencing project leaders in the Genomic
Exchange era we have entered.

    When the Augustinian monk Gregor Mendel was conducting his pea experi-
ments and laying the foundation for modern genetics, the city in which he labored,
Brno, was gaining fame as the "Manchester of the Hapsburg Monarchy" for the
rapid growth of its manufacturing industries. Today Brno in the Czech Republic
is a city of nearly 400,000 residents and eight universities that attract students and
skilled workers from around the region. The scientific seeds that Mendel planted
have sprouted from the soil in the form of scientific and innovation communities
like Medipark, a $200 million life science campus developed at Masaryk University
in Brno. Home to more than 1,000 PhD students, Medipark is focusing its innovation efforts in the fields of genomics, gene therapy, pharmaceutics, and, of course,
plant biotechnology. "We are trying to connect industry, education and infrastructure to make it easier for companies to come here to create an environment that
suits biotech companies best," said the city's mayor. Brno has ambitions to be central Europe's biotech hub.

It is a short walk from Medipark to the garden of the fourteenth-century Abbey of St. Thomas where Mendel set about in 1856 to see if he could throw some light on the laws of inheritance by cross-breeding pea plants. In all, he cataloged the traits of some 28,000 pea plants over eight years. He classified them by seed color (yellow or green), seed shape (round or wrinkled), pod color (yellow or green), pod shape (inflated or pinched), flower color (purple or white), flower position (axial or terminal), and stem height (tall or dwarf). With mathematical precision uncommon for his day, Mendel charted the complexities of the process by which genetic traits travel from one generation to the next. His reported findings tended to be neatly one thing or the other (as in the binary digital world of zeros and ones), not reflecting intermediate features. But Mendel's laws of segregation and independent assortment remain intact, and his systematic analysis and statistical methods were unparalleled for his time. They laid a theoretical foundation for the discrete information unit of what came to be called the gene and the gene's role in heredity. A little more than a century after Mendel published his findings, the gene's isolation and transfer from one organism to another led to the founding of Genentech.

Mendel labored in his garden as the Industrial Revolution was transforming Brno into a center of the textile and machine industries and making it a key link in the Austro-Hungarian Empire thanks to new transportation and communications channels. In 1836 Brno became the first city to be connected to Vienna by railway. In 1847, just three years after Samuel F. B. Morse dispatched the first telegraphic message "What hath God wrought?" over an experimental line, Brno became the first city to be connected to Vienna by telegraph lines. But the telegraph did not make much headway, at least not early on, in distributing ideas and discoveries within the high priesthood of science. The same year the first transatlantic telegraph cable was put into service, in 1866, Mendel published the results of his experiments in the *Proceedings of the Natural History Society of Brno* after giving two lectures to the Society the previous year. The *Proceedings* with Mendel's paper "Experiments in Plant Hybridization" were mailed out to the libraries of some 120 institutions, including the Royal and Linnean Society of Great Britain. Mendel himself had forty additional reprints at his disposal, which he sent to leading biologists of Europe.

Despite stories to the contrary, there is no firm evidence that Charles Darwin ever read Mendel's paper. If he had, says evolutionary biologist Richard Dawkins, the history of biology would have been very different. Mendel's paper escaped the notice of his peers and of the scientific community generally, even though the scientific world was eagerly awaiting the discovery of the laws of heredity following the 1859 publication of Darwin's *Origin*, a book Mendel read in the German translation and annotated. Mendel's is perhaps the best-known case of a now-famous paper finding its way to obscurity almost before the ink was dry. By one account, the heavy use of mathematics by the "father of genetics" put off his scientific contemporaries who were not accustomed to looking at the world of botany and plant breeding that way. Mendel's paper revealed a level of scientific understanding and methodology that was quite uncommon among botanists in the mid-nineteenth century. In the decades after it was published, Mendel's paper was referenced very little in the scientific literature. Not until the turn of the twentieth century was the significance of his work recognized independently by the Dutch botanist Hugo de Vries, the German botanist Carl Correns, and the Austrian botanist Erich von Tschermak. Mendel's keen insights were not embraced, wrote technophile Kevin Kelly, because "they did not explain the problems biologists had at the time, nor did his explanation operate by known mechanisms, so his discoveries were out of reach even for the early adopters." Mendel was simply a man ahead of his time.

Mendel and the Transmission of Scientific Information

Mendel's discovery of the laws by which genetic information is transmitted through succeeding generations together with Darwin's theory of evolution by natural selection and paleontological evidence formed the modern evolutionary synthesis, the paradigm for evolutionary biology. Yet his brilliance lay dormant for half a century in part because the way information was transmitted among scientific peers in his day was still in an embryonic stage. During the century following the rediscovery of his genius by de Vries, Correns, and von Tschermak, the world changed.

Today, the transforming power of mass electrification, electronics, digital communications, and wireless undergird the post-industrial order the way that the steam engine captured the rise of industry in Britain two centuries ago and the electrical dynamo powered America into global manufacturing leadership a century later. That power is increasingly distributive rather than hierarchically structured. It is being yoked to the task of bringing basic discovery in the biosciences and genetics to the attention of innovators—scientific innovators in research laboratories, entrepreneurs in regional bioclusters, economic development and planning experts, and other players in the emerging bioeconomy. Biology, as Kevin Kelly observed, is "the biggest science," with the most scientists, the most funding, the most scientific results, the most ethical significance, and where there is the most to learn given its billions of years of experimental results involving self-replicating organisms.

Could future Mendels be missed, given the open-access world we are entering and the incredible search tools we have? Will the sheer mass of information end up inundating the inquirer? The person most responsible for setting in motion the American federal scientific research and development enterprise was the engineer and presidential science advisor Vannevar Bush, dubbed the "patron saint of American science." His report to President Truman, "Science, The Endless Frontier," brought government into the process of research and discovery as never before. During World War II and the Cold War, Bush championed technological innovation and entrepreneurship, viewing these activities as critical for both economic and geopolitical security. He viewed communication as central to the task of innovation. In July 1945, when he was director of the Office of Scientific Research and Development, Bush published an article "As We May Think" in the *Atlantic Monthly*. The article describes how machine technology could help to connect disparate fields of knowledge thereby spurring the innovation process. The "memory extender" or Memex was a "microfilm rapid selector" that enabled the machine's user to follow trails connecting related subjects, a form of analog computer.

In his article, Bush issued a cautionary note about the growing mountain of research spurred by specialization. Because specialization is "increasingly necessary for progress," the effort to bridge the information flow between and among disciplines becomes a critical task for innovation in science and industry. That represented to Bush a major communications challenge. Mendel's laws of inheritance were "lost to the world for a generation because his publication did not reach the few who were capable of grasping and extending it; and this sort of catastrophe is undoubtedly being repeated all about us, as truly significant attainments become lost in the mass of the inconsequential."

In Mendel's case, say some scholars, it was a matter of bad timing. Geneticist Gunther Stent argued that Mendel's research on peas, which relied heavily on mathematics, was not a good "fit" for the canon of botanical knowledge at the time. The monk's statistical methodology was essentially foreign to the botanical community, and thus his research was an example of "scientific prematurity." This was not the first time nor would it be the last that a transformational discovery in biology lay dormant because scientists of the time were not prepared to consider a novel concept or mechanism. The American pathologist Peyton Rous discovered the first tumor virus, a transmissible agent in chickens, in 1910. But it took decades before Rous's discovery was embraced. Rous's research paved the way for a better understanding of the cause of cervical cancer, human papilloma virus, which decades later led to a preventative vaccine. In the intervening years between Mendel's publication and the rediscovery of his work, Theodore Boveri discovered chromosomes in the nucleus of the cell; mitosis and meiosis were witnessed; and Friedrich Miescher discovered what he called "nuclein" and what we today call nucleic acids, the molecules that carry genetic information. These discoveries, providing as they did a structural basis for the transmission of genetic traits in the nucleus of the cell, set the stage for the rediscovery of Mendel's laws of inheritance by de Vries, Correns, and von Tschermak, who brought the Austrian monk back from obscurity.

Whether or not recognition of Mendel's research was delayed due to a failure of significant distribution of Mendel's paper or its "scientific prematurity," the digital revolution that began in the late twentieth century has reduced dramatically the possibility of distributive failure and condensed the vetting time frame for a novel idea or discovery. Digital technologies are accelerating the pace of research discoveries in biology as well as in physics, engineering, and other fields and directing these discoveries to innovators. At the end of his presidency George W. Bush (no relation to Vannevar Bush) signed an appropriations bill that mandated public access to the results of research funded by the National Institutes of Health (NIH). It was the first open-access mandate adopted by the US government. NIH grantees are obligated to deposit a copy of their research articles once accepted for publication into the National Library of Medicine's PubMed Central database from where it is or will be freely available to the worldwide research community as well as the public. Given that NIH funding results in an estimated 80,000 published articles annually, open access "will introduce more authors to self-archiving—posting one's own research results online for free access—than any single event to date," said an open-access advocate. Within three years of Bush's signature, the trends were showing the success of open access. The number of papers in freely accessible journals was growing 20 percent per year. In 2012 the UK government-commissioned Finch Report advocated a transition to an open-access academic ecosystem in which authors pay commercial publishers directly to make their articles free to read. The European Commission acknowledged that open access is essential for Europe's competitiveness and is taking steps to implement a mandate. In the United States the Obama administration directed federal agencies with more than $100 million in R&D expenditures to develop plans to make the published results and digital data from federally funded research freely available to the public within one year of publication. Within weeks of that 2013 directive, mandatory budget cutbacks at federal funding agencies threatened a decline in US scientific journal publication output, projected to be 8 percent by one estimate.

Open access is a movement that incubated in the 1990s and emerged in the early 2000s advocating free access to and use of digital scientific and scholarly material. By the end of 2007, the Massachusetts Institute of Technology (MIT) had made available online core teaching materials from nearly all of its 1,800 courses through its OpenCourseWare program. Harvard became the first university in the United States to mandate open access to its faculty members' research publications. In the space of two months, three open-access mandates were passed—those of the NIH, the European Research Council, and Harvard University—paving the way for other universities and colleges in the United States to adopt similar policies. A decade after MIT launched OpenCourseWare in 2002, top American universities, including MIT, were participating in massive open online courses. Online startups with names like Coursera, Udacity, EdX, 2tor, and Udemy aim to bring higher education services, full degree programs among them, to the burgeoning middle class around the world. EdX offers "The Secret of Life," an introductory biology course taught by MIT genomics pioneer Eric Lander.

The fears of Vannevar Bush that specialization and restricted access to research findings would hamstring innovation—fears that he believed lost Mendel's historic contribution for a generation—no longer hold sway. It is digital rather than analog technology (the technology Bush championed) that is revolutionizing education, research, discovery, and innovation. Yet when it comes to data storage, silicon may not have the last word. In 2012 Harvard researchers used next-generation DNA synthesis and sequencing technologies to encode and store digital information. They reported storing 700 terabytes—equivalent to 14,000 50 gigabyte Blu-ray discs—in a single gram of DNA, representing 1,000-fold increase over the previous DNA data storage density record. A British research team followed up with a comparable system. In a feat of self-reference, the team encoded into DNA Watson and Crick's 1953 1,000-word research paper in *Nature* describing DNA's double helix. It is only a matter of time before someone encodes Mendel's 13,000-word paper into DNA, the very agent of his laws of inheritance.

Cultivating Mendel's Garden with Computational Tools and Webs

The mathematics of Mendel's laws finds its contemporary expression in the fields of bioinformatics and computational biology. These fields involve the use of applied mathematics, informatics, statistics, computer science, artificial intelligence, chemistry, and biochemistry to solve biological and genetic problems. Bioinformatics is, as one might expect, increasingly an activity of free access to computer source codes for anyone to tinker with and, perhaps, improve. Bioinformatics is one of many fields benefiting from the open-source movement, an offshoot of the free software movement in the computer industry. Whether by improving source code for tracking genes or fostering gene transfer technologies, both bioinformatics and genetic technologies are featured in lectures delivered in the Mendel Museum of Genetics (Mendelmuseum.muni.cz) housed in the old abbey where the fundamental laws of inheritance were penned.

The rise of broadband technology has enabled what is sometimes called Web 2.0, which involves the emergence of social networking media like Facebook and Twitter; wikis like Wikipedia; maps and mashups; video via YouTube; weblogs and blogs; and mobile phones that support voice, video, and data transmission. All facilitate creativity, collaboration, and sharing among users, including users with an interest in science and innovation. Similar activities include crowdsourcing, citizen science activities, and virtual 3D worlds like Second Life in which resident participants have their own distinct digital persona or "avatar" that can explore virtual domains, interact with other avatars, and build virtual objects. In Second Life, science, technology, engineering and math students can immerse themselves in Mendelian genetics in Mendel's garden on "Genome Island" and do Mendel's breeding experiments with green and yellow peas. These features together with the open-access movement in research and education do not guarantee that geniuses like Mendel won't find their work ignored, but they dramatically reduce that likelihood. They are part and parcel of the biosciences in the twenty-first century, a manifestation of what Harvard law professor, Creative Commons founder, and best-selling author Lawrence Lessig calls a "hybrid economy" in which information sharing creates virtual communities *and* commercial value.

Online collaboration is growing apace within multinational firms as well as among universities and research centers. Take IBM Brno, a Silicon Valley-styled building of steel, slate, and glass highlighted by design motifs in nitro green and electric orange for example. IBM Brno is a global delivery and outsourcing center for some of the world's largest firms. It is outfitted with a cutting-edge digital infrastructure and populated by 2,500 employees representing nearly seventy nationalities speaking almost forty different languages. Web 2.0 technologies such as blogs, wikis, podcasts, social networking tools like Facebook and Twitter, and virtual worlds such as Second Life enable it to engage in brainstorming sessions with team members in IBM branches in places as far-flung as Manila, Buenos Aires, and São Paolo.

IBM debuted the "IBM Virtual Healthcare Island" in Second Life in 2008, the fiftieth Second Life island IBM had created at the time. IBM described the island as "a unique, three-dimensional representation of the challenges facing today's healthcare industry and the role information technology will play in transforming global healthcare-delivery to meet patient needs." Visitors can walk, fly, or use transporters to visit the various island stations, including the Patient's Home, the Laboratory, the Clinic, the Pharmacy, the Hospital, and the Emergency Room. The IBM island's tour guide explains to visitors the company's plans to transform health care in the years ahead.

Data Mining and Gene Mining Join Forces

The Mayo Clinic in Rochester, Minnesota, one of the most famous medical institutions in the world, launched "Mayo Clinic Island" on Second Life's alternate universe in 2010. The outpost's virtual hospital, outdoor conference center, and other features are meant to explore new avenues in medical education, research, and practice. Mayo cardiologist Paul Friedman sees Second Life as a way making his talks on cardiology available to the millions of users of the service.

Just as Mendel pioneered the study of the laws of inheritance in Brno with his gardening, his mathematics, and his information processing and analysis, in the United States IBM is pioneering a data-mining project that some call the Holy Grail of medicine: individualized patient treatment. Its partner is the Mayo Clinic. In 2004 IBM and Mayo teamed up to, as Mayo put it, to "tap the power of IBM's Blue Gene supercomputer" to harness "the explosion of valuable, yet untapped healthcare data that is emerging due to breakthroughs in genomics, proteomics and molecular modeling, and the digitization of patient records." They have also forged ahead in biomedical imaging, exploiting parallel computer architecture and memory bandwidth to dramatically speed up the processing of 3D medical images via collaboration at a medical imaging informatics innovation center in Rochester.

Czech biotechnology and medical genomics got a boost when the Mayo–IBM collaboration expanded to Brno's new International Clinical Research Center, one of the European Union's largest initiatives in the biomedical field. In announcing the venture in 2006, the Czech Republic said the "new research cluster" will strengthen European–US collaboration in medical research and education and bring information-based medicine to the diagnosis and treatment of cardiovascular and neurovascular diseases, internal medicine, and oncology. The European Union awarded the International Clinical Research Center $220 million in 2011 to foster the collaboration, with $10 million of the EU funds going directly to the Mayo Clinic.

Both information-based medicine and bioscience and information-enabled research collaboration are joined in the city where Mendel described what genes do even though he did not know what genes are. Brno-based researchers are among many working to develop biochips. "Lab on a chip" technologies join DNA microarray technologies, both with roots in Silicon Valley, as cornerstones of the life sciences revolution. They are located at the swelling crossroad where biology and technology meet—where materials science, chip design and manufacturing, signal processing, and software take aboard molecular biology and genomics. It is a busy crossroad involving submicroscopic laboratories, Internet-driven experiments, and virtual seminars in Second Life. It is a crossroad of exploration conducted on silicon wafers and over fiber optics cables and via satellite-beamed images as well as in conventional Earth space. It is where genome-wide association studies of far-flung populations and whole genome sequencing of individual patients can identify genetic factors that influence health and disease.

Gene expression analysis of *Pisum sativum*, Mendel's garden pea, enabled geneticists to locate the gene that controls the color of the peas' seeds, called "Mendel's green cotyledon gene." The gene, *sgr* (for "staygreen") was originally identified in a variant of prairie grass that stayed green in drought because of its inability to break down green pigment. The gene has also been found in rice and in the mustard plant *Arabidopsis*, whose sequence was reported in 2000, the first complete DNA sequence of a plant. Mendel selected particular traits in his pea plants through careful observation and meticulous cross-breeding experiments. Crop scientists today use plant breeding guided by DNA marker-assisted selection and targeted mutation-selection to produce plants with desirable traits such as improved productivity; disease and pest resistance; and ability to grow in saline soil and deal with stress, climate change, and drought. Many if not most varieties of large market crops including corn (maize), soybeans, wheat, rice, potatoes, rapeseed, common beans, cassava (manioc), cotton, and rubber—all of which figured prominently in the Columbian Exchange—will be produced in the future using knowledge from their sequenced genomes (Table 3.1). Massive computing power is guiding crop plant improvements in the Genomic Exchange era.

The nexus of genetics and technology, particularly information technology and analysis, tracks an intellectual current beginning in Brno with Mendel's notebooks. It runs through the work of the English evolutionary biologist, geneticist and statistician R. A. Fisher, a leading figure in the modern synthesis reconciling Mendelian genetics with evolution by means of natural selection. Fisher conducted genetic studies of multifactorial disease susceptibility in human populations. He introduced the concept that the role of genes in diseases like diabetes and heart disease is not a straightforward and predictable pattern of Mendelian inheritance found in genetic diseases like Huntington's disease and cystic fibrosis but a complex distribution. A pioneer of the principles of experimental design, Fisher was a sharp critic of Mendel's pea experiments. Today modern computers do the biostatistical analysis Mendel did in his head and Fisher did using his "Millionaire" mechanical desk calculator. Computers perform trillions of calculations a second, run on powerful bioinformatics software, and generate cumulative genomic data measured in petabytes with each petabyte constituting just over a million gigabytes. The storage capacity of a DVD is 4.7 gigabytes.

What would Mendel think were he alive today? "I am certain—as certain as one can be about such things—that Mendel would have been delighted with a whole genome project and the idea of having his own sequenced," wrote biologist Simon Mawer, Mendel's biographer and author of the novel *Mendel's Dwarf.* "It would have appealed to the mathematician in him, and the chess-player." Such information to Mendel would constitute just another step in the long journey of discovery. Yet steps on the journey through the endless frontier of biological science are proceeding at an accelerating clip because of data mining and gene mining. At way stations along the journey today, entrepreneurs and scientists team up with investors to create economic value.

In 1998, an international team of scientists identified and cloned the first gene known to control the production of the most abundant organic compound in Mendel's garden and on Earth, cellulose. Cellulose fibers give hundreds of thousands of plant species their strength by wrapping around the cells like the metal hoops around barrels. Cellulose is the raw material for making paper and holds promise for many high-fiber products. But cellulose also locks up the vast potential of plants as a source for renewable energy. As scientists announced their discovery of the first cellulose gene, the private firm Mendel Biotechnology, Inc. was sprouting in Hayward, California, on the edge of Silicon Valley. A self-described functional genomics company, Mendel Biotechnology is one of a growing number of companies focused on identifying and patenting genes that control certain aspects of plant growth and development, the key components of plant genome regulatory networks. Its goal is to use its inventions to develop or co-develop new, more productive plant varieties and to find ways to produce ethanol from the sugar and starch locked up in the plant by breaking cellulose down with enzymes.

Mendel Biotechnology formed a partnership with the biotech giant Monsanto in 2008 to develop high-yielding Chinese silvergrass, *Miscanthus sinensis*, as a twenty-first century biofuel. The company has developed valuable lines of *Miscanthus sinensis* seed and is working with Chinese researchers to develop additional lines. The oil company BP as well as Monsanto are significant shareholders in Mendel Biotechnology. BP Biofuels commenced field trials in 2012 with Mendel Biotechnology's trademarked PowerCane Miscanthus developed by its subsidiary Mendel BioEnergy Seeds to evaluate its performance as a feedstock for cellulosic biofuel production. Scaled-up cellulosic ethanol production from woody waste may be commercially viable in the near future. Renewable plant biomass has potential as a next-generation cellulosic biofuel despite the enzymatic challenge of inducing plant biomass to release its energy potential eons before it becomes fossil fuel.

Biology, Technology, and Planetary Concerns

A century and a half after Mendel cultivated his garden at St. Thomas Abbey and Colonel Edwin Drake struck "black gold" in Titusville, Pennsylvania, the genetic revolution Mendel seeded is being harnessed to find renewable sources of energy that can help replace the fuel from fossilized plants on which the world has become dependent. The man whose name is synonymous with the beginning of genetics and biomathematics today lives on in entrepreneurial companies as well as in research laboratories, field biology stations, and high-school classrooms. He lives in virtual worlds that spur innovators around the globe eager to compete in the race for health and wealth, high-tech economic growth, and scientific prestige that the future of the sciences of life have promised for decades.

A four-letter chemical code is responsible for producing the different traits Mendel painstakingly catalogued in his pea plant experiments. The code of life and our growing ability to produce it cheaply will come into play as society confronts such immense challenges as growing energy demand and climate change, which will have profound effects on its future growth. Biological and digital technologies will be critical for mitigating the effects of climate change. The migration and destruction of species and the shifting patterns of food–cropland use in response to recurrent flooding, drought, and crop-specific growing seasons are among those effects.

Climate change "is the greatest and widest-ranging market failure ever seen," wrote Sir Nicholas Stern in his heralded report on the economics of climate change published in 2006. A team of economists at HM Treasury prepared the report in consultation with academic experts. One of the documents Stern's team received was from the UK's Tyndall Centre, an interdisciplinary climate change research center. The center was named after the nineteenth-century Irish physicist John Tyndall, a friend and correspondent of Charles Darwin who championed Darwin's theory of evolution by natural selection. Tyndall demonstrated the atmospheric greenhouse effect experimentally in 1859, the year Darwin published *On the Origin of Species*.

Technologies that manipulate information, organisms, and materials, which we know as, respectively, information technologies, biotechnologies, and nanotechnologies, are critical tools we have to address shortfalls of natural resources in a world of increasing demand for them. The Tyndall group predicted that digital technology will facilitate growth of the service sector, which will constitute more than 80 percent of world gross domestic product (GDP) by 2100, mainly due to massive IT investment and diffusion. "There is an inextricable link between technological change, industrial dynamics and environmental impact over the long term." By 2050, the adoption and diffusion of biotechnologies and nanotechnologies could change industrial structure. Biotechnology has the potential to create "generic platforms" with pervasive effects across a wide range of industries, including pharmaceuticals, health care diagnostics, agriculture, food, forestry, materials technology, energy, and environmental monitoring. Industrial biotechnology processes in the bulk chemical industry will go a long way toward reducing inputs of electricity. A new advanced materials industry stemming from biotechnology and nanotechnology could also reduce current fossil fuel power inputs as well as petroleum-derived synthetic fabrics, packaging, and other materials.

Because intensive agriculture accounts for as much as 80 percent of water use and makes heavy use of nitrogen-based fertilizers that emit nitrous oxide, a greenhouse gas, some think agriculture may have to be reinvented. New approaches, new methods, new technology—"indeed perhaps even new crops and new agricultural systems"—will be necessary, and soon, an international team of agricultural scientists writing in *Science* concluded. There is no escaping the need for biological technologies to overcome "declining feedstock yields due to global climate change, natural resource depletion, and an increasing demand for limited water and land resources."

The complex, costly, and time-intensive regulatory system in the United States discourages public-sector researchers from using molecular methods to improve crops for farmers. Thus genetically modified (GM) crops have been limited largely to those for which there is a large seed market. GM soybeans, corn, and cotton account for more than 80 percent of those crops grown in the United States. Worldwide, of the 1.5 billion hectares of arable land, about 12 percent were planted with GM seed in 2012. Nearly all were planted with GM soybeans, corn, cotton, and canola in five countries: the United States, Brazil, Argentina, Canada, and India. Eighty percent of the soybean and cotton crops and one-third of the corn crop around the world were planted with GM seed in 2012. "Without broader research programmes outside the seed industry," editorialized *Nature*, "developments will continue to be profit-driven, limiting the chance for many of the advances that were promised 30 years ago," among them feeding the planet's growing population in a sustainable way and reducing agriculture's environmental footprint.

The benefits of agricultural biotechnology, whether through marker-assisted selection, targeted mutation-selection, genetic modification, precise genetic editing, or other technologies, have not been realized for the vast majority of food crops, though transgenic rice, cassava, bananas, apples, and plums may make their debut soon. The exchange of crop plant whole genome sequence information through open-source initiatives could accelerate the process. Yet even with the powerful tools of bioscience, it is not clear that current crops can be pushed to produce as well as they do now at expected higher temperatures and with less water. Indeed, researchers studying yield trends of four key crops from 1961 to 2008 found that more than a quarter of corn, rice, wheat, and soybean cropland areas worldwide are stagnating or in production decline.

Yet institutional, regulatory, and proprietary obstacles as well as public apprehension have constrained the knowledge revolution that began in Mendel's garden from being more widely applied to produce plant-based food in ways that mitigate environmental damage. The search for agricultural solutions "should remain technology neutral" say environmental and crop scientists in "Solutions for a Cultivated Planet" published in *Nature*. Because the paths to improving the production, food security, and environmental performance of agriculture are many, "we should not be locked into a single approach a priori, whether it be conventional agriculture, genetic modification or organic farming."

Growth in crop yields proceeded very slowly until the eighteenth century and "really took off only after Mendelian genetics opened new opportunities," says the Czech-born Canadian environmental scientist Vaclav Smil. Higher harvest indexes, denser planting, optimum nutrient supply, and herbicide and pesticide application boosted cereal yields with national averages often more than doubling during the twentieth century. At the beginning of the twenty-first century, roughly half of the food, feed, and fiber crops—together with their residues and forage crops—was consumed by animals. That consumption produced nearly 300 million tons of meat, some 700 million tons of milk, and 65 million tons of eggs per year by Smil's calculation in "Harvesting the Biosphere." In addition to the fact that feeding grain crops to animals always entails a loss of potential food output, meat production is also a major source of greenhouse gases when all inputs and outputs are taken into account. What Smil calls human appropriation of photosynthetic products is undertaken today at such a vast scale that natural biogeochemical cycles involving carbon, nitrogen, and other elements are distorted (Figure 4.1). The only way out, in his view, is to reduce consumption rates and use resources more efficiently.

Can advances in animal genetics play a role as well? Livestock animal scientists think they can. They argue that precision genetics with the information-intensive tools of genomic selection and genetic engineering will be essential given population growth and growing demand for animal protein. Pressure for solutions will mount when feedstock yields fall as a result of climate change and the demand for water and land resources continues to grow. Besides contributing to more efficient food production and resource use, genetically modified animals and genomic technologies also can enhance livestock health, for example by elevating the expression of antibacterial proteins so that the use of antibiotics in livestock feed, a widespread practice with serious public health consequences, can be reduced or eliminated. Genomic technologies could also be used to create similar-sized animals to facilitate meat processing.

Today we live in what some environmental scientists call the "Anthropocene" due to the impact of human activity on the biosphere. Most of them place the genesis of the Anthropocene in the Industrial Revolution, though it accelerated exponentially following World War II as a result of atmospheric nuclear bomb testing coinciding with oil- and coal-fired growth. It is a historical curiosity that James Watt's fellow Scotsman and Lunar Society member William Small offered

FIGURE 4.1 Bioscience and the biosphere. Human activity is altering nature's biogeochemical cycles including the global carbon cycle illustrated here in a simplified form. Genomic science is being applied to detect, monitor, and remediate some of the biological consequences of these changes.

Source: Carbon Cycling and Biosequestration: Report from the March 2008 Workshop, DOE/SC-108, United States Department of Energy Office of Science. Genomic Science Program, http://genomicscience.energy.gov.

an opinion on what he regarded as the perils of future climate change. Around 1770, a period of global cooling had set in, the second of three "minima" that characterized the Little Ice Age. In correspondence with Watt while Watt was in the midst of his Caledonian canal survey and encountering plenty of challenges in the barren, rocky Scottish Highlands, Small observed that it is no wonder people leave Scotland for America with its bountiful land and forests. Small reasoned that Scotland suffered from the southward spread of cold climate. Like a good Lunar Man, he had an idea for mitigating the problem. He proposed corrective geoengineering before the entire planet went into the deep freeze. Small's big idea was to blow up polar ice.

The Boulton-Watt engine was an innovation no less brilliant for the fact that burning its fuel had unforeseen consequences that today demand innovation of at least comparable brilliance to counter. At the time Small described to Watt his geoengineering scheme, in 1773, the carbon dioxide content of the earth's atmosphere was about 280 parts per million (ppm). In the mid-nineteenth century Mendel's pea plants grew in atmospheric CO_2 at less than 300 parts per million. Today's pea plants grow in atmospheric CO_2 at 400 parts per million thanks mostly to fossil fuel combustion, which has added more than half a trillion tons of carbon to the atmosphere. The international team that is studying the genome of *Pisum sativum* notes that the pea is a major food legume that "has the capacity for enhanced nitrogen fixation and CO_2 capture, which may partially offset growth reduction associated with higher temperature, shorter growing season, and periods of drought." Like most major crops, the pea has been bioengineered to resist disease. Could bioengineered crops mitigate the concentration of atmospheric CO_2 by enhancing its capture? Could they ease the atmospheric burden of carbon by reducing fossil fuel use for herbicide or insecticide applications and soil cultivation, as has been proposed?

However those questions are answered in the years ahead, genes are responding to the power of climate. By tracking the evolution of wild cereals over nearly three decades of global warming in Israel, not far from the Fertile Crescent where they were domesticated, scientists found clear evidence of adaptive phenotypic and genotypic changes, namely earlier plant flowering times and loss of alleles in wild emmer wheat and wild barley. No factors other than global warming in Israel "would induct such essentially similar behaviors for many natural populations." The Israeli scientists advocate a "red light" and "green light" approach to the problem. The red light is a warning that the threat of climate change to wild cereals is real. These important germplasms and genetic resources must be conserved for future crop improvement. Cultivated wheat, the world's most important food staple, has been "genetically impoverished" by long-term selective breeding. The green light is to advance "the green revolution to the gene and genomic revolution" through tools such as marker-assisted selection to enhance the introduction of beneficial and adaptive alleles of genes from wild emmer wheat into cultivated wheat. Shotgun sequencing of the bread wheat genome in 2012 produced seventeen DNA gigabases, more than five times the size of the human genome, and nearly 100,000 genes. Genomic information of bread wheat will help crop scientists identify useful variants in genes of wheat landraces and progenitor species, increase genetic diversity, and provide markers to guide selection.

Bioscientists are not waiting for a calculation of how much their field can contribute to ameliorating the harmful effects of our agricultural practices and photosynthetic harvests. Biologists and bioentrepreneurs are generating novel biomass materials and modifying microorganisms to eat carbon dioxide or generate biofuels that do not require arable land. They are modifying crops to tolerate marginal or harsh growing conditions and to take up nitrogen more efficiently so that use of nitrogen-based fertilizer, which contributes to greenhouse gas emissions by adding reactive nitrogen to the atmosphere, can be minimized.

Innovation in the biosciences will be critical in the search for energy efficiency improvements, cost reductions, sustainable development, and the preservation of the planet's diverse life forms. Introducing and applying new technologies and systems expeditiously to mitigate the pace of global warming and its consequences will command large capital investments over many years. National policymakers and international agencies are exploring new paradigms designed to stimulate innovation and investment in low carbon technologies.

The digital revolution has largely met and overcome the "communications challenge" that Vannevar Bush believed caused Mendel's laws of genetics to be lost to the world for a generation—and just in time. The mind-numbing problems posed by dirty energy derived from fossil fuels and growing carbon dioxide emissions from their combustion require that all channels to deep knowledge and collaboration be pried wide open. Today it is accepted wisdom that technological innovation is most likely to occur at disciplinary boundaries of scientific research (interfacial innovation) and the organizational boundaries of public and private enterprises. It is also becoming accepted wisdom that to proceed relying solely on the familiar pattern of incremental progress to solve problems of planetary scope is a risky gambit for our future.

That is why federal agencies like the Department of Energy through its Advanced Research Projects Agency—Energy program and the NIH through its New Innovator Award program borrow from the Defense Advanced Research Projects Agency with its heralded and successful "high-risk, high-reward" approach to innovation. The former's $130 million program called Plants Engineered to Replace Oil, which involves universities, federal laboratories, and private enterprise, aims to optimize the ability of plants to capture solar energy and carbon dioxide. As we saw earlier, the Defense Advanced Research Projects Agency through its "Living Foundries" program is seeking to convert cells into factories for producing renewable fuels. Microbial photosynthetic biorefineries that use atmospheric CO_2 and sunlight to make energy-rich molecules that, in turn, can be converted to fuels, chemicals, and other bio-based products are now within the realm of possibility.

Shotgun metagenomic analysis—the study of genetic material recovered directly from environmental samples such as water and soil as well as animal microbiomes without the need for cell cultures—is unearthing new species of microorganism as well as new genes from around the world. Some of these microorganisms/genes are involved in recycling carbon and CO_2; in degrading plant cell walls to produce renewable biofuels such as ethanol, butanol, and methane; and in generating electricity in microbial fuel cells. Bioengineers have made notable advances toward the goal of turning cells into batteries by genetically fine-tuning their electron transport systems, harnessing their energy, and transforming living cells into electrical conduits that can interact directly with conventional electronics. Communication between living and nonliving systems could also open up new avenues in bioenergy conversion and biosensing.

As we have seen, digital technology is empowering academic and industrial research networks where technologies are converging. Whether in research institutions, large industrial research laboratories, or small, high-tech startups, networked science will loom ever larger as humanity seeks to address problems affecting the entire planet. The accelerating trend of multi-authorship of scientific papers across national boundaries is destined to continue through shared global priorities in health, energy, climate, and social structures.

The Anthropocene marks the emergence of what the *Economist* called "a form of intelligence that allows new ways of being to be imagined and, through co-operation and innovation, to be achieved." The magazine ran a cover story that focused on how humans have remade the living environment through the domestication of plants and animals and disrupted key biogeochemical cycles through industry. "On a planetary scale, intelligence is something new and powerful." In a decade or two, we will know whether or not the *Economist*'s observation was prescient.

Health Technologies Are Getting Personal

Innovation, in its broadest sense, made the modern world. Today innovation is being brought to bear in perhaps the ultimate exercise of self-discovery by mapping our genomes, tracking our genetic history from our earliest days as a separate species, calculating statistical probabilities about our health and risk of disease based on our genes and their interactions with our environment, and giving us power to shape our own evolution.

It has been nearly half a century since Intel cofounder Gordon Moore described the trend that bears his name. Moore's Law is based on the observation that, over the history of computing hardware, the number of transistors on integrated circuits doubles approximately every two years. The unfolding of Moore's Law has reshaped how we interact across the surface of the earth through a web of connectivity. It has enabled us to visualize worlds below us and above us from the earth's core to distant stars. Now a powerful new trend is in the making, a more potent technological accelerator than the one Moore envisioned. It concerns living things as bearers of encoded information that brought them into being, manages their development and internal affairs, and makes arrangements for them to create new life consistent with the laws of inheritance that Mendel discovered and codified. Its focus is the chemical sequence of the code of life.

Next-generation DNA sequencing (NGS) advances have already surpassed the price-performance pace of computer chips (Figure 4.2). Next-generation DNA synthesis is not far behind. As we noted earlier in this chapter, researchers using next-generation DNA synthesis and sequencing technologies encoded and stored hundreds of terabytes of digital information in a single gram of DNA, which was accomplished "for ~one hundred thousand-fold less cost than first generation encodings." The speed of technological change that we see in the networks and devices all around us is being ramped up. Soon we will be flooded with DNA sequence data from our own healthy and diseased cells and from the cells of life forms that we host, from the cells of organisms that transmit disease, from the cells of organisms that nourish and sustain us, and from the cells of other creations that

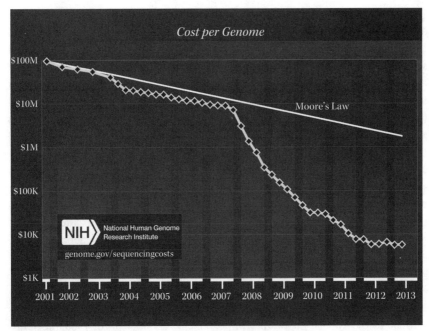

FIGURE 4.2 Cost trend of sequencing a human-sized genome and Moore's Law 2001–2013.

Source: Kris A. Wetterstrand, DNA Sequencing Costs: Data from the National Human Genome Research Institute's Genome Sequencing Program, http://www.genome.gov/sequencingcosts.

constitute inhabit what we call the biosphere. The challenge will be, as the statistician, blogger, and writer Nate Silver would put it, to find the signal in the noise.

We have come a long way from Mendel's experiments in a historically short time. The first complete sequence of the human genome was announced in 2003. Since then biotechnology and information technology have formed an ever-tighter symbiotic relationship around the world. Many advances in biology have been made possible by rapidly growing computational hardware, software, and distributed systems. Biomedical and health technology firms are compelled to adjust their business models for the digitally enabled world because these firms arise at the confluence of biology, computer science, engineering, and material science The blending of these technologies brings together people and business practices in the industry's pharmaceutical, medical device, diagnostics, and biotechnology segments. The rise of predictive, preventive, and personalized medicine enabled by digital technologies will force players in the field to form networks with the health care delivery sector, forging entirely new business models for biomedical innovation. These networks, in turn, will contribute to precision medicine, what the National Research Council describes as the ability to classify individuals into subpopulations that differ in their susceptibility to a particular disease, in the biology or prognosis (or both) of those diseases they may develop, or in their response to a specific treatment.

In the years following completion of the Human Genome Project, the rise of direct-to-consumer (DTC) personalized genomics companies with names like Navigenics, 23andMe, deCODEme, Knome, and Pathway Genomics heralded what some called a new era of "Genomes for the Masses" powered by the proliferation and plummeting cost of DNA genotyping and sequencing. These companies offer services that search for genetic variants or markers associated with disease and then assign statistical probabilities of developing disease based on an individual's genetic variant profile. The number of disease-associated genetic variants identified by more than one study grew from 6 in 2003 to 3,000 in 2013. DTC genetic testing can provide individuals with health information based on statistical probabilities such as their risk for developing heart disease, cancer, diabetes, Alzheimer's disease, mental illness, and a variety of other disease conditions as well as genetic factors that may affect how they react to certain foods, drugs, alcohol, and caffeine.

Although studies show significant public interest in acquiring genetic risk information, many geneticists and public health officials question the value of DTC genetic testing in assessing disease risk, particularly for chronic diseases. These diseases are polygenic, and their associated genetic variants typically have low penetrance. In addition, though most DTC companies run their tests in laboratories certified by the Clinical Laboratory Improvement Amendments of 1988, the industry lacks uniform testing standards. Different companies can produce markedly different results from testing the same individual. Unless they have recently completed medical school and residency training, very few primary-care physicians are prepared to respond to questions their patients may have about DTC test results, which critics say may raise anxiety among consumers. In addition, there is the possibility that their privacy may be compromised, which is an ongoing concern of the Presidential Commission for the Study of Bioethical Issues. The tension between human genome sequencing and privacy was highlighted when researchers using a sophisticated algorithm and data from public genealogical records were able to identify individuals who had volunteered to have their genomes sequenced as part of a genomics study. Genomes, transcriptomes, and microbiomes constitute unique personal signatures. The personal privacy of "de-identified" public registry data derived from these studies cannot be guaranteed.

All of this has led federal government agencies and professional genetics associations to cast a jaundiced eye on the nascent DTC industry. An official from the US Government Accountability Office testifying before Congress described DTC genetic testing services to be "misleading and of little or no practical use." The US Food and Drug Administration claims jurisdiction to regulate commercial genetic testing including DTC companies, and in 2013 it laid down a marker. The agency issued a warning letter to 23andMe for marketing its Saliva Collection Kit and Personal Genome Service (PGS) "without marketing clearance or approval." The FDA wrote that the company "markets the PGS for providing 'health reports on 254 diseases and conditions,' including categories such as 'carrier status,' 'health risks,' and 'drug response,' and specifically as a 'first step in prevention' that enables users to 'take steps toward mitigating serious diseases' such as diabetes, coronary heart disease, and breast cancer." Though FDA warning letters are not legally binding, the government can take companies to court if steps are not taken to resolve FDA concerns about the safety and effectiveness of drugs and medical devices, which include genetic testing kits used for diagnosing disease. The *Wall Street Journal* recommended that 23andMe consider moving offshore: "The scientists and entrepreneurs helping to lead medicine into the genomic era have little need to operate inside the U.S. if that means begging the government for a hall pass every time they want to do something new and potentially life-saving."

Innovation in the biodigital age backed by market demand from individuals wanting access to their own genetic disease-risk information is likely to erode regulatory barriers meant to protect consumers from inaccurate tests and misinterpretation of test results. The Genetics and Public Policy Center at Johns Hopkins University identified twenty-seven DTC genetic testing companies (DTC or DTC through a physician) in the United States in 2011. Consumer genomics serves more than a million customers worldwide, including genealogical enthusiasts through Ancestry.com, FamilyTreeDNA.com, and similar services as well as through DTC genetic testing companies. The market is estimated to be worth tens of millions dollars with projections that it could reach hundreds of millions of dollars in the years ahead. Some of these companies collaborate with academic researchers by making genetic variation and phenotypic data available for study with the consent of their customers.

Consumer genomics based on genotyping of genetic variations is just the warm-up act for whole genome sequencing and whole exome sequencing, the sequencing of the protein-coding parts of genes (exomes). These technologies are poised to transform health care in the long run both for healthy people through disease prevention and for people with disease through early diagnosis and targeted therapies based on an individual's or a tumor's genetic profile. A complete genomic or exomic sequence typically comes loaded with confounding information ranging from single nucleotide polymorphisms (SNPs), to sequence inversions and deletions (indels), to copy number variations of chromosomal segments (CNVs). Sophisticated bioinformatics tools and thousands of whole genome sequences from diverse ethnic populations (e.g., the 1000 Genomes Project) are required to catalogue the immense variation that the human genome harbors. As the authors of a review on human genome sequencing in health and disease stated, the challenge for personal genomics is to identify disease-causing mutations among the approximately 3.0 million to 3.5 million simple nucleotide variations (SNVs, which include SNPs and indels) and the approximately 1,000 large CNVs, on average, in a given human diploid genome. More accurate sequencing, deep resequencing, and improved algorithms are needed to meet that challenge. Meanwhile, whole genome sequencing and whole exome sequencing have proved effective for studying diseases, among them dominant and recessive Mendelian disorders (principally through exome sequencing); congenital, metabolic, autoimmune and bone disorders; neuropathies, mental retardation, autism, leukemias, and carcinomas. Yet with all the gene identification tools we now have on hand, some 20,000 of the estimated 23,000 protein-coding human genes have no assigned function or trait association.

Genome sequencing technologies and their application to thousands of ethnically diverse human genomes are driving the development of pharmacogenomics by identifying genetic variants associated with how an individual metabolizes a given drug. That information allows clinicians to adjust dosage to maximize the drug's effectiveness and minimize its side effects. In the decade from 2003 to 2013, the number of prescription drugs with pharmacogenomic information on the label grew from 46 to more than 120. In the case of cancer, whole genome sequencing has enabled clinicians to discover that seemingly unrelated tumors can share specific genetic features and associated cellular signaling pathways known to be responsive to already existing drug therapies. Rare and common DNA variants important in pharmacogenomics are typically found in genes but also in nongenic regions of the genome. More than 98 percent of the human genome is nongenic. Included in that 98 percent of noncoding sequences are nearly 90 percent of single-letter DNA variants (SNPs) associated with disease risk found through genome-wide association studies and used extensively by DTC gene testing companies.

Yet the fact that these sequences do not code for proteins does not mean they are not functional. That cautionary note was highlighted by the Encyclopedia of DNA Elements (ENCODE) Project. The collaborative project systematically mapped regions of gene transcription (the transcriptome), chromatin structure, and histone modification as well as the epigenome (compounds called methyl groups mark the genome and play a role in turning genes off or on), enabling an international scientific team to assign biochemical functions for genomic sequences previously described as "junk DNA." Results from the massive data set the project produced, reported in 2012, showed that putative genetic regulatory elements, many of them non-protein-coding RNA transcripts (long noncoding RNA), can reside in the genome far away from the genes they regulate. Sometimes they are tens or hundreds of thousands of bases away. Sometimes they are on different chromosomes. Many of these distant switch-like elements may regulate gene expression in a given cell type. That, in turn, may determine whether a gene is functioning normally or, as in the case of disease and disease susceptibility, abnormally. If genetic function of these distant regulatory elements is firmly established, the extra time and expense of whole genome sequencing can be justified over sequencing just exomes (1.2 percent of the genome). The ENCODE Project challenges gene-centric conventions, causing Thomas Gingeras of Cold Spring Harbor Laboratory to declare: "New definitions of a gene are needed." Indeed, ENCODE prompts reexamination of the widely held view that a gene is a specific intact sequence at a specific chromosomal location activated by neighboring promoters and enhancers. Rather, chromosomes with their chromatin threads may behave like Pilobolus dancers as they flex and curl and connect to put remote regulatory elements in contact with the genes they regulate. Genomes, like natural ecosystems, innovation clusters, and other nonlinear adaptive systems, are more than the sum of their parts.

The ENCODE Project was able to scale up dramatically from its pilot study reported in 2007—producing five times the data it originally anticipated at no extra cost—thanks to massively parallel NGS technologies. That year executives from 454 Life Sciences, an NGS pioneer, showed seventy-eight-year-old James Watson, co-discoverer of the structure of DNA, a first draft of his own genome it had sequenced at a cost of $2 million. Whole genome sequencing raises the possibility of locating key DNA markers for health and disease by looking at our entire code of As, Ts, Gs, and Cs—all 3 billion letters in a haploid chromosome set and 6 billion letters in a diploid chromosome set. The technology is rapidly becoming a practical reality. The reason is unusually rapid technological innovation at the nanoscale level. From 2001 to 2007, the cost per human-sized genome sequenced and the cost per DNA megabase of genome sequenced tracked the computer chip price trend of Moore's Law very closely, showing a gradual decline from $100 million to $10 million for an entire sequence and $8,000 to $800 for a sequenced DNA megabase (1 million base pairs). Around 2007 the trend in sequencing costs parted company with Moore's Law, dramatically (Figure 4.2). By 2012, the cost of sequencing a genome had dropped to less than $8,000, and the cost of sequencing a DNA megabase had dropped to one cent. Both represented exponential improvements over Moore's Law. Had DNA sequencing costs continued to trend with Moore's Law, the cost of sequencing a human-sized genome in 2012 would have been $4 million rather than $10,000, and the cost of sequencing a DNA megabase would have been $200 rather than a penny.

When Gordon Moore made his prediction in 1965, an integrated circuit contained thirty transistors; in 2012 a microprocessor could contain more than 2 billion transistors. If DNA sequencing succeeds in establishing a new law remotely approaching the half-century durability of Moore's Law, it is possible to imagine that decoding any species could eventually be as routine as an ATM transaction.

Data Overload and Lingering Doubts

Six generations after Mendel began systematically recording genetic data from his pea experiments, spectacular advances in DNA sequencing are producing a data overload problem. For all the advances Moore's Law has made possible, computers are scrambling to keep up. Making sense of raw data is why Mendel is one of five biological researchers listed in scientist and polymath Stephen Wolfram's timeline of systematic data and the development of computable knowledge dating from 20,000 BC to the present. The other four are the ancient Roman physician Galen for organizing anatomy and physiology, the Swedish taxonomist Carl Linnaeus for systematizing the classification of living organisms, and James Watson and Francis Crick for discovering that DNA contains "a digital genetic code." It took two decades from the time Watson and Crick discerned the structure of DNA for scientists to develop DNA sequencing methods and another three decades for the human genome to be sequenced. That feat is also represented in Wolfram's timeline because it reflects the growing systematization of data and knowledge and the ability of such systems to provide "core infrastructure for the world." In short, knowledge systems enable future innovations and inventions.

Mendel's data notebooks have been lost, but there is little doubt that if they existed they could easily be captured and stored on a standard CD-ROM compact disc. The 1000 Genomes Project is described as one of the largest distributed data collection and analysis projects ever undertaken in biology. It amassed more than 260 terabytes of data from 1,092 human subjects from fourteen geographically diverse ethnic populations, the rough storage capacity of 400,000 compact discs. The University of California Santa Cruz hosts the world's largest cancer-genomics data hub (CGHub) to store and make available the data from US cancer-sequencing projects. CGHub is planning to store a terabyte of data from each of 10,000 tumors from patients that will be sequenced as part of a project that is cataloging genetic mutations responsible for cancer, the Cancer Genome Atlas. CGHub in partnership with the Silicon Valley startup Annai Systems launched with a storage capacity of five petabytes (one petabyte is roughly 1,000 terabytes) and can scale up to twenty petabytes.

The nascent industry that is responsible for flooding genetic researchers and databanks with information, outstripping their ability to store, transmit, and analyze the data, makes high-throughput sequencing machines and systems with names like GS20, HiSeq, Proton, SMRT, Polonator, and SOLiD. NGS companies—Illumina, 454 (Roche), Ion Torrent (Life Technologies), Oxford Nanopore, Pacific Biosciences, and Applied Biosystems (Life Technologies)—manufacture these machines. These companies are racing to provide ever-faster, ever-cheaper, and more accurate sequencing capabilities, with longer sequence read lengths and shorter run times, based on strikingly different technologies. Oxford Nanopore Technologies' revolutionary MinION device, a sequencing tool the size of a memory stick that can be plugged directly into a laptop's USB port, was made available to participants in an early access program in 2013. Do these companies represent the DNA sequencing counterparts to firms that pioneered the information technology behind Moore's Law—Fairchild Semiconductor, Texas Instruments, National Semiconductor, Intel, and Advanced Micro Devices?

However that question is answered, their machines are populating sequencing laboratories around the world, exploring human and animal health, plant genetics, conservation genetics, bioremediation, bioenergy, and other fields. As of 2013, the number of high-throughput sequencing machines in North America numbered more than 1,000 (the great majority in the United States), more than 800 in Europe, and nearly 600 in Asia, followed by Australia, South America, and Africa. By one estimate, made in 2011, these machines together are producing thirty-three terabytes of DNA sequence data per day and twelve petabytes per year, an output

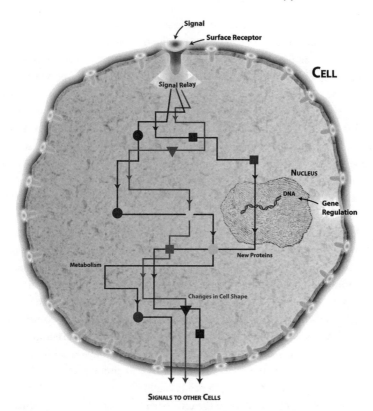

FIGURE 4.3 A simplified illustration of a biological signaling pathway in a human cell. Hundreds of signaling pathways regulate gene expression, protein production, metabolism, cell shape, and cell-to-cell interactions. They are keys to understanding health and disease and for developing drug and cell regenerative therapies. By designing and enhancing circuits and pathways in mammalian and plant as well as microbial cells, scientists and engineers aim to make the cell a bioproduction center for the twenty-first century economy.

Source: The National Human Genome Research Institute, http://www.genome.gov/27530687.

that is growing five-fold annually. Knome, DNAnexus, Five3 Genomics, Real Time Genomics, Gene By Gene, Genophen, Foundation Medicine, Personalis, Bina Technologies, Curoverse, and Illumina's BaseSpace platform are players in a growing commercial market for providing whole genome sequence processing and analysis services for genomic data. With petabytes of storage capacity for patient genomic data, Foundation Medicine has lured big pharma and big names like Bill Gates to invest in its diagnostic test and analytics of the 200 or more genes suspected of driving or influencing cancer growth. Cancer genome sequencing and other technologies have revealed that these genes operate through one or more of a dozen cellular signaling pathways in malignant tumors that regulate three core cellular processes: cell fate, cell survival, and genome maintenance (Figure 4.3). A cancer's genetic profile, not the tissue or organ where it originates, marks its true character.

At the beginning of the industrial era, James Watt and his employee John Southern invented a device for automatically tracing variation of pressure with volume in a steam engine and used today in cardiovascular and respiratory physiology (pressure-volume or indicator diagram). That is why Stephen Wolfram included Watt in his timeline of how our civilization "has systematized more and more areas of knowledge—collected the data associated with them, and gradually made them amenable to automation." The biomathematician Mendel would surely agree. It was his ability to see distinct patterns in his data and make deductions that earned him his place in history. In our age of deep automation, the challenge for bioinformatics is to develop improved algorithms for sequence assembly, pattern recognition, and data compression. The problem is we cannot count on algorithmic breakthroughs sufficient to deal with the vast amounts of data being generated. A more practical alternative, as Cold Spring Harbor Laboratory's Michael Schatz and his bioinformatics colleagues see it, is to find ways that make better use of multiple computers and processors that take apart tasks and perform them in parallel. Valuable bioinformatics resources have been developed for cloud computing, the computational resources made available by vendors over the Internet like Amazon Web Services for processing large data sets. The key is to know what to process and what to discard in a raw data set, which can be enormous and unwieldy. In theory, when the human reference sequence is established on the basis of millions of genomes, we will be able to set aside the estimated 99.5 percent of the sequence that is shared among all humans and focus only on the differences of a given genome with the reference genome. In the end, those differences will probably constitute a few gigabytes of data and thus could be carried on a handy data storage device such as a thumb drive.

A decade after President Bill Clinton announced in 2000 that the first draft of the human genome sequence was complete, critics argued that the human genomics field had yet to live up to its billing as a predictor of disease let alone as a foundation for a new generation of therapies. Linking common genetic variations tightly with disease risk proved devilishly difficult. The tight linkage of disease risk seemed to be reserved for rare genetic variations. Only by sequencing all of an individual's genes—indeed all an individual's DNA based on results from the ENCODE Project—are rare variations revealed that could determine his or her genetic risk.

The number of gene mutations found in rare genetic disorders doubled to nearly 3,000 in the decade following publication of the human genome in 2003. NGS technologies are expected to identify the mutations responsible for the remaining 4,000 Mendelian or monogenic diseases affecting millions of people worldwide by 2020, if not before. Precise genetic editing technologies such as the CRISPR-Cas system mean that such genetic mutations could potentially be corrected in patients who have them, an idea that has launched startup companies. Geneticists gave a preview of the power of personal genomics in 2010 when they reported that whole genome sequencing enabled them to pinpoint rare mutations that cause recessive Mendelian disorders in families. The findings suggested to geneticists that it is possible to sequence the entire genome of a patient "at reasonable cost and with sufficient accuracy to be of practical use to medical researchers," wrote the veteran science journalist and author Nicholas Wade in *The New York Times*. One of the studies, conducted by a team that included innovator Leroy Hood's Institute for Systems Biology and the sequencing company Complete Genomics (purchased by Shenzhen-based BGI in 2013), laid out a vision of where genomics enabled by digital technology is taking us just a century and a half after Mendel published his famous paper: "As our knowledge of gene function increases, we will be able to use the power of family genome analysis rapidly to identify disease-gene candidates." Together with relevant environmental and medical information, they wrote, such data will characterize the integrated medical records we can expect to see as a standard feature of future medical care.

Not everyone sees a future in which whole genome sequencing revolutionizes personal health and medical practice. Despite its indisputable benefits, whole genome sequencing "will not be the dominant determinant of patient care and will not be a substitute for preventative medicine strategies incorporating routine checkups and risk management based on the history, physical status, and life-style of the patient." So concluded a study, published in *Science Translational Medicine* in 2012, on the predictive capacity of personal genome sequencing. The authors used a mathematical model of monozygotic twin pairs with "identical genetic risk factors" to determine the capacity of whole genome sequencing to identify individuals at clinically significant risk for twenty-four different diseases. Most of the tested individuals would test negative for most diseases. The predictive value of the negative tests would generally be small "because the total risk for acquiring the disease in an individual testing negative would be similar to that of the general population." A clinically meaningful risk for at least one disease in most patients is the best that could be expected from whole genome sequencing.

The study drew a great deal of news media attention, including front-page coverage in *The New York Times* under the headline "DNA's Power to Predict Illness Is Limited." The Scripps Research Institute's Eric Topol, a leading advocate for genomic medicine, countered in a letter to the journal that, mathematical models aside, the predictive value of whole genome sequencing cannot be known until a large number of individuals with like conditions have been sequenced. The study's authors, including Bert Vogelstein of Johns Hopkins University, responded that Topol undervalued the utility of twin studies. The data they used "are the result of tenacious, decades-long efforts by epidemiologists around the world." Monozygotic twins, they argued, have been studied throughout their lifetimes and "are internally controlled at the genetic level: One twin is a perfect genetic control for the other."

Yet evidence from genomic analysis suggests that monozygotic twins are not by definition genetically identical and do not have "identical genetic risk factors." Sometimes the presence of CNVs—the duplication and deletion of chromosomal segments—in the somatic cells of some twins makes them genetically different from their fellow twin. As we noted above, CNVs are identified through whole genome genotyping and sequencing. They are believed to be associated with disease risk in some instances and may be causative factors, including in monozygotic twins. That monozygotic twins are genetically identical has been a bedrock assumption in biomedicine and the behavioral sciences for decades. Conventional wisdom holds that studies of identical twins help disentangle environmental and genetic influences on individual traits and behaviors. Twin studies are considered to be the "gold standard" for weighing the relative importance of genetic susceptibility and environmental influences to the development of complex traits and diseases in humans.

Differences in the genetic profiles of monozygotic twins are telling. Dutch and Australian researchers who reviewed the utility of twin studies "in the omics era" concluded that the study of discordant monozygotic twins—twins who derive from a single fertilized egg cell but who are dissimilar for a certain characteristic or disease—is "a powerful method to identify DNA sequence variants, epigenetic variation and metabolites that are associated with disease." Indeed, twin registries offer a valuable resource for large-scale molecular studies. NGS across many different tissues and cell types will help us find genome-wide SNPs, CNVs, and epigenetic variation in discordant twins "at an unprecedented scale," meaning that twins are likely to continue to make an exceptional contribution to our understanding of how heredity, environment, and disease are linked.

Laboratory Medicine Meets Large-Scale Genomics

Most clinical decision making, as much as 70 percent by one account, is based on laboratory data even though generating that data accounts for a relatively small fraction of the $2.6 trillion Americans spend each year on health care. Large-scale genomics now constitutes a major activity in laboratory medicine and pathology, one that is certain to grow. The integration of genomic data in laboratory testing is part of the drive toward bringing evidence-based medicine to the hospital and clinic and improving treatment outcomes while reducing costs. Whether genomic testing can make a contribution toward fashioning "a more nimble health care system that is consistently reliable and that constantly, systematically, and seamlessly improves," a goal recommended by the Institute of Medicine, remains to be seen. Health care, the Institute contends, should look to other industries for guidance on "how to better meet specific needs, expand choices, and shave costs." It reported that about one-third of annual health care expenditures—$750 billion—is wasted, including more than $200 billion spent in the United States in 2009 on unnecessary services such as needless tests or repeated tests. As Bill Gates has observed, most innovations in the health care field today actually increase the net costs of the health care system. Health information technology has yet to make significant inroads into that dim reality. McKinsey & Co. analysts suggest that if early successes are brought to scale, big-data applications "could eventually strip more than $300 billion in costs from the nation's health care system and improve transparency to drive better patient outcomes."

Genomic testing is still at an early stage. As it is systematically incorporated by clinical laboratories, it can be expected to have an important economic impact as well as improving medical outcomes and quality of life. In an analysis for the American Clinical Laboratory Association, the Battelle Memorial Institute found that genetic testing—disease diagnosis, genotyping, and presymptomatic testing to determine disease risk, pharmacogenomic testing to determine drug safety and efficacy, identity and forensic testing, and carrier and newborn screening—is a major enterprise. It already produces more than $16 billion annually in national economic output and supported more than 100,000 jobs in the United States in 2009. Such testing is "at the heart of a new paradigm of medicine that is evidence-based and rooted in quantitative science." These increases in economic activity around clinical testing flow directly from the Human Genome Project and subsequent research and development, which Battelle estimated in 2011 had produced nearly $800 billion in economic output and more than 310,000 jobs, though the figures were disputed by some economists. In a follow-up study published in 2013 Battelle found that the $14.5 billion in federal investment in genomics between 1988 and 2012 produced nearly a trillion dollars in economic output in fields ranging from biomedicine to energy and agriculture. The investment contributed $31 billion to the US GDP in 2012 alone, according to the report. "Every dollar we invested to map the human genome returned $140 to our economy—every dollar," said US President Barack Obama in his 2013 State of the Union address, drawing directly from Battelle figures. But the greater value of genomic science resides in its contribution to human, animal, and environmental health rather than its contribution to GDP or job creation. As genetic and genomic tests become more specific, precise, and accurate, and as the analytics needed to interpret them correctly are advanced, the chances of misdiagnoses and unnecessary treatments will be reduced, patient outcomes will improve, and the burden of inefficient health care on economic productivity and growth will be eased.

The College of American Pathologists (CAP), which accredits thousands of clinical laboratories in the United States and abroad, published a revised version of its molecular pathology checklist in 2012 that addressed NGS. CAP checklists are used in the accreditation process to help laboratories meet the Centers for Medicare & Medicaid Services requirements for human clinical laboratory testing, which the agency regulates through Clinical Laboratory Improvement Amendments (CLIA) certification. The revised checklist includes NGS library creation, sample preparation, barcoding, sequencing, bioinformatics, variant identification, annotation, and the final patient report describing the methods, reagents, and bioinformatics analysis used. Although many clinical labs were already using NGS tests, until CAP revised its checklist there had been no standardized framework established that could be used to validate them and document their performance. The CAP NGS working group that recommended the standards, which included members from the Association for Molecular Pathology and the American College of Medical Genetics and Genomics, also recommended proficiency testing guidelines. The CAP standardization and accreditation of genomic testing is critical. It is designed to assure the highest quality use of this emerging technology for patient care. Patients and their physicians and representatives will demand it. In the European Union, the European Society for Human Genetics recommended guidelines for whole genome sequencing in health care, including diagnostic testing, clinical data, counseling, and informed consent. Step by step, the chemical sequence underpinning Mendel's laws is changing laboratory medicine and pathology and, as a consequence, how physicians make decisions about how to treat their patients and how public health officials make decisions about emerging disease threats.

In the future, genomic information will be integrated into electronic medical records in a systematic way. The Electronic Medical Records and Genomics (eMERGE) Network, organized by the National Human Genome Research Institute, is one such project. eMERGE is a national consortium of medical research institutions formed in 2007. Its goal is to develop, disseminate, and apply approaches to research that combine DNA biorepositories with electronic medical record systems for large-scale, high-throughput genetic research. The consortium has shown the feasibility of using electronic medical record systems to investigate gene–disease relationships between targeted genotypes and multiple phenotypes (phenome-wide association studies). Such investigations are also being advanced through initiatives such as the Genetic Testing Registry (GTR) and ClinVar, both developed by the National Center for Biotechnology Information, part of the NIH. GTR is a free online tool in which health care providers type in a disease or gene to get a list of available genetic tests. Companies that develop such tests are invited to share detailed information about them and explain how the information improves health for patients and families. ClinVar integrates numerous existing genetic variant databases and provides repository services, giving clinical testing laboratories access to aggregated information about gene variations linked to human health—a one-stop shop for disease genes. MutaDATABASE, an international open-access online DNA variation database, is another example of the effort to normalize and share genomic information and link genetic variants with human health and disease. The entry of genomic data into health systems was highlighted when Britain's National Health Service was allocated £100 million ($161 million) in 2012 to undertake whole genome sequencing of 100,000 patients. "By unlocking the power of DNA data, the NHS will lead the global race for better tests, better drugs and above all better care," said British Prime Minister David Cameron. In the United States, government grants, foundations, and health care organizations currently provide most of the funding for clinical genomics. Clear clinical utility of whole genome sequencing will need to be demonstrated before health insurance companies will cover its costs.

At the individual patient level, the NGS firm Illumina launched an individual sequencing service in 2012 that can provide a physician with his or her patient's whole genome sequence at thirty-fold coverage (meaning the genome has been sequenced thirty times to eliminate errors) in two weeks. The service, which is CLIA certified and CAP accredited, is designed to assist clinicians with diagnosis and treatment decisions. It is available only through a physician's order, and the information is given only to the physician and not the patient. For patients with cancer, the genomes of both healthy and tumor cells are sequenced. The information derived is transmitted to the physician's iPhone or iPad together with a reference genome and bioinformatics tools for identifying structural variations within, say, a tumor gene or anomalies identified in the patient's genome. The health care provider with the assistance of a geneticist or pathologist translates the information into a standard of care for the patient with the goal of delivering optimal treatment.

Services like Illumina's TruGenome individual sequencing service not only have the potential to change health care delivery, says one health care analyst, "but can be a catalyst for shaking up existing life science business models." The National Human Genome Research Institute's eMERGE consortium, the National Center for Biotechnology Information's ClinVar, and Illumina's individual sequencing service are expressions of the effort to align science and informatics, patient–clinician partnerships, and incentives and instill a culture of continuous improvement. They reflect a systematic approach to health care that the Institute of Medicine says can improve outcomes and reduce costs.

Digital Technology Recodes the Genetics of Communications

Plant photosynthesis is the foundational natural process for life on Earth. Sequencing the half a million species of flowering plants or angiosperms, 100 of which supply 90 percent of the calories humans consume, is a long-term goal for plant science. The eyes of an Augustinian monk were the observational tools employed to establish the first genetic laws. Flow and image cytometry, laser-based technologies that count, sort, image and classify cells, are examples of observational tools used in the laboratory today. Such instruments give biologists the ability to take technological "snapshots" of almost any attribute of a living organism whether a DNA sequence, a genetic mutation, a protein, an ion channel, or a cellular signaling pathway, says plant scientist David Galbraith. He measures gene expression in individual cells and the nuclear DNA content of angiosperms, which on average have genomes twice the size of humans. Data acquisition is no longer a problem. "We are experiencing a revolution, underpinned and empowered by Moore's famous law and the computational power enabled by that law, that has largely eliminated the traditional scientific bottle-neck associated with data acquisition."

Zooming out from the research laboratory to the landscape of health systems and competitive global industries, the picture is also one of technological empowerment and with it disruptive change. The inexorable unfolding of Moore's Law is forcing the health care and bioscience industries to adapt. Players failing to evolve and participate in the industry's transformation will not succeed and ultimately will thwart patients seeking to reap the benefits of the new technologies, say economists Alicia Löffler and Scott Stern in describing the future of the biomedical industry in an era of globalization. Winners in the personalized medicine market will be "highly profitable, highly specialized small and medium-size companies, mainly in the biotech sector, focusing on research and health delivery innovation." The number of companies forging alliances with non-traditional groups, away from pharmaceutical companies, can be expected to grow given the dynamics of the field.

As the code of life enters both the clinic and the commercial space for purchase by consumers interested in personal characteristics, personal health, and family history, digital technology is expanding its presence in biopharmaceutical discovery, development, and production. Battelle estimates that the biopharmaceutical industry in the United States has an economic impact of nearly $1 trillion annually. Three-fourths of the industry is made up of small and midsized firms. Overall it depends on continuing advances in the power of microprocessors, software, and computer networks and has identified a number of key technology infrastructure needs. Among them are bioimaging, biomarkers, bioinformatics, gene expression tools, commercial manufacturing standardization, and postmarket surveillance protocols for adverse event data. All depend to one degree or another on digital technology. Because such data often travels over high-speed networks, a sound communications infrastructure is a fundamentally enabling precondition for the adoption of standardized protocols and data-sharing systems in the biopharmaceutical industry.

The tools of two-way communication used by the founders of the biotechnology in the late 1970s and early 1980s were the telephone, the fax machine, snail mail, printed material, and, most important of all, face-to-face meetings in restaurants, coffee shops, bars, and more mundane establishments such as offices. They were the communications DNA of the new industry. Through their use ideas were exchanged, intellectual property was secured, investment was procured, people were hired, and businesses were built. Because biotechnology is a science-based enterprise, the businesses tended to be built in the neighborhood of research centers where face-to-face communication between scientists, entrepreneurs, and venture capitalists was easy to facilitate: Genentech near Stanford, Berkeley, and the University of California San Francisco; Biogen near Harvard and MIT; and Hybritech near the University of California San Diego and the Salk and Scripps Institutes. Social scientists have developed interorganizational linkage diagrams and "family tree" models for how regional biotech industries evolved from anchor firms and their university partners. No one can track precisely how social networks in cities and regions fostered by the communications tools noted above led to thriving bioscience clusters. Such phenomena have kept analysts and policymakers busy for three decades.

Today we have a bundle of new networking instruments and mobile communications tools. People called users operate them, and users are transforming how innovation is done. "Ingenious leading-edge users—not everyday consumers or profit-focused products—are becoming the economic engines that drive innovation," wrote Eric von Hippel and Michael Schrage in 2007 just as smartphones moved into the mainstream. "In sectors as diverse as software, biotechnology, medical instrumentation, telecommunications and sports equipment, users are spurring growth." The democratization of innovation brought about by users is advancing traditional bioscience fields like biochemistry and emerging fields like genomics and synthetic biology. By being "democratized," says Lawrence Lessig, a founder of Creative Commons, the right to innovate is no longer dependent on "a special status or hierarchy within some company or government." Innovation is more open and participatory than it has ever been. Perhaps nothing better illustrates the change than the crowdsourcing of science through online puzzle video games like those for genetic disease research (phylo.cs.mcgill.ca), protein folding (Foldit.com) and RNA design (EteRNA.cmu.edu). These games attract hundreds of thousands of players and yield valuable insights into biomolecular design and function.

With the relentless insinuation of the new communications tools into everyday life, doctors themselves are users and clinics are being converted into their own innovation communities. Over time more patients will become users, follow their care, evaluate their physician, and understand the value of the care they are receiving. Consider the growing use of mobile health devices such as medical smartphones and "smart" Band-Aids that can transmit physiological information from the patient to his or her doctor's office or a clinical laboratory. Real-time information on patient health means that doctors have the opportunity to shift from treating illness and disease to working with patients to maintain their health and well-being.

These technologies give everyone a digital voice—doctors, patients, researchers, inventors, financiers, economic developers and city planners, and ordinary citizens. User-generated content and feedback can lead to novel insights and new products and service delivery systems. For these and other reasons, digital technologies are disruptive to familiar structures and ways of doing things. They are disruptive not just in industries like music and entertainment and journalism but in deeply entrenched and government-regulated systems like health care, with its three decades of cost inflation in the United States, as "disruptive innovation" scholar Clayton Christensen and his coauthors describe in their book *The Innovator's Prescription*. Digital communications networks and devices and their users are inexorably altering the model of biotechnology industrial development that we have seen for more than three decades.

Digital technology in its many and varied manifestations will grow in importance to bioscience innovation, from the laboratory to the incubator to the bioscience firm that brings a drug, a device, a diagnostic, or a service to market. Unlike almost everything else, costs associated with computing power continue to fall. As air and ground transportation grow more costly, digital communications technologies from videoconferencing, desktop, laptop, and notebook computers to smartphones and electronic tablets will continue to decline in relative cost per performance capability.

These technologies are accelerating the pace of biological discovery through smarter instruments, networked science, and open access. They are also accelerating the rate of understanding, adoption, and use of biology and biological technologies.

The field of genetics, where we began this chapter, is encountering what some call the "app economy," the creation and proliferation of useful programs for smartphones and tablet computers. iCut DNA, for example, makes available a restriction enzyme database on a mobile device that enables researchers and graduate students to look up the recognition site where an enzyme cuts DNA. Genetic Decoder gives them ready access to information about amino acid molecules either through listings or images. A mobile phone version was released for bioGPS, a web-based gene annotation portal for learning about gene and protein function. A number of educational apps allow users to explore Mendel's laws of inheritance and even "swap DNA" between devices to create novel images based on the laws.

Then there is the matter of the human genome meeting clinical medicine. DNA sequence information transmitted by mobile devices has debuted in the clinic. Nothing better illustrates the changing communication of genetics and the changing genetics of communication in our expanding information ecosystem than that. Nothing better illustrates how far we have come since an Augustinian monk published a scientific paper reporting the results of his hybridization experiments with the garden pea.

References

Listed sequentially based on their order in the chapter.

Frank Salomon, "Testimonies: The Making and Reading of Native South American Historical Sources," in *The Cambridge History of the Native Peoples of the Americas*, Vol. 3, Bruce G. Trigger et al., eds. (Cambridge UK: Cambridge University Press, 1999), 33.

Petr Smýkal et al., "Pea (*Pisum sativum* L.) in the Genomic Era," *Agronomy* 2 (2012):74–115.

Michael Khan, "Czech City Bids to be Global Biotech Hub," Reuters, September 3, 2008.

Ernst Mayr, *The Growth of Biological Thought: Diversity, Evolution, and Inheritance* (Cambridge, MA: Harvard University Press, 1982).

William D. Stansfield, "Mendel's Search for True-Breeding Hybrids," *Journal of Heredity* 100.1 (2009): 2–6.

David Galton, "Did Darwin Read Mendel?" *QJM: An International Journal of Medicine* 102.8 (2009): 587–589.

Richard Dawkins, *The Greatest Show on Earth: The Evidence for Evolution* (New York: Free Press, 2009).

MendelWeb, accessed February 3, 2010, http://www.mendelweb.org.

Kevin Kelly, *What Technology Wants* (New York: Viking Press, 2010).

Mendel and the Transmission of Scientific Information

"The Next 100 Years of Science: Long-term Trends in the Scientific Method." Talk by Kevin Kelly, Long Now Foundation, March 10, 2006, accessed February 3, 2010, http://foratv.org.

G. Pascal Zachary, *Endless Frontier: Vannevar Bush, Engineering of the American Century* (New York: Free Press, 1997).

"Science, The Endless Frontier," Report to the President by Vannevar Bush, Director of the Office of Scientific Research and Development, July 1945 (Washington, DC: US Government Printing Office, 1945).

Vannevar Bush, "As We May Think," *The Atlantic Monthly* (July 1945): 101–108.

Gunther S. Stent, "Prematurity and Uniqueness in Scientific Discovery," In *Paradoxes of Progress* (New York: W. H. Freeman, 1978), 94–114.

"President Bush Signs Omnibus Appropriations Bill, Including National Institutes of Health Research Access Provision," *Medical News Today*, December 29, 2007, accessed February 3, 2010, http://www.medicalnewstoday.com.

Gavin Baker, "Public Science: NIH's New Open Access Policy Can Benefit Everyone," *Science Progress*, January 28, 2008, accessed February 3, 2010, http://www.scienceprogress.org.

John Whitfield, "Open Access Comes of Age," *Nature* 274 (2011):428.

Mikael Laakso et al., "The Development of Open Access Journal Publishing from 1993 to 2009," *PLoS One* 6.6 (2011), accessed September 3, 2012, doi: 10.1371/journal.pone.0020961.

Report of the Working Group on Expanding Access to Published Research Findings—The *Finch* Group, "Accessibility, Sustainability, Excellence: How to Expand Access to Research Publications," June 2012, accessed September 13, 2012, http://www.researchin-fonet.org/publish/finch/

Mike Taylor, "Open Access Means a Bright Future for Scientific Research," *The Guardian*, July 17, 2012.

Richard Van Noorden, "Britain Aims for Broad Access," *Nature News* 486 (June 19, 2012): 302–303.

Horizon 2020: The EU Framework Programme for Research and Innovation, accessed September 13, 2012, http://ec.europa.eu/research/horizon2020/index_en.cfm

J. P. Holdren, Public Access Memorandum From the Office of Science and Technology Policy, the White House, February 22, 2013, accessed December 3, 2013, http://www.whitehouse.gov.

Michael Stebbins, "Expanding Public Access to the Results of Federally Funded Research," White House Office of Science and Technology Policy, February 22, 2013, accessed February 24, 2013, http://www.whitehouse.gov.

Justin Hicks and Robert D. Atkinson, "Eroding Our Foundation: Sequestration, R&D, Innovation and U.S. Economic Growth," Information Technology & Innovation Foundation, September 2012, accessed March 4, 2013, http://www2.itif.org/2012-eroding-foundation.pdf.

Andrew Odlyzko, "Tragic Loss or Good Riddance? The Impending Demise of Traditional Scholarly Journals." *International Journal of Human-Computer Studies* 42.1 (1995):71–122.

"MIT Marks OpenCourseWare Milestone," *MIT Tech Talk* 52.10 (November 28, 2007).

"Harvard Faculty Adopts Open-Access Requirement," *Chronicle of Higher Education*, February 12, 2008.

Jeffrey R. Young, "Coursera Hits 1 Million Students, with Udacity Close Behind," *Chronicle of Higher Education*, August 12, 2012, accessed September 16, 2012, http://chronicle.com/blogs/wiredcampus/coursera-hits-1-million-students-with-udacity-close-behind/38801

George M. Church, Yuan Gao, and Sriram Kosuri, "Next-Generation Digital Information Storage in DNA," *Science* 337 (2012):1628, accessed September 14, 2012, doi: 10.1126/science.1226355.

Nick Goldman et al., "Towards Practical, High-Capacity, Low-Maintenance Information Storage in Synthesized DNA," *Nature* 494 (2013):77–80.

Sebastian Anthony, "Harvard Cracks DNA Storage, Crams 700 Terabytes of Data into a Single Gram," ExtremeTech.com, August 17, 2012, accessed September 14, 2012, http://www.extremetech.com.

Cultivating Mendel's Garden with Computational Tools and Webs

Second Life Website, accessed November 20, 2012, http://www.secondlife.com.,

Lawrence Lessig, *Remix: Making Art and Commerce Thrive in the Hybrid Economy* (New York: Penguin Press HC, 2008).

"The Life of a Global Delivery Center," *Industry Week*, October 22, 2007.

"IBM Opens New 3D Virtual Healthcare Island on Second Life," News Release, IBM Corp., February 24, 2008.

Data Mining and Gene Mining Join Forces

Maura Lerner, "Mayo Clinic Lands Own Fantasy Island," *Star Tribune* (Minneapolis), August 11, 2010.

"Mayo Clinic, IBM Aim to Drive Medical Breakthroughs." News Release, IBM Corp, August 4, 2004, accessed January 9, 2012, http://www-03.ibm.com/press/us/en/pressrelease/7255.wss.

"EU's Largest Platform for Medical, Biotech Research to Open In Czech Republic," Market Wire (press release), March 21, 2006, accessed January 9, 2012, http://www.marketwire.com/press-release/-691352.htm.

Jeff Hansel, "Mayo to Collaborate with Czech Research Center," *Post Bulletin* (Rochester, Minnesota), September 26, 2011.

Ian Armstead et al., "Cross-Species Identification of Mendel's *I* Locus," *Science* 315 (2007):73.

Arabidopsis Genome Initiative, "Analysis of the Genome Sequence of the Flowering Plant Arabidopsis Thaliana," *Nature* 408 (2000):796–815.

R. A. Fisher, "Has Mendel's Work Been Rediscovered?" *Annals of Science* 1 (1936):115–137.

Simon Mawer quoted by Jason Bobe at Personal Genome Website, January 2008, accessed February 10, 2010, http://thepersonalgenome.com/2008/01/.

Simon Mawer, *Mendel's Dwarf* (New York: Harmony Books, 1998).

Tony Arioli et al., "Molecular Analysis of Cellulose Biosynthesis in Arabidopsis," *Science* 279 (1998):717–720.

"Monsanto, Others Invest in Mendel Biotechnology," *San Francisco Business Times*, October 15, 2007.

Bryan Sims, "Mendel Biotechnology, BP Biofuels to Conduct Miscanthus Trials," *Biomass Magazine*, March 21, 2012.

Matthew L. Wald, "Fuel From Waste, Poised as a Milestone," *The New York Times*, November 13, 2012.

Biology, Technology, and Planetary Concerns

Nicholas Stern, *The Economics of Climate Change: The Stern Review* (Cambridge UK: Cambridge University Press, 2007).

John Tyndall, "The Belfast Address," in *Fragments of Science: A Series of Detached Essays, Addresses and Reviews* (New York: D. Appleton & Co., 1884), 506.

Mike Hulme, "On the Origin of 'the Greenhouse Effect': John Tyndall's 1859 Interrogation of Nature," *Weather* 64.5 (2009):121–123.

"The Stern Review on the Economics of Climate Change," Tyndall Centre for Climate Change Research, Submitted to HM Treasury, December 9, 2005, accessed February 4, 2010, http://www.tyndall.ac.uk.

Nina V. Fedoroff et al., "Radically Rethinking Agriculture for the 21st Century," *Science* 327 (2010):327–328.

"GM Crops: A Story in Numbers," *Nature* 497 (2013):22–23.

"Fields of Gold," *Nature* 497 (2013):5–6.

Daniel Cressey, "Transgenics: A New Breed," *Nature* 497 (2013):27–29.

Deepak K. Ray et al., "Recent Patterns of Crop Yield Growth and Stagnation," *Nature Communications* 3.1293 (2012), accessed January 8, 2012, doi: 10.1028/ncomms2296.

Jonathan Foley et al., "Solutions for a Cultivated Planet," *Nature* 478 (2011):337–342.

Vaclav Smil, "Harvesting the Biosphere: The Human Impact," *Population and Development Review* 37.4 (2011): 613–36.

Vaclav Smil, *Harvesting the Biosphere: What We Have Taken From Nature* (Cambridge, MA: MIT University Press, 2012).

Vaclav Smil, "Eating Meat: Evolution, Patterns, and Consequences," *Population and Development Review* 28.4 (2002): 599–639.

S. C. Fahrenkrug et al., "Precision Genetics for Complex Objectives in Animal Agriculture," *Journal of Animal Sciences* 88 (2010):2530–2539.

Paul Voosen, "Geologists Drive Golden Spike Toward Anthropocene's Base," *Greenwire*, September 17, 2012, accessed September 24, 2012, http://eenews.net/public/Greenwire/2012/09/17/1.

Patrick Muirhead, *The Life of James Watt: With Selections From His Correspondence* (London: John Murray, Albemarle Street, 1859), 212.

Historical CO_2 Records from the Law Dome DE08, DE08-2, and DSS Ice Cores, Carbon Dioxide Information Analysis Center, Oak Ridge National Laboratory, accessed September 23, 2012, http://cdiac.ornl.gov.

Myles R. Allen et al., "Warming Caused by Cumulative Carbon Emissions Towards the Trillionth Tonne," *Nature* 458 (2009):1163–1166.

Smýkal et al., "Pea (*Pisum sativum* L.) in the Genomic Era."

Awah Anna Amian, Jutta Papenbrock, Hans-Jörg Jacobsen and Fathi Hassan, "Enhancing Transgenic Pea (*Pisum sativum* L.) Resistance Against Fungal Diseases Through Stacking of Two Antifungal Genes (Chitinase and Glucanase)," *GM Crops* 2.2, (2011): 104–109.

Graham Brookes and Peter Barfoot, "GM Crops: The First Ten Years—Global Socio-Economic and Environmental Impacts," *ISAAA Brief* 36 (2006), accessed May 3, 2013, http://www.isaaa.org/resources/publications/briefs/.

Graham Brookes and Peter Barfoot, "Key Environmental Impacts of Global Genetically Modified (GM) Crop Use 1996–2011," *GM Crops & Food* 4.2 (2013), accessed May 3, 2013, http://dx.doi.org/10.4161/gmcr.24459.

Eviatar Nevo et al., "Evolution of Wild Cereals During 28 Years of Global Warming in Israel," *Proceedings of the National Academy of Sciences USA*, 109.9 (2012): 3412–3415.

Rachel Brenchley et al., "Analysis of the Bread Wheat Genome Using Whole-genome Shotgun Sequencing," *Nature* 491 (2012):705–710.

Robert Sanders, "Fertilizer Use Responsible for Increase in Nitrous Oxide in Atmosphere," UC Berkeley News Center, April 2, 2012.

Bush, "As We May Think."

Plants Engineered to Replace Oil, Advanced Research Projects Agency—Energy, US Department of Energy, accessed September 22, 2012, http://arpa-e.energy.gov/ProgramsProjects/PETRO.aspx.

"Photosynthetic Biorefineries: Emerging Frontiers in Research and Innovation 2013" (EFRI-2013), National Science Foundation, Program Solicitation NSF 12-583, July 11, 2012, accessed September 22, 2012, http://www.nsf.gov/pubs/2012/nsf12583/nsf12583.htm.

Basit Yousuf, Jitendra Keshri, Avinash Mishra, and Bavanath Jha, "Application of Targeted Metagenomics to Explore Abundance and Diversity of CO_2-fixing Bacterial Community Using cbbL Gene From the Rhizosphere of Arachis Hypogaea," *Gene* 506.1 (2012):18–24.

Alexander M. Cardoso et al., "Metagenomic Analysis of the Microbiota from the Crop of an Invasive Snail Reveals a Rich Reservoir of Novel Genes," *PLoS ONE* 7.11 (2012): e48505, accessed February 25, 2013, doi: 10.1371/journal.pone.0048505.

Sebastian Jaenicke et al., "Comparative and Joint Analysis of Two Metagenomic Datasets from a Biogas Fermenter Obtained by 454-Pyrosequencing," *PLoS ONE* 6.1 (2011): e14519, accessed February 25, 2013, doi: 10.1371/journal.pone.0014519.

Cheryl P. Goldbeck et al., "Tuning Promoter Strengths for Improved Synthesis and Function of Electron Conduits in *Escherichia coli*," *ACS Synthetic Biology* 2.3 (2013): 150–159.

Michael Nielsen, *Reinventing Discovery: The New Era of Networked Science* (Princeton, NJ: Princeton University Press, 2012).

"A Man-Made World," *The Economist*, May 26, 2011, accessed September 24, 2011, http://www.economist.com/node/18741749.

Health Technologies Are Getting Personal

Moore's Law Timeline. Intel Corp., accessed December 13, 2012, http://www.intel.com.

Church, Gao, and Kosuri, "Next-Generation Digital Information Storage in DNA."

Alicia Löffler and Scott Stern, "The Future of the Biomedical Industry in an Era of Globalization," White Paper, Kellogg Center for Biotechnology, Kellogg School of Management, January 2008, accessed February 6, 2009, http://www.kellogg.northwestern.edu/biotech/faculty/articles/future_biomedical_industry.pdf.

George Weinstock, "Genomes for the Masses," *Technology Review*, May 2007.

National Research Council, *Toward Precision Medicine: Building a Knowledge Network for Biomedical Research and a New Taxonomy of Disease* (Washington, DC: National Academies Press, 2012), 8.

Eric Green, "Welcome and Opening Remarks: HGP10: The Genomics Landscape a Decade after the Human Genome Project," National Human Genome Institute, April 25, 2013, accessed May 21, 2013, http://www.genome.gov/27552257.

Timothy Caulfield and Amy L. McGuire, "Direct-to-Consumer Genetic Testing: Perceptions, Problems, and Policy Responses." *Annual Review of Medicine* 63 (2012):23–33.

"Privacy and Progress in Whole Genome Sequencing," Presidential Commission for the Study of Bioethical Issues, October 2012, accessed November 2, 2012, http://www.bioethics.gov/cms/node/764.

Melissa Gymrek et al., "Identifying Personal Genomes by Surname Inference," *Science* 339 (2013):321–324.

Gregory Kutz, "Direct-to-Consumer Genetic Tests: Misleading Test Results Are Further Complicated by Deceptive Marketing and Other Questionable Practices," Rep. US Government Accountability Office, July 22, 2010, accessed October, 29, 2012, http://www.gao.gov/new.items/d10847t.pdf

U.S. Food and Drug Administration, Warning Letter, 23andMe, Inc., November 22, 2013, accessed November 25, 2013, http://www.fda.gov/ICECI/EnforcementActions/WarningLetters/2013/ucm376296.htm.

"The FDA and Thee," *The Wall Street Journal*, November 25, 2013.

"GPPC Releases Updated List of DTC genetic Testing Companies," Genetics & Public Policy Center, The Johns Hopkins University, August 11, 2011, accessed November 29, 2013, http://www.dnapolicy.org.

Caroline F. Wright and Shelley Gregory-Jones, "Size of the Direct-to-Consumer Genomic Testing Market," *Genetics in Medicine* 12 (2010):594.

Razib Khan and David Mittelman, "Rumors of the Death of Consumer Genomics Are Greatly Exaggerated," *Genome Biology* 14 (2013):139, accessed November 30, 2013, doi: 10.1186/gb4141.

Nicholas Eriksson at al., "Web-Based, Participant-Driven Studies Yield Novel Genetic Associations for Common Traits," *PLoS Genetics* 6.6 (2010): e1000993, accessed February 3, 2013, doi: 10.1371/journal.pgen.1000993.

Claudia Gonzaga-Jauregui, James R. Lupksi, and Richard A. Gibbs, "Human Genome Sequencing in Health and Disease," *Annual Review of Medicine* 63 (2012):35–61.

Li Ding et al., "Clonal Evolution in Relapsed Acute Myeloid Leukaemia Revealed by Whole-Genome Sequencing," *Nature* 481 (2012):506–510.

Green, "Welcome and Opening Remarks."

Brendan Maher, "The Human Encyclopaedia," *Nature* 489 (2012):46–48.

ENCODE Project Consortium, "An Integrated Encyclopedia of DNA Elements in the Human Genome," *Nature* 489.57 (2012): 57–74.

"In Massive Genome Analysis ENCODE Data Suggests 'Gene' Redefinition," *Science Daily*, September 5, 2012, accessed November 2, 2012, http://www.sciencedaily.com/releases/2012/09/120905135012.htm.

"DNA Sequencing Costs," National Human Genome Research Institute, accessed October 31, 2012, http://www.genome.gov/sequencingcosts/.

Moore's Law Timeline.

Data Overload and Lingering Doubts

Stephen Wolfram, "Timeline of Systematic Data and the Development of Computable Knowledge," WolframAlpha, accessed November 4, 2012, http://www.wolframalpha.com/docs/timeline/.

1000 Genomes Project Consortium, "An Integrated Map of Genetic Variation From 1,092 Human Genomes," *Nature* 491 (2012):58–65.

Laura Clarke et al., "The 1000 Genomes Project: Data Management and Community Access," *Nature Methods* 9.5 (2012): 1–4.

Erika Check Hayden, "US Cancer-Genome Repository Hopes to Speed Research," *Nature News* blog, May 2, 2012.

Nicholas J. Loman and James Hadfield, "World Map of High-Throughput Sequencers," accessed November 20, 2013, http://omicsmaps.com.

Kevin Davies, "Laying the Foundation for Next-Gen Cancer Diagnostics," *Bio-IT World*, February 8, 2012, accessed January 19, 2013, http://www.bio-itworld.com/2012/02/08/laying-foundation-next-gen-cancer-diagnostics.html

Michael Schatz, "Cloud Computing and the DNA Data Race: Emerging Computational Methods for the Life Sciences," HPDC 2011, San Jose, California, June 8, 2011, accessed November 4, 2012, http://schatzlab.cshl.edu/presentations/.

Bert Vogelstein et al., "Cancer Genome Landscapes," *Science* 339 (2013):1546–1558.

Stephen Wolfram, "Advance of the Data Civilization: A Timeline," WolframAlpha blog, August 16, 2011, accessed November 4, 2011, http://blog.wolframalpha.com/2011/08/16/advance-of-the-data-civilization-a-timeline/.

Michael Schatz, Ben Langmead, and Steven L. Salzberg, "Cloud Computing and the DNA Data Race," *Nature Biotechnology* 28 (2012):691–693.

Green, "Welcome and Opening Remarks."

Kym M. Boycott, Megan R. Vanstone, Dennis E. Bulman, and Alex E. MacKenzie. "Rare-disease Genetics in the Era of Next-generation Sequencing: Discovery to Translation," *Nature Reviews Genetics* 14 (2013):681–691.

Zhonggang Hou et al., "Efficient Genome Engineering in Human Pluripotent Stem Cells Using Cas9 from *Neisseria meningitides.*" *Proceedings of the National Academy of Sciences USA* 110.39 (2013):15644–15649.

Jared C. Roach et al., "Analysis of Genetic Inheritance in a Family Quartet by Whole-Genome Sequencing," *Science* 328 (2010):636–639.

Nicholas Wade, "Disease Cause is Pinpointed with Genome," *The New York Times*, March 10, 2010.

Nicholas J. Roberts et al., "The Predictive Capacity of Human Genome Sequencing," *Science Translational Medicine* 4.133 (2012): 133ra58, doi: 10.1126/scitranslmed.3003380.

Eric J. Topol, "Comment on 'The Predictive Capacity of Personal Genome Sequencing,'" *Science Translational Medicine* 4.135 (2012): 135lr3, doi: 10.1126/scitranslmed.3004126.

Bert Vogelstein et al., "Response to Comments on 'The Predictive Capacity of Personal Genome Sequencing,'" *Science Translational Medicine* 4.135 (2012): 135, doi: 10.1126/scitranslmed.3004246.

Carl E. G. Bruder et al., "Phenotypically Concordant and Discordant Monozygotic Twins Display Different DNA Copy-Number-Variation Profiles," *American Journal of Human Genetics* 82.2 (2008): 763–771.

J. Breckpot et al., "Differences in Copy Number Variation Between Discordant Monozygotic Twins as a Model for Exploring Chromosomal Mosaicism in Congenital Heart Defects," *Molecular Syndromology* 2.2 (2012): 81–87.

E. Schorry et al., "Copy Number Variants in Monozygotic Twins with Neurofibromatosis 1," Paper presented at the annual meeting of the American Society of Human Genetics, November 7, 2012, accessed November 2, 2012, http://www.ashg.org/2012meeting/abstracts/fulltext/f120123496.htm.

Britney L. Grayson et al., "Genome-Wide Analysis of Copy Number Variation in Type 1 Diabetes," *PLoS ONE* 5.11 (2012): e15393, accessed November 2, 2012, doi: 10.1371/journal.pone.0015393.

Feng Zhang et al., "Copy Number Variation in Human Health, Disease, and Evolution," *Annual Review of Genomics and Human Genetics* 10 (2012):451–481.

Jenny van Dongen et al., "The Continuing Value of Twin Studies in the Omics Era," *Nature Reviews Genetics* 13 (2012):640–653.

Laboratory Medicine Meets Large-Scale Genomics

UK Department of Health Pathology Modernisation Team. Modernising Pathology Services, 2004, p. 7, accessed November 5, 2012, http://www.dh.gov.uk.

Mike J. Hallworth, "The '70% Claim': What Is the Evidence Base?" *Annals of Clinical Biochemistry* 48 (2012):487–488.

"Best Care at Lower Cost: The Path to Continuously Learning Health Care in America," Institute of Medicine, Report Brief, September 2012, accessed November 5, 2012, http://www.iom.edu/bestcare.

Ezra Klein, "Bill Gates: 'Death Is Something We Really Understand Extremely Well,'" Wonkblog, *Washington Post*, May 17, 2013, accessed May 19, 2013, http://www.washingtonpost.com/blogs/wonkblog/

Basel Kayyali, David Knott, and Steve Van Kuiken, "How Big Data is Shaping US Health Care," *McKinsey Quarterly*, May 2013.

Simon Tripp, Martin Grueber, and Deborah Cummings, "The Economic and Functional Impacts of Genetic and Genomic Clinical Laboratory Testing in the United States," Battelle Technology Partnership Practice, prepared for the American Clinical Laboratory Association, January 2012, accessed November 5, 2012, http://www.labresultsforlife.org/news/Battelle_Impact_Report.pdf.

"Economic Impact of the Human Genome Project," Battelle Technology Partnership Practice, Battelle Memorial Institute, May 2011, accessed January 9, 2012, http://www.battelle.org.

Nadia Drake, "What is the Human Genome Worth?" *Nature News*, May 11, 2012, accessed November 5, 2012, doi: 10.1038/news.2011.281.

"The Impact of Genomics on the U.S. Economy," Battelle Technology Partnership Practice, Battelle Memorial Institute, June 2013, prepared for United for Medical Research, accessed June 12, 2013, http://www.unitedformedicalresearch.com.

Barack Obama, "Remarks by the President in the State of the Union Address," February 12, 2013, accessed February 13, 2013, http://www.whitehouse.gov.

"CAP Checklist a First for Next Generation Sequencing Laboratory Standards," Press Release, College of American Pathologists, July 31, 2012, accessed November 5, 2012, http://www.cap.org.

Carla G van El et al., "Whole-Genome Sequencing in Health Care: Recommendations of the European Society of Human Genetics," *European Journal of Human Genetics* 21 (2013):580–584.

Electronic Medical Records and Genomics Network, National Human Genome Research Institute, accessed November 6, 2012, http://www.genome.gov/27540473.

Joshua C. Denny et al, "Systematic Comparison of Phenome-wide Association Study of Electronic Medical Record Data and Genome-wide Association Study Data," *Nature Biotechnology*,(2013):1102–1111.

Francis Collins, "An Evolving App for Genetic Tests," NIH Director's Blog, April 30, 2013, accessed November 18, 2013, http://directorsblog.nih.gov/.

Monya Baker, "One-stop Shop for Disease Genes," *Nature News* 491 (2012):171, accessed November 7, 2012, doi: 10.1038/491171a.

Sherri Bale et al., "MutaDATABASE: A Centralized and Standardized DNA Variation Database," *Nature Biotechnology* 29 (2011):17–18.

"DNA Tests to Revolutionise Fight Against Cancer and Help 100,000 NHS Patients," British Prime Minister's Office, December 10, 2012, accessed March 13, 2013, http://www.number10.gov.uk/news/dna-tests-to-fight-cancer/.

Jim Golden, "Your Genome's on an iPhone (Trying to Call Home...)," *Forbes*, November 1, 2012.

Digital Technology Recodes the Genetics of Communications

David William Galbraith, "Frontiers in Genomic Assay Technologies: The Grand Challenges in Enabling Data-Intensive Biological Research," *Frontiers in Genetics* 2 (2011):26, accessed November 6, 2012, doi: 10.3389/fgene.2011.00026.

Löffler and Stern, "The Future of the Biomedical Industry in an Era of Globalization."

"The U.S. Biopharmaceuticals Sector: Economic Contribution to the Nation," Battelle Technology Partnership Practice, prepared for Pharmaceutical Research and Manufacturers of America, July 2011, accessed November 8, 2012, http://www.phrma.org/research/publications/profiles-reports.

"Economic Analysis of the Technology Infrastructure Needs of the U.S. Biopharmaceutical Industry," National Institute of Standards and Technology. Prepared for Gregory Tassey, NIST, by RTI International, November 2007, accessed November 6, 2012, http://www.nist.gov/director/planning/upload/report07-1.pdf.

Eric von Hippel and Michael Schrage, "Users Are Transforming Innovation," *Financial Times*, July 11, 2007.

Lessig, *Remix.*

Clayton Christensen, Jerome H. Grossman, and Jason Hwang, *The Innovator's Prescription* (New York: McGraw-Hill, 2008).

Michael Mandel, "Where the Jobs Are: The App Economy," Research by South Mountain Economics, LLC for TechNet, February 7, 2012, accessed November 11, 2012, http://bit.ly/AlGe4l.

5

Toning Up Universities for Regional Growth

Needless to say, no institution lasts nine centuries without
adapting.

—*The Economist*

Like many if not most modern universities, the University of Bologna offers prospective students a virtual tour of the campus and the city as well as the school's
academic programs from its website. Bologna is the capital of the Emilia Romagna
region in northern Italy. Consisting of nine provinces and 4 million people in the
Po River Valley, Emilia Romagna takes its name from the ancient Roman road that
marks the line between the northern plain and the mountains to the south. Bologna
is a highly touted tourist destination. The city offers an unbroken thread of mag-
nificent architecture and design going back more than 500 years when the current
city was built.

But perhaps Bologna's greatest claim to fame today, a time when economies must
tap into new knowledge to grow out of what has been called the "Great Stagnation"
in job-producing technological innovation, was what happened there nearly a millennium ago. Sometime toward the end of the eleventh century—some historians
put the date at 1088—Bologna was the birthplace of the university in the West. For
nearly 1,000 years the university has been an organization of higher education, and
for the past 200 years, since Wilhelm von Humboldt established the University of
Berlin in 1810, it has been a place where faculty conduct research.

Bologna saw the first Universitas scholarium, a free association between teachers and students, with the students collectively having far more power than they do
today. And unlike the University of Paris with its ties to ecclesiastical authority,
the University of Bologna had a practical aim. Its academic focus was the study
of the legal system, an effort to bring order to bear on the confusing mass of laws
city officials and merchants had to deal with. That was important because Bologna
was a largely independent free city and one of the main commercial trade centers
in Europe thanks to a system of canals that allowed large ships to enter and depart.

The university helped provide a legal foundation for the economic and political development of the city.

During its "golden era" in the seventeenth century, the University of Bologna was a leader in the emerging fields of science and medicine. The Pole Nicolaus Copernicus began his observation of the stars while a law student there. Marcello Malpighi pioneered the use of the microscope, founding comparative physiology and microscopic anatomy. When the Industrial Revolution arrived in the eighteenth century, the university seized the moment to promote scientific and technological development. One of Bologna's academic luminaries was Luigi Galvani. Together with fellow Italian Alessandro Volta, Benjamin Franklin in America, and Henry Cavendish in England, Galvani laid the foundation for modern electrical studies. It was in Bologna where heirs to the mythical Prometheus brought the "vital force" of electricity to cells and tissues. And it was in Galvani's laboratory with its electrostatic generators and frog parts scattered about where bioelectrical experiments animated the human imagination and, in the hands of writers like Mary Shelley and film stars like Boris Karloff, fueled dreams of immortality and dread of a future gone awry.

Like many modern universities, the University of Bologna is also vested in the economic development of its surrounding space, the dynamic Emilia Romagna region. Indeed, the links between universities and their local and regional economies have become critical channels for knowledge transfer and economic growth, diffusing the insular ethos that defined campus life for nearly a millennium. At a time of fiscal austerity and budget woes, the University of Bologna, like research universities around the world, is more eager than ever to stress such external interactions as part of its mission, recruiting entrepreneurial scientists, transferring knowledge and technology, setting up university–industry collaborative research centers, pursuing patenting and licensing, and creating spin-off companies from its research.

Figuring out how to widen and deepen these knowledge transfer channels to enhance innovation and productivity in the face of global competition is a key challenge for regional development. It will require rigorous experimentation at every level of the regional economy just as mutation, recombination, and selection drive adaptation to change in a natural ecosystem. By the time the University of Bologna celebrates the millennium of its founding in or around 2088, experiments now underway, under consideration, and yet to be conceived are likely to make the way new knowledge is produced and put to work different exercises than what we see today. The territory between basic discovery and commercial application constitutes an experimental field plot for new ideas, new arrangements, and new structures designed to accelerate knowledge flow from universities to industry and society and from productive enterprises back to university educators, scientists, scholars, and students.

The University of Bologna is well situated for such experimentation thanks in part to what happened in its neighborhood during the previous century. Emilia

Romagna is well known internationally for its conglomeration of independent artisans or "artigianati" and flexible manufacturing networks. These networks of small producers make up a diversified competitive cluster of manufacturing strength that was featured in *The Competitive Advantage of Nations*, the 1990 classic by Harvard Business School strategist Michael Porter. The region lists more than 400,000 business enterprises. The 90,000 small manufacturers in Emilia Romagna with an average of fewer than ten employees rival multinational corporations in their collective productivity. They include manufacturers in diverse fields: packaging machinery, electrical components, textiles, clothing, furniture, ceramics, leather goods, and footwear. The region also hosts the famed automobile manufacturers Ferrari, Maserati, and Lamborghini. Per capita gross domestic product in Emilia Romagna is nearly a third higher than that of Italy as a whole. The region is one of Europe's leaders in digital infrastructure, which is enabling the formation of business webs and data sharing.

Emilia Romagna has become one of the most productive regions in Europe and a global model for successful entrepreneurship. It is also a model for the study of global business on regional productivity processes and the unique role these productive regions play in the information economy. The region has a seed-capital investment fund to support the creation and growth of companies and cooperatives and a solid civic foundation on which to build. Harvard social scientist and best-selling author Robert Putnam highlighted Emilia Romagna as Italy's most civic and best-governed region in his book *Making Democracy Work* (1993). Putnam found a strong connection between participation in civic and community associations, social cooperatives, and the performance of civic institutions. Just as the density of business links is an indicator of productive economic activity in a given geographic space, the density of civic organizations and cultural and recreational associations is an indicator of civic cooperation and community health. The evolution of the cooperative system in Emilia Romagna with its manufacturing networks and industrial districts has become a model not just for the European Union but indeed for economic development around the world. Based on his survey of Emilia Romagna, Putnam coined the term *social capital*, a term that has since entered the lexicon of regional innovation and economic development. Social capital is one of the ingredients that makes the knowledge economy tick.

If Emilia Romagna is to succeed in the twenty-first century as a model for a regional innovation architecture, it needs to move production further up the value chain by systematic innovation partnerships with universities, pursuing the necessary finance and world-class quality and environmental standards. That is the view of Philip Cooke who has analyzed industrial districts, innovation clusters, and technology development around the world for decades, including in the life sciences. The Emilia Romagna research system has five universities in eleven locations plus two American universities that all together enroll more than 150,000 students. The system employs more than 6,000 professors and researchers, two-thirds of whom are in the science and engineering fields.

Italy's National Research Council has twelve institutes in Emilia Romagna and ten centers within the region's universities. These institutes employ nearly 1,000 researchers, technicians, and students with grants and PhD students. According to the Organisation for Economic Co-operation and Development (OECD), the Paris-based international economic organization of advanced industrial countries, Emilia Romagna accounts for 15 percent of Italy's national scientific production from just 6 percent of the national research investment, a critical factor as Italy struggles to climb out from under its debt burden. The region is also the clear leader in Italian academic entrepreneurship based on university patents, spin-off companies, and university–industry collaborations.

The University of Bologna, at the heart of Emilia Romagna's capital city, enrolls 100,000 students in its twenty-three faculties and sixty-eight departments. It has long been involved in educating and training the Emilia Romagna workforce. Take Bologna's ASTER Science and Technology Enterprise, a public–private consortium involving government agencies, enterprises, and universities established in 1985. Its mission is to boost the competitiveness of the Emilia Romagna regional production system by promoting research, technology transfer, and innovation, with a focus on small to medium-sized enterprises.

ASTER lists environmental science; food and health; motors and materials; information technologies, including e-business and industrial automation; life science; biomedical sensors; microsystems; and nanotechnologies as key research and tech transfer areas. These targeted high-tech fields are a far cry from traditional technologies used by Emilia Romagna's entrepreneurs and small manufacturers. The reasons for focusing on them are evident. Emilia Romagna cannot compete with low-cost producers, especially China, in fields like textiles, clothing, ceramics, manufacturing equipment and machine tools, and components assembly.

How does a historically dynamic region like Emilia Romagna cope with the inexorable forces of globalization? Will it be compelled to take a leap into the economic unknown? A leap strategy is less likely to succeed than one that incorporates new technology and services into traditional economic sectors, in the view expressed by Massachusetts Institute of Technology's (MIT's) Suzanne Berger and Richard Locke. From her scholarship, colloquia, and factory visits, Berger is as well informed as anyone about the changes in manufacturing systems, including those used by life sciences companies. She cochaired MIT's Production in the Innovation Economy project (2011–2013), asserting that a renewal of American manufacturing "is not just desirable but possible, if only we can learn more about how technological innovations fuel productivity." With respect to Emilia Romagna, Berger and Locke say that integrating great manufacturing and design with new information technologies would create valuable products. The region's firms need to stay close to home "for the same reasons that information technology firms stick to Silicon Valley or new biotech firms cluster around universities: to gain access to information that is only transmitted through social relationships, to incorporate this knowledge into new high-value-added products, and to find a highly skilled workforce."

But even in regional manufacturing centers that are lower tech and more labor intensive like Emilia Romagna, lower labor costs in Asia are outweighed by the advantages of remaining located where new ideas emerge and are debated, and where the experimentation of others constantly offers lessons. Dynamic regions like Emilia Romagna will not disappear with globalization but will be transformed, assisted by new knowledge, know-how, and technologies flowing from their native universities, research institutes, health centers, and public–private development agencies. That transformation will be reflected in new products and services, health technologies and computer-assisted manufactured devices among them.

Like its sister universities in Emilia Romagna, in Europe, and around the world, the University of Bologna today sees its role in dual terms: traditional academic excellence and outreach to its regional companies, cooperatives, and communities through public engagement. Given the high level of civic engagement already existing in Emilia Romagna, which Robert Putnam sees as a key factor for human and social capital formation, the University of Bologna's public engagement mission is made that much easier as the end of the millennium since the university's founding approaches.

The Triple Helix Then and Now

The term *ecosystems of entrepreneurship* is an example of how biological metaphors help us to understand regional systems of innovation and their adaptability to change. Another example is *the triple helix*. In early 1953, Linus Pauling and colleague Robert Corey published a proposed structure for DNA, a triple helix. Their proposed DNA structure had three spirals, with three sugar-phosphate backbones on the inside and the nucleotide bases sticking out. They shared their findings in advance of publication with James Watson and Francis Crick. Watson and Crick received the manuscript in February and immediately saw that a triple helix could not be right. Just two months later, on April 25, 1953, *Nature* published their proposal that DNA was a double helix, not a triple helix.

If a triple helix was the wrong model for the structure of DNA, the chemistry of life, it seems to be a good model for regional innovation in the life sciences. Henry Etzkowitz of the State University of New York and Loet Leydesdorff of the Amsterdam School of Communications introduced the model in *Universities and the Global Knowledge Economy: A Triple Helix of University–Industry–Government Relations*, a book they edited that was published in 1997. Their research gave a new look to the role of the sciences in society and particularly in economic development. The scholars saw what they call a "neo-evolutionary" system of knowledge production and wealth creation, drawing on Darwin's ideas in the natural world and applying them to a social dynamic.

It was once the case that government, universities, and firms operated largely within their own spheres. But today, the boundaries between public and private, science and technology, university and industry are less clear. Indeed, they are in flux. Leydesdorff observed that in contrast to the double helix, which is a stable molecular form, the triple helix is not as stable. A knowledge-based regime of innovations arising from university–industry–government interactions "can be expected to remain in transition" as one would expect of a dynamic system. As the university crosses traditional boundaries in developing new linkages to industry, it looks for ways to make teaching, research, and economic development efficient in ways that are compatible with its traditional mission.

More than ever before, firms are looking beyond their confines for knowledge and technology. "Companies increasingly look to universities, as well as other firms and government laboratories, as a potential source of useful knowledge and technology, especially in biotechnology and software," as Etzkowitz and Leydesdorff saw it, writing in the late 1990s. Cooperative initiatives involving academia and industry, often encouraged by government, sprang up at the regional, national, and international levels. The triple helix of innovation was advanced as the essential code for both growth and sustainability, with the helical strand representing the university more important than ever to the future of innovation, as important as the strands representing the private sector and government research and innovation funding. Reflecting on the bankruptcy of Rochester, New York-based Eastman Kodak, University of Rochester physicist Adam Frank said that behind the headlines of corporate demise "there's something else going on in here and it's happening in small cities all over the country. Much of the role that industry used to play in innovation has now shifted to the universities." Dramatic expansion of the University of Rochester's medical research complex has helped to make the university the area's largest employer, with some of the research finding its way into intellectual property, new companies, and new economic activity. Such developments would appear to vindicate the argument of triple helix theorists that the university is key to future innovation:

> From its medieval origins, whether a student- or faculty-led foundation as in Paris or Bologna, the university has proved to be a flexible and evolving organizational form capable of reconciling conservatism with innovation. The passage of student generations combined with relative faculty autonomy is the basis on which these two institutional characteristics co-exist in a creative tension.

A third mission for the university, in addition to teaching and research and considered part of service mission, has emerged across the globe. It goes under the label "technology transfer." Today activities that go under that name can be found at the oldest universities in the West in Bologna and Paris, at the classic English universities at Oxford and Cambridge, and across the American academic landscape. More recently Japan has gotten into the technology transfer act, as have the rapidly rising economic powers of Asia such as China, where the triple helix theory of economic development has been embraced if not yet implemented.

The impetus for the wave of interest in technology transfer was a bill passed by a lame duck session of Congress and signed into law by President Jimmy Carter on December 12, 1980. The Bayh–Dole Act—Public Law 96-517, the Patent and Trademark Act of 1980—had a long gestation. The new law that turbocharged universities as agents of technology transfer actually had roots in the founding of the American Republic and in legislation crafted and signed into law in the early days of a war that nearly split the nation apart.

Education and Knowledge Flow Early On

Institutions of higher education have been in the business of knowledge transfer for many centuries. Indeed, it is their raison d'être. Education is fundamentally all about knowledge transfer, from teacher to student and from student to society at large as an educated member of society. In the medieval university, like the universities in Bologna, Paris, and Oxford, that was pretty much as far as it went. Universities and clerics controlled the flow of information by controlling the means of that flow, the textbooks handwritten by their stationers.

But the rise of printing changed everything. As Elizabeth Eisenstein observed in her classic book about the printing press as an agent of change, the advent of printing went hand in hand with progress of science and technology. Printing created a "bridge over the gap between town and gown." The shift from script to print "helps to explain why old theories were found wanting and new ones devised even before telescopes, microscopes, and scientific societies appeared," she wrote. "Professors and printers began to engage in fruitful collaboration almost as soon as the new presses were installed." The revolt was on against clerical elites steeped in Latin. Students allied themselves with those professors who pursued the new communications technology and indeed began demanding printed learning materials, just as they demand electronic information services today.

But printing was slow to break down social barriers between the learned, the privileged, and the unschooled masses. It took several centuries for science, technology, and innovation to break out into society at large. When we talk today about innovation systems and knowledge clusters in manufacturing, finance, aerospace engineering, energy, telecommunications, and bioscience, we are talking about contemporary expressions of economic forces that had their roots in the eighteenth century, when the modern age of economics took distinctive shape thanks to luminaries of the Scottish Enlightenment like Adam Smith. Commerce and trade were central to their view of an organic society. Social improvements went hand in hand with new ideas and new technologies for the creation of wealth. Without the economic productivity that results from a culture of invention and innovation, a nation can increase its wealth only by the zero-sum game of mercantilism, that is, by seeking a trading advantage through encouraging exports and discouraging imports. Or it can seize the natural resources of other lands through conquest and colonization.

Indeed, the writers of the US Constitution working toward the end of the Scottish Enlightenment recognized what the Scots had wrought and incorporated their ideas into the document. James Madison, a graduate of what is today Princeton University, foresaw social change in similar evolutionary terms, from the simple to the more complicated. Even after new "arts of life" such as manufacturing was introduced, change was naturally resisted. "The manufacturer readily exchanges the loom for the plough, in opposition often to his own interest, as well as to that of his country," Madison said in a presidential address to a Virginia agricultural society in 1818. "Where do we behold a march to the opposite direction? the hunter becoming the herdsman; the latter a follower of the plough; and the last repairing to the manufactory or workshop." Regressing to earlier forms of livelihood, however idealized, was not an option for the nascent American Republic.

In addition to protecting the rights of authors and inventors, Madison believed that growth of the professions, voluntary associations, scientific societies, and educational institutions should be encouraged to help ease social transitions. The advancement and diffusion of knowledge was critical to the success of the American experiment.

> The American people owe it to themselves, and to the cause of free Government, to prove by their establishments for the advancement and diffusion of Knowledge, that their political Institutions, which are attracting observation from every quarter, and are respected as Models, by the new-born States in our own Hemisphere, are as favorable to the intellectual and moral improvement of Man as they are conformable to his individual & social Rights.

Technological development in America for the last years of the eighteenth century and the first half of the nineteenth century was driven by private enterprise. The factory system was a new organizational phenomenon on the postcolonial landscape, drawing workers from surrounding rural districts and thereby, as in Europe, launching the first massive rural-to-urban migration since the dawn of agriculture. But even more disruption was in store, namely from the building of canal, rail, and telegraph systems; the massive capital investment they required; and the separation of ownership from management.

The new economic system that began to emerge in the new American Republic by the early decades of the nineteenth century involved government, scientific societies, professional associations, and higher education in addition to private enterprise. Together they constituted a new dynamic, a kind of self-organizing system that economic historian Joel Mokyr describes as "one of the most powerful and influential ideas of the modern age and perhaps the most important element in Adam Smith's thought." Self-organization appears throughout our social system, including language, science, technology, arts and manners, and indeed by what Alexis de Tocqueville described as the American characteristic of volunteerism. "Americans of all ages, all conditions, and all dispositions, constantly form associations," he wrote in *Democracy in America*. Surveying the American experience in the early 1830s, he observed that in democratic countries knowledge of how to combine is the wellspring of all other forms of knowledge.

The joining of biology, technology, and economics on the contemporary American economic landscape has a taproot in two events that occurred in consecutive days in July 1862. On July 1, Abraham Lincoln signed into law the Pacific Railway Act, enabling the transcontinental reach of the railroad and, as a precursor to the networked economy, the telegraph. Planning for the railroad across the American West had been conducted by the Pacific Railroad Surveys undertaken during the 1850s just as Darwin was putting his ideas of evolution by natural selection on paper in his Down House study in England and Gregor Mendel commenced his studies of the inheritance patterns in pea plants. The Pacific Railroad Surveys turned out to be a sort of early version of what we know today as environmental impact studies. The naturalists employed by the federal government returned vast quantities of specimens to the Smithsonian Institution where they formed the foundation for its zoological and botanical collections.

The day after Lincoln signed the Pacific Railway Act he signed the Land-Grant College Act, better known as the Morrill Act. He thereby joined in time the concept of exchanging land for progress: progress in transportation, commerce, and communications through the Pacific Railway Act and progress in education by equipping America's youth, its students, with the knowledge and skills for work in an agricultural and manufacturing economy. The Pacific Railway Act and the Morrill Act secured the industrial economy of the nineteenth century and set the United States on its present path of global economic, scientific, and military dominance. They were a clear expression of active, energetic government laying the groundwork for long-term economic growth in the tradition of Alexander Hamilton and Henry Clay. It was the Morrill Act that first championed the critical role of knowledge transfer from institutions of higher education to the economy at large. That dimension of what we call today technology transfer was part and parcel of the American experience. Land-grant colleges sprouted up all over the country in the decades following the Morrill Act until, eventually, every state had one.

A distant forerunner of the Bayh–Dole Act as a vehicle for knowledge transfer, the Morrill Act was the first substantial legislation joining the federal government, higher education, and the American economy. It was the first attempt by the federal government to support states in their efforts to prepare students for participating in the local economy rather than endowing them with classical learning. Traditional agriculture benefited enormously from that prescient legislation and from its successor laws: the Hatch Act of 1887; the Second Morrill Act of 1890; and the Smith-Lever Act of 1914 that established the system of cooperative extension services to bring people the benefits of current developments in crop management, natural resource and waste management, small business development, home economics, and other related subjects. By the 1920s extension services across the country were helping to introduce hybrid seed, the product of a genetic revolution a century ago.

The second Industrial Revolution—the second phase of the economic process James Watt and Matthew Boulton set loose with their manufactory and Adam Smith with his postmercantilist laissez-faire theory—saw mechanization and production on a mass scale. It witnessed the rise of the knowledge-based industries of the time, the chemical, electrical, petroleum, and steel industries. Universities played a role in the evolution of these industries through the diffusion of knowledge and in some cases key inventions in mechanical, electrical, and chemical engineering. Faculty participated as expert consultants in research and development (R&D) activities, especially after the rise of industrial research laboratories toward the century's end. University graduates supplied educated workers for the new industries.

At the dawn of what some have termed the third Industrial Revolution, satellite, cable, broadband, mobile telephony, and wireless networks constitute the same "disruptive innovation" that the rise of printing and the telegraph once did. They challenge not only the way conventional business is conducted and society interacts but notably the way new knowledge is discovered and distributed. Indeed, they are accelerating exponentially the speed of knowledge transfer around the world. Knowledge systems have not been the exclusive province of institutions of higher learning for half a millennium now. The rise of printing ended their monopoly. Yet universities remain at the heart of the transfer of knowledge, technology, and culture, particularly in the biosciences. Their role was amplified by war and its aftermath.

Tapping into the "Endless Frontier" for Innovation

Knowledge flowed in a steady if narrow stream between universities and industry in the United States until a huge channel was built in the wake of World War II. During the war President Franklin D. Roosevelt saw the need to bring the intellectual firepower of university researchers into the war effort in an organized way. He established the Office of Scientific Research and Development. Under the direction of Vannevar Bush, it was tasked with coordinating weapons development research. Bush chose Harvard University, MIT, and the University of California (UC) at Berkeley as lead partners in an effort to engage American universities in scientific and technological innovation for the Armed Forces. The Office of Scientific Research and Development laid the foundation for postwar legislation that launched the National Science Foundation.

That was beginning of the federal largess for university research and its contribution to military and civilian innovation. Its foundational document was Vannevar Bush's "Science, The Endless Frontier," a report to President Truman submitted in July 1945. Bush argued that basic research was "the pacemaker of technological progress." New products and new processes "do not appear full-grown" like Athena from the forehead of Zeus. "They are founded on new principles and new conceptions, which in turn are painstakingly developed by research in the purest realms of science!"

Bush was a strong advocate of patents to provide the essential incentives to spur private-sector investment. He also embraced uncritically the linear idea of technological progress. In his view, advances in pure science lead to advances in technology and not the reverse. Today we appreciate that technology provides basic science with ever-more powerful tools that enable us to ask questions of nature that we might not ask without these tools to prompt them. Plus we understand today that technological innovation has more in common with nonlinear adaptive systems or ecological communities than with logical linear progress. But in Bush's time, the image of technology as a "handmaiden to science" was widely adopted by the popular press, presumably because it is easier to understand than a more accurate explanation of how technology is developed and transferred.

University scientists and engineers in the United States had a long cooperative relationship with the agricultural, industrial, and public sectors. Some universities patented faculty inventions. Such patenting tended to be uncommon despite the development of patent offices and formal patenting procedures at universities, particularly after World War II. Before the war, universities had misgivings about getting involved in the patenting business. They feared their nonprofit tax status could be jeopardized. That was particularly true in the biomedical field. Universities that did develop internal patenting policies typically prohibited medical schools from patenting the inventions of their faculty. Other universities outsourced their patenting activity.

One of the first organizations to provide patenting services was Research Corporation, founded in 1912. It was the brainchild of UC Berkeley scientist and inventor and then philanthropist Frederick Gardner Cottrell with the assistance of Charles Doolittle Walcott, secretary of the Smithsonian Institution. Like Vannevar Bush in a later era, Cottrell was passionate about the advancement of science, guided by Progressive Era ideals that entailed cooperation between academic and industry scientists to further the public interest. Research Corporation was launched with funds derived from proceeds of Cottrell's invention, the electrostatic precipitator for controlling air pollution. The foundation's management of university patents and licensing expanded during the 1920s and 1930s as funding for research increased and universities linked up with industry.

The university–industry linkage that Research Corporation sought to foster entered a new era with the rise of venture capital, a key force in drawing ideas from the academy into the economy and putting them to work. Though birth of the venture capital industry is usually traced back to the late 1970s, it was putting out feelers decades earlier. Following the economic trials and tribulations of the Great Depression and World War II, efforts were undertaken to secure capital for funding fledgling companies, some based on new technologies being produced by university engineering schools and research institutes. The American Research and Development Corporation was a venture capital and private equity firm founded by a former dean of the Harvard Business School and today sometimes referred to as the "father of venture capitalism," Georges Doriot. Doriot understood that entrepreneurs and corporate managers had different mindsets, once observing that the moment of critical danger for a company was the moment of success because that was when it stops innovating.

Together with Ralph Flanders, then head of the Federal Reserve Bank of Boston, and Karl Compton, a former president of MIT, Doriot established the American Research and Development Corporation in 1946. The company was unique for its time in that it looked beyond approaching wealthy individual investors and tapped into institutional and pension funds. Its most notable success in the twenty-six years of its independent existence (it merged with Textron in 1972 after having invested in over 150 companies) was its investment of $70,000 in 1957 to found Digital Equipment Corporation. Ken Olsen and Harlan Anderson, formerly engineers at MIT's Lincoln Laboratory, built Digital Equipment Corporation into an anchor firm of what came to be known as Boston's Route 128 high-tech corridor, a location second only to Silicon Valley for the development of computer technology in the United States.

A second experimental initiative, this one from government, was the Small Business Investment Companies (SBICs) program established through the Small Business Investment Act of 1958. The act allowed the Small Business Administration to license private SBICs—a professionally managed private equity fund formed, say, through partnerships or consortia of small banks—to help the financing and management of small entrepreneurial businesses in the United States. Its larger purpose was to help the country regain its technological edge in the wake of Sputnik and growing Cold War competition with the Soviet Union. By enabling SBIC companies to borrow half their capital from the federal government and enjoy an assortment of tax incentives, the program was the first major initiative by the federal government to provide long-term funding for growth-oriented small businesses and laid the groundwork for the future growth of the venture capital industry.

To be sure, the SBIC program has had its flaws and critics over the years, particularly with regard to its rules regulating securities and financing, but it helped create a national mindset that saw entrepreneurial investment as critical for the nation's economic well-being. The SBIC legacy (financings totaled $1.8 billion in fiscal year 2009) and the rise of the venture capital industry ($18.7 billion invested in 2009) and angel investing ($17.7 billion invested in 2009) are intertwined, serving to foster the rise of many venture-backed companies predominantly in technology nurseries like Silicon Valley, Boston–Cambridge, the San Diego–La Jolla coastal cluster, and Research Triangle Park in North Carolina.

Among public universities looking to patent the intellectual property of its faculty and seek investment in it through licensing, Wisconsin was a pioneer. It established the Wisconsin Alumni Research Foundation (WARF). WARF was founded in 1925 by Harry Steenbock, inventor of a process for using ultraviolet radiation to add vitamin D to milk and other foods. Steenbock did something rare in those days, given the prevailing ethos that university scientists mixing it up with industry was frowned upon by academia. He commercialized the intellectual property and used the proceeds to fund research. It didn't take long for the WARF to show that such an organization could yield significant returns, both monetary and for public health and the public good. As the story goes, a Wisconsin farmer showed up at the School of Agriculture with a milk can full of blood from a heifer that would not coagulate. From it Karl Paul Link successfully isolated an anticoagulant factor that first found commercial application as a rodent killer and then in medicine for treating blood clots or thrombosis. The factor is known today as Warfarin (Coumadin), named after WARF. The story of its discovery and subsequent development is emblematic of the "Wisconsin Idea," the principle that education "should influence and improve people's lives beyond the university classroom."

Patent Rights and Patent Fights

In the two decades before the Bayh–Dole Act was signed into law in 1980, many US universities continued to avoid direct involvement in patent filing and commercialization. Most people who work in the technology transfer offices that are a conspicuous feature of American research universities today would not recognize their office as it existed before 1980. Historically, university patent offices were small-time affairs, often occupied by a lone and solitary individual, the patent officer.

But the tide was turning. By the 1960s, traditional "institutional ambivalence" in patenting and licensing activities, as UC Berkeley's David Mowery described it, was no longer so common. Federal funding of science had surged in the post-Sputnik years. Baby boomers, weaned in postwar optimism about the power of science and technology for war, space exploration, and mass consumption, were flocking to colleges and universities. As laboratories of basic research and new technology expanded, and as the psychological and even the physical barriers to cooperative university–industry activities loosened, universities began to look anew at the resources patenting and licensing could potentially provide to their institutions.

The decade of the 1970s was a watershed in patenting and licensing at US universities. Not only did universities dramatically expand their patenting activities, they did so especially in the previously verboten field of biomedical research. Steady growth of National Institutes of Health (NIH) research dollars and massive federal spending on research infrastructure during the 1960s supported universities in their dreams of constructing laboratory meccas on their campuses. As the research meccas were established and new faculty were hired, university demand for research dollars, from government, industry, and private philanthropy, naturally grew. When the federal dollar flow began to narrow beginning in 1970, research universities increasingly looked to the possibilities of new revenue produced through patenting and licensing, a trend that grew steadily in the following decades and began to change the face of the research university.

As the federal research infrastructure took shape after World War II, federal funding agencies were invested with patent rights over intellectual property developed through federally funded research. When universities began to show interest, federal agencies began to loosen their control, grant waivers, and enter into contractual relationships with universities, or their agents such as Research Corporation, that allowed limited licensing of research advances. In Mowery's view, the development of new agency waivers for patent rights was a response to greater demands for such rights by universities and a contributing factor in the growth of university patenting during the 1970s.

The Department of Defense was first to allow universities to keep patent title to inventions of their agency-funded faculty, followed by the Department of Health, Education and Welfare (HEW), home of the NIH, and then the National Science Foundation. An extensive study of government patent policy by Harbridge House published in 1968 found that the policy stifled cooperation between NIH-supported university investigators and the pharmaceutical industry in testing compounds synthesized under NIH grants. In response, HEW established Institutional Patent Agreements that gave universities the right to retain title to agency-funded research leading to patents as long as they had an internal technology transfer process approved by HEW. Patent attorney Norman Latker, a strong proponent of inventor's rights who joined HEW the year Harbridge House published its report, revised HEW's patent policies and was fired after criticizing a subsequent review. Latker favored waiving title to universities if they requested it, telling Congress that "a licensable patent right is probably a primary factor in the successful transfer of a university innovation to industry and the marketplace, and failure to protect such right may fatally affect a transfer of a major health innovation." A civil service review board overturned Latker's dismissal, and he was reinstated. Meanwhile, the idea of awarding patent title to university investigators rather than the federal agency supporting their research was winning the fight over the status quo. The tide was turning.

Bayh–Dole on the Fast Track

What put the impetus for Bayh–Dole on the fast track was a perfect storm of frustration shared by Congress, industry, and universities about the lack of consistency and uniformity in determining intellectual property rights resulting from federally funded research. "When your innovative idea gets tied up by piles of paperwork and months of delay as Washington dawdles over whether to let you market the thing or not," William Broad wrote in *Science* in 1979, "nasty thoughts about the U.S. patent policy are never far off." Broad described a steady stream of inventors who had been showing up at Congressional hearings to complain about the bureaucratic knots that tied up the transfer of patents derived from federally funded research. Their goal was to boost new legislation, "and it seems to be working."

A Senate bill, the University and Small Business Patent Procedures Act, coauthored by Birch Bayh (D-Indiana) and Robert Dole (R-Kansas), would give patent rights from federally sponsored research to universities and small businesses. The idea was that federal funds could be used to make money. Universities and their faculty would have an incentive to generate new ideas that could be "monetized" through the patenting and licensing processes. No longer would they be required to petition a federal funding agency and then wait to hear whether their application was approved. Past criticism of such legislation as the giveaway of public funds largely melted away. What Bayh–Dole implied was that just about any university can manage its own inventions better than the federal government can.

A profound shift was taking place, a shift that would place universities and their research at the heart of economic innovation. Before the 1970s lawmakers saw universities as "providing the fundamental science that firms would draw upon as needed to solve industrial problems and make technical advances," wrote sociologist Elizabeth Popp Berman in her book *Creating the Market University*. That was the way Vannevar Bush envisioned their role in the economy. By the early 1980s, universities, like policymakers before them, were starting to see science as having the potential "to actively drive economic growth by serving as a fount of innovation that could launch new industries or transform old ones beyond recognition." University administrators, following the lead of federal policymakers, came to accept the idea that their university could actually serve as an economic *engine* rather than merely an economic *resource* for local, regional, and national economic development. They began to talk of the social value of their institutions in terms of innovation, productivity, and competitiveness rather than their value in "meeting national needs," which was how their value was traditionally framed. In Washington, D.C., it was a change of thinking that revealed a deep anxiety about the future of an economy that had been the world's largest since just after the Civil War and its most productive in the twentieth century.

Bayh–Dole was born with a rising tide of free-market thinking and calls for deregulation and privatization. These were new times, and not particularly good ones in the United States. The impressive growth rate of the American economy following World War II and the revenue it provided for an activist government was coming to an end. Stagnate growth, double-digit inflation, a second oil shock in 1978 following the 1973 oil crisis, and rising concern about the lagging pace of innovation were combining to alter the country's mood. As a result, the relative balance of big government, the limits of regulation, and opportunities through free enterprise shifted. The government's collection of tens of thousands of patents had not yielded much in the way of economic returns. Newsweeklies ran headlines like "Has America Lost its Edge?" that contributed to as well as reflected a shift the winds of public opinion. Good public policy might include incentives for private gain when the alternative is patent portfolios that, as Broad put it then and as free-market advocates reflexively repeat today, gather dust on shelves.

Universities, for their part, had operated largely under what Berman calls the "logic of science" since Vannevar Bush championed the practice of pure science as the beginning of the road to technological progress and material prosperity, a serendipitous road with no shortcuts. That remained their operational logic despite their "experiments with market logic" in the way of university–industry research centers, industrial affiliates programs, industrial extension offices, and research parks. Perhaps more than anything else, what reflected universities' adoption of the "logic of the market" was the rise in academic entrepreneurship in the biosciences. "Emerging from disciplines with almost no tradition of industry relations, [biotechnology] ignored the old model entirely in favor of a university–firm relationship that was both better integrated and blurrier," as Berman described it. She cites three policy developments in the late 1970s as responsible for the rapid growth of bioscience entrepreneurship involving university faculty: a decision by Congress not to regulate recombinant DNA research, Congressional passage of a large capital gains tax cut, and a regulatory clarification by the Department of Labor allowing pension funds to invest in venture capital. Then, in 1980, the Supreme Court ruled in *Diamond v. Chakrabarty* that genetically modified organisms could be patented. It was a remarkable confluence of developments. The federal government tuned in to the role of innovation in economic growth, enlisted research institutions to help make it happen, and acted legislatively to deal with what neoclassical economics informs us causes self-interested economic agents to act: incentives.

Bayh–Dole passed by a thin margin in a lame duck session of the 96th Congress and became law on December 12, 1980, with President Jimmy Carter's signature. "Given Bayh–Dole's success," wrote Ashley J. Stevens in his study of the law's enactment, "it is surprising that there is not more general awareness of how fragile the coalition was that passed Bayh–Dole and indeed that it almost didn't get passed at all." Carter himself had favored an alternative bill co-sponsored by Senators Adlai Stevenson (D-Illinois) and Harrison Schmitt (R-New Mexico) that reflected the Hamiltonian philosophy of a strong central governmental role as opposed to the Jeffersonian individualist philosophy embedded in Bayh–Dole. He was not convinced that universities and small businesses could contribute nearly as much to the economy as big companies working with government agencies. Carter was lobbied to let it suffer a pocket veto. In the end, however, he relented and signed the bill on the last possible day.

The act amended Title 35 of the United States Code through the addition of a new chapter titled "Patents." Bayh–Dole was an example of revenue sharing through royalty sharing. Universities and other nonprofit institutions must share royalties with the inventor. The net income is to be used strictly for scientific research and education. This requirement means that universities must establish an in-house facility to manage technology transfer and the licensing of inventions. Though technology transfer offices began to grow in number beginning in the 1960s, passage of Bayh–Dole saw a wave of new offices established. Today the Association of University Technology Managers (AUTM) has more than 350 institutional members in the United States, including universities, research institutes, and teaching hospitals.

The Bayh–Dole Act was an expression of broader shift in US policy toward free markets, deregulation, and stronger intellectual property rights. As Berman chronicles in her book, the trend began years before the arrival of the conservative Reagan administration in January 1981, an administration that commenced nearly three decades of a strong property rights and deregulatory free-market ethos that continued until the financial crisis of 2008 and the severe global recession that followed led to a pull-back. Two years after Bayh–Dole became law, Congress established the Court of Appeals for the Federal Circuit to handle intellectual property disputes. The Court emerged as a strong champion of patent holder rights.

Other legislation in addition to Bayh–Dole designed to spur the transfer of technology to the private sector included the Stevenson-Wydler Act (1980) amended by the Federal Technology Transfer Act (1986) that encouraged joint research projects through cooperative R&D agreements (CRADAs). CRADAs are contracts between a government agency and a private company to work together on R&D. At the outset, CRADAs applied primarily to government-owned, contractor-operated laboratories and third-party laboratories. The National Competitiveness Technology Transfer Act (1989), another amendment to the Stevenson-Wydler Act, made technology transfer a mission of government-owned, contractor-operated facilities. It clarified the manner in which CRADAs are implemented, a matter further clarified in 1995 by the National Technology Transfer and Advancement Act that laid out guidelines to how CRADAs may be expedited. The Technology Transfer and Commercialization Act of 2000 sought to further improve the ability of federal agencies to license federally owned inventions and to make technology transfer "industry friendly" in light of rising global competitiveness.

Focusing on Innovation: Universities in the Balance

As crucibles of discovery, representative of what James Madison called "establishments for the advancement and diffusion of Knowledge," universities were expected to play an active role in the innovation process as the twentieth century came to an end. They were seen as vital components in the national drive to achieve the "commanding heights" of economic growth in the new global economy. Bayh–Dole entered the campus lexicon at universities across the United States as the free-market ethos replaced one of government controls and intervention that characterized the Progressive Era, the New Deal, and the Great Society.

In the late twentieth and early twenty-first centuries, universities paid ever more attention to what Adam Smith called "the current opinions of the world" in *The Wealth of Nations*, bristling at the fact that the elite universities of his day ignored such opinions. They are expected to do more than merely analyze the human propensity to "truck, barter, and exchange one thing for another" by developing complex econometric models. They are expected to be players in the "political economy." They are expected to be involved more directly in creating value for the commonwealth in addition to their basic missions of educating students, conducting research, and engaging the public. That said, the modern university, to the careful observer, remains easily recognized as the direct descendant of the University of Bologna, the first university in the West. "That is by any standard a formidable success: something to bear in mind whenever yet another book or essay pronounces— as this one will not—that the academy is 'in crisis,'" *The Economist* editorialized in a special edition "Inside the Knowledge Factory" issued during the dot-com boom of the late 1990s.

The wisdom of the Bayh–Dole Act is more actively debated today than during its first decades. In "Innovation's Golden Goose," *The Economist* editorialized that Bayh–Dole is possibly "the most inspired piece of legislation to be enacted in America over the past half-century." By the time of Bayh–Dole's twenty-fifth anniversary in 2005, several thousand US patents were being issued to universities annually compared to less than 250 in 1980, the year the act was passed and signed into law. During the first quarter-century of Bayh–Dole, these patents yielded universities and not-for-profit research institutions and hospitals an estimated $1.4 billion from 5,000 licenses, about 70 percent of them in the biomedical and life sciences. "There have been more than 1,500 new products are in the marketplace developed as a result of technology transfer in the past five years, demonstrating impact and benefit to the public," AUTM president Wriston Mark Crowell observed in a speech marking the anniversary. Academic technology transfer had added more than $40 billion to the US economy in its first two decades by one estimate.

Bayh–Dole has also spurred growth in company startups and spin-offs from university discoveries. In 2010, some 650 companies were started based on discoveries made in academic and nonprofit research institutes, the great majority of whose primary business was located in the licensing institution's home state. That is in line with surveys that show about 75 percent of companies spawned by university technologies locate near the university that produce them. In many cases, researchers and their graduate students work with start-up managers, venture capitalists, and industrial partners as members of an innovation team. Start-ups that scale up successfully can generate many jobs and bring dollars to their region.

Like the concentration of new machines and their inventors and operators in Birmingham, England, in the late eighteenth century and the rise of small manufacturers and artisans in Emilia Romagna in the twentieth century, life science companies tend to cluster in specific regions. In recent years, some two-thirds of US investment in biotechnology has gone to firms based in northern California, San Diego, and Boston. That means states like California and Massachusetts but also New Jersey, North Carolina, Pennsylvania, Maryland, and several others where biotechnology and pharmaceutical firms are currently concentrated stand to benefit disproportionately from the Bayh–Dole Act in terms of bioscience startups. If economists are right that technology hubs will be responsible for a growing share of national economic output, the spatial diffusion dimension of Bayh–Dole is likely to loom larger in the years ahead. The pull of local and regional market advantage could become a more important factor when universities negotiate licenses for technologies used by local industry.

In the United States, entrepreneurial universities such as MIT and Stanford University have contributed enormously to cultivating a culture of innovation in their neighborhoods. OECD launched the "Bologna Process" in 2000 to foster the growth of small and medium-sized enterprises. Among other things, the charter called for supporting the creation of "linkages between enterprises and education systems, and between industry and public and university research." The fact that the OECD chose to issue the program's charter at the oldest university in the West was meant to signal the key role of university research in regional innovation and that universities need to adapt to change and pay more attention to the world around them, as Adam Smith prescribed. Technology transfer efficiency in Europe is poor compared to the United States.

University licenses typically have no locational restrictions, but entrepreneurship is almost always local. Entrepreneurs who start companies tend not to relocate but instead stay close to the source of their perceived competitive advantage, typically the organization where the founder was previously employed, says Maryann Feldman and her colleagues who study the entrepreneurial university and technology transfer. In brief, what happens between Stanford University and Silicon Valley tends to happen elsewhere: University spin-offs stay close to home. That fact points up the tension between federal government policy, the interests of a state's Congressional delegation, and the tendency of certain regions, locales, and universities to benefit disproportionately from the nature of innovation itself. If Congress had been aware of how much innovation spurred by the Bayh–Dole Act would be in the biosciences and how much location matters in the field, it is possible to imagine that some members would have looked for ways of spreading the benefits of entrepreneurial activity resulting from the act more evenly among the states, perhaps giving a new twist to the term *political economy*.

The tension between faculty autonomy characteristic of the traditional university and institutional participation in economic development through the establishment of technology transfer offices led to a backlash against Bayh–Dole in its third decade. The backlash is reflected in books with titles like *Universities in the Marketplace: The Commercialization of Higher Education* and *Science in the Private Interest: Has the Lure of Profits Corrupted Biomedical Research?* Perhaps the sharpest criticism stemmed from Bayh–Dole's effect in the biomedical arena. "While Bayh–Dole was clearly a bonanza for big pharma and the biotech industry, whether its enactment was net benefit to the public is arguable," wrote Harvard physician and former *New England Journal of Medicine* editor-in-chief Marcia Angell. A frequent critic of big pharma, Angell views the entire era of Bayh–Dole, its incubation during the Carter administration, and its rollout during the Reagan administration as inherently destructive of American medical science and a major contributor to the high cost of drugs. Bayh–Dole, in her view, also transformed the ethos of medical schools and teaching hospitals because they "started to see themselves as 'partners' of industry, and they became just as enthusiastic as any entrepreneur about the opportunities to parlay their discoveries into financial gain."

Angell is right about one thing: It is hard to exaggerate the importance of Bayh–Dole to the emergence of the biotechnology industry in the United States. From the industry's beginning with the founding of Genentech in 1976, attempts were made to limit the influence of industry in public science. One of the first such attempts occurred early in the Reagan administration when then-Congressman Al Gore proposed exempting any inventions having to do with biotechnology from Bayh–Dole, arguing that this was far too important an area of technology to be left to universities to manage. Gore's concerns were echoed later by Michael Crichton, author of numerous science-fiction thrillers and later Gore's archnemesis in the global warming debate. At the end of his book *Next,* in which a chimp's egg is fertilized by a postdoctoral fellow's sperm yielding a "humanzee," Crichton argued we should, among other things, "rescind the Bayh–Dole Act (which allows universities to patent and make money from their research)." Crichton, a Harvard-trained physician who died in 2008, argued in a television interview that Bayh–Dole has altered the nature of the research university. "What this has done is [to] enormously stimulate the commercial aspect of university life," he said, adding: "From my point of view, [Bayh–Dole] was a great idea that hasn't really turned out the way anybody expected." Some universities are getting into the early-stage drug development business instead of licensing the intellectual property to entrepreneurs, giving universities a commercial atmosphere. Changing the law won't be easy because "universities are very strongly wedded to it" despite what Crichton saw as questionable evidence of its benefit to universities.

At the annual meeting of the Biotechnology Industry Organization in 2005 on the twenty-fifth anniversary of the legislation that bears his name, Birch Bayh appeared before a microphone. Looking over multitude of biotech boomers and boosters at the Philadelphia Convention Center, he expressed surprise by the consequences of his narrow victory in driving the legislation that bears his name through the Senate. "I don't think any of us who worked so hard to get that passed had the slightest idea we'd be sitting here twenty-five years later in the birthplace of our country, with all these people created by it," Bayh said, recalling the environment that led to the legislation. Before the legislation, thousands of government-owned patents, the proprietary yield of the billions spent by federal agencies to fund academic research, sat idle on the shelf, he said. The longer they sat, the lower their potential value as research invariably raced ahead, making the inventions obsolete. "Nobody would spend thousands or millions to get it into your medicine cabinet."

As Bayh spoke, the reaction against the pioneering legislation was in full swing, a fact that was tacitly acknowledged by a Biotechnology Industry Organization panel. University-industry collaborations fostered by the statute could go off the ethical track, damaging the independence of academic science. Does the growth of the number and size of technology transfer offices on campuses all over the United States make bias, misdirection, and abuse more likely, even inevitable? Is the culture of business inserting itself into the culture of academia—into the way basic science is done—in ways that raise the natural tension between the two cultures, erecting obstacles to the free flow of information and discovery? University administrations are on the spot as they struggle to find that right balance at a time when they also need to cover the costs of a large and often expanding research enterprise.

Academic critics of the new world of technology transfer, a world characterized by endless forms, legal experts, prolonged negotiations, bureaucratic logjams, misunderstandings and, of course, money, grew in number and stridency through the first decade of the new century. Even Bayh–Dole backers in industry complained. Venture capitalist Steven Burrill, who has put bioscience on the global investment map for more than two decades, accused some universities of being greedy and holding up the innovation process. "I think at many schools there's been an increasing push from the provost, who's asking about the royalty streams," Burrill said. "And that's a barrier to guys like me, who want to pull out [technology] at the lowest possible cost. As a result, we're spending less time dealing with tech transfer." Other critics complain that technology transfer offices still don't understand the needs of businesses, some of which are going overseas to do their idea mining at foreign universities as well as to establish new R&D facilities on foreign soil.

The consequences of Bayh–Dole to the traditional values of higher education continue to be assessed and reassessed. *The Economist* itself, which had lauded Bayh–Dole in 2002, acknowledged the increasing costs of the landmark legislation as proponents were celebrating its twenty-fifth anniversary. "Many scientists, economists and lawyers believe the act distorts the mission of universities, diverting them from the pursuit of basic knowledge, which is freely disseminated, to a focused search for results that have practical and industrial purposes," the magazine editorialized. "What is not in dispute is that it makes American academic institutions behave more like businesses than neutral arbiters of truth."

Studies have shown that licensing is just part of a "flurry of activities" that is consistent with fundamental discoveries from basic research and has not distorted faculty research agendas. But does faculty patenting and licensing really produce the economic punch that supporters of Bayh–Dole point to? Bayh–Dole scholar David Mowery argues convincing evidence is lacking that the legislation substantially increased the contributions of US universities to economic growth and innovation during the 1980s and 1990s. The nature of the contributions and the channels through which they have been realized before and after Bayh–Dole "have been complex and have included much more than patenting and licensing."

The arguments of critics like Mowery were given added weight with the mounting evidence that even as higher education expands around the world, university patenting activity is actually waning. At the conclusion of her book *Creating the*

Market University (2012), Berman notes that the number of patents issued to universities peaked in 1999 and industry funding for academic R&D has declined in real dollars, suggesting to her that the "logic of the market" in academia has reached its apogee. The dynamics of "economic rationalization" that led universities to view themselves as machines for the production of knowledge and educated workers (rather than, say, gardens for nurturing the mind) "are also constraining."

Scholars of the triple helix theory of regional economic development noted the decline in university patenting and spin-off activity in most advanced economies during the first decade of the 2000s. They found that the incentives provided through Bayh–Dole are no longer as strong as incentives for engaging in other activities. Chief among the alternative activities is the focus on university rankings, which typically do not take into account patenting and licensing activities. The nature of the competition among universities nationally and internationally is changing. Rising economic powers like China are investing heavily in their higher education infrastructure. More students in developing countries than ever before are enrolled in online courses offered free of charge by universities in the West. With growing international competition to achieve higher academic excellence standings and thus attract higher caliber students, the argument goes, the incentives for faculty to patent their discoveries are no longer as strong as they once were.

With the rapid rise of "networked science" and the data-sharing, problem-solving ethos that pervades online communities, international collaborations and coauthorships have become more important in research assessment than university–industry relations. University administrators today have to concern themselves more with changing academic ecosystems than innovation ecosystems. In addition, public universities and publicly funded science are common goods. At a time of social and economic stress and projected slow growth, that value may come to the fore and serve as a reminder that the best agent of knowledge transfer to society and the economy is a well-educated student who becomes a valuable employee or an entrepreneur.

Experimentation Underway in Knowledge Ecosystems

Has the luster of Bayh–Dole indeed started to fade? It is too soon to know. An analysis in 2011 by the Congressional Research Service, the public policy research arm of the US Congress, acknowledged "areas of concern" that Congress may decide to pursue. With respect to pharmaceuticals and biotechnology, some argue that under the Bayh–Dole Act companies are receiving too many benefits at the expense of the public particularly given the price of many drugs. Others observe that it is not the act itself but university policies designed to maximize institutional benefits that are the problem. What is not in dispute is that the tenure of the act has coincided with an expansion of academic life sciences R&D expenditures from approximately 40 percent of total research spending at universities in 1980 to 52 percent today.

Overall, the Congressional Research Service analysis found that the positive impact of the legislation "is still seen as significant."

By the time of its thirtieth anniversary of Bayh–Dole in 2010, according to AUTM, the number of patents granted to universities had grown ten-fold, to more than 3,000 a year (though it peaked at 3,698 in 1999), adding nearly $2 billion to their coffers every year. Even as more questions about the effectiveness and viability of Bayh–Dole are raised in the United States, its reputation outside the country is growing. Universities in Europe, Japan, Australia, Singapore, and other advanced economies, not to mention universities in economies rapidly on the rise such as those of China and India, are looking to emulate the American model of technology transfer.

Government leaders from the national, state, provincial, and local levels in countries around the world are undoubtedly proud of their Nobel laureates and scholarly heroes, but they are banking on economic growth. The role of universities in economic development tends to be what brings them to campus laboratories, not academic accolades. MIT, the quintessential technology transfer institution of higher education, is what nations and consortia of nations such as the European Union want their own universities to look like. But that fact does not automatically mean that Bayh–Dole is the best model for their technology transfer needs. "Developing countries should not follow the United States in enacting policies that undermine traditional ways of commercializing research output," says Bhaven Sampat, who studies the intersection between health and innovation policy. These countries should pay notice to the problems that have arisen with the act, including overly restrictive patenting and licensing, and craft their own legislation to avoid them.

Ideas for expediting the transfer of technology from university labs to the commercial sector are being proposed and explored in the context of Bayh–Dole. The Kauffman Center for Entrepreneurship, for example, called for unleashing America's academic entrepreneurs by revamping the technology transfer process. Kauffman scholars urge adoption of the open, flexible licensing approach to replace what they called the current "monopolistic model." It could be accomplished, they argue, by having the US Department of Commerce and possibly the Small Business Administration amend the rules of Bayh–Dole in a way that would allow university inventors to take their invention to any technology licensing office to serve as their licensing agent and not be restricted to the office of their resident university. Faculty would negotiate licenses themselves and then turn over a portion of their profits to the university. Universities within a region might form consortia and contract independently with companies. The Internet could be used to bring together companies and professors. "Let's stop penalizing professors who come up with new ideas and the universities they work for," wrote Kauffman's Robert Litan and Lesa Mitchell in 2010. Earlier Bayh–Dole critics complained of government patents languishing on shelves. Litan and Mitchell targeted *university* patents: "Most important, let's not keep the world waiting for new products and services—some of them lifesaving—while valuable ideas languish on university shelves."

The academic technology transfer community regards the ideas put forth by Kauffman as unworkable and in any event unnecessary because technology transfer offices are performing very well. Settling for the status quo does not appeal to innovation activists. From the business perspective, universities are seen as slow to act and do not appreciate that much of the technology they possess is very early stage, unproven, and often not all that valuable. But change is afoot. As it is throughout the economy, automation is making inroads into technology transfer management. A growing number of universities, research institutions, hospitals, and national laboratories are employing proprietary web-based software platforms and customized search engines to join available technologies with potential licensees, streamlining the flow of intellectual property to market application and experimentation. Such deal-flow tools may turn out to be a useful disruptive innovation to traditional university–industry technology transfer during the Bayh–Dole Act's fourth decade.

Federal governments are acting on an unmistakable feature of the evolving the modern economy: clustering, which almost invariably involves technology transfer from universities and research institutions. "We're all familiar with clusters like Silicon Valley," Barack Obama said at a small business forum in Cleveland in 2011. "When you get a group of people together, and industries together, and institutions like universities together around particular industries, then the synergies that develop from all those different facets coming together can make the whole greater than the sum of its parts." In recent years the US government has spent several hundred million dollars on pilot studies and other programs designed to cultivate new technologies and innovation clusters. As we have seen, university–industry linkages within regions drive bioscience innovation.

The case for federal cluster support is bolstered, in the eyes of some, by the federal government's rescue of the domestic automobile industry in 2009. General Motors, Chrysler, and Ford and their massive supply chains exist largely within a regional industrial ecosystem. That ecosystem was kept alive and intact by the bailout, setting the stage for the industrial resurgence that followed. The resurgence culminated with General Motors overtaking competitors as the top-selling automaker just two years after it was rescued. The auto bailout was an extreme case of direct government intervention in the private sector. Federally funded experimentation through pilot studies of how regional innovation clusters arise and can be sustained and what role universities play in their vitality should not be as controversial.

The Small Business Innovation Research (SBIR) and Small Business Technology Transfer (STTR) programs, which address funding issues at the earliest stages for small technology companies, also serve to build companies from university discoveries. The National Academy of Sciences found SBIR to be a highly successful program when it comes to linking small firms with university technology and expertise. The biotech pioneers Genentech, Biogen, and Amgen received SBIR support at an early stage of their development. In its bioeconomy blueprint report released in 2012, the Obama administration cited the NIH's SBIR Technology Transfer Program, which now offers exclusive short-term licensing agreements (it had previously offered only nonexclusive licenses) for start-up companies interested in discoveries generated through NIH's intramural research efforts. NIH SBIR program modifications designed to accelerate the commercialization of biomedical discoveries are underway following the establishment of the NIH National Center for Advancing Translational Sciences. The STTR program, which operates like SBIR but focuses on collaborative projects between small businesses and a university, nonprofit research organization, or federal R&D laboratory, is a particularly effective approach for building and supporting an innovation ecosystem because it fosters the functioning and growth of collaborative networks.

Congress reauthorized the SBIR and STTR programs for six years in 2011. In so doing it increased the overall allocation for the programs to $2.8 billion, trimmed application processing time to ninety days, and raised the venture capital/hedge fund/private equity fund participation for agency awards to 25 percent for the NIH, the Department of Energy, and the National Science Foundation, and 15 percent for eight other participating federal agencies. Allowing companies that are primarily funded through venture capital to compete for SBIR and STTR grants, the most important change from the standpoint of capital-intensive biotechnology, "will increase the number of new medical discoveries and innovations available to patients," said Biotechnology Industry Organization president and former Pennsylvania Congressman James Greenwood. In another federal initiative, the National Science Foundation established the Innovation Corps program to encourage collaboration between universities and industry and speed up the pace of technology-based startups. Innovation Corps–funded entrepreneurial teams, many of which are working on bioscience projects, typically consist of a principle investigator, a lead entrepreneurial graduate student or postdoctoral research associate, and a mentor from the business world.

Universities, with their research expenditures estimated to be $65 billion annually nationwide and their ability to educate some 3 million students each year for the workforce, are at the heart of the innovation ecosystem in the United States. In addition to forming multiyear, multimillion-dollar collaborations and partnerships with industry, they are trying out a number of new ideas, sometimes with private support, to spur innovation and economic activity. One such idea is that of offering start-up services within the institutions themselves. University venture centers or "proof-of-concept centers" introduce faculty and their technologies to prospective investors, typically at an earlier stage than incubators do. Such centers characteristically trumpet their successful start-up companies as evidence that investment in academic science pays dividends to public stakeholders and the regional economy, though their role in bridging distinct cultures and the "market gap" is still evolving.

Pittsburgh-based Carnegie Mellon University has a long history of turning its discoveries into businesses, particularly in Pennsylvania, where its spin-offs constitute a third of total university technology-based companies created in the state in recent years. Carnegie Mellon launched a "Greenlighting Startups" initiative, a portfolio of five new and existing campus incubators designed to build on its entrepreneurial success. The initiative is part of the university's "Five Percent, Go in Peace" spin-off model, which limits university equity in faculty-based startup to 5 percent and establishes clear royalty guidelines. Other universities are experimenting with innovation partnership incentives, for example, by granting companies exclusive licenses to technologies that result from research they sponsor in university laboratories. As noted in chapter 2, the Broad Institute's partnership with Bayer Healthcare grants Bayer an option to an exclusive license to potential drug candidates arising from their joint research in cancer genomics.

Universities are also participating in a variety of state programs and privately funded initiatives to speed the flow of technology from academia into the marketplace. The Maryland Innovation Initiative, a $6 million technology transfer program funded by the state legislature with contributions from five Maryland universities, brings together the state's public and private universities, federal research labs, and its quasi-public technology agency to create new jobs and companies. The fact that Maryland is home to the NIH helps.

Another approach is to tap into the brainpower of students before they graduate. Perhaps the boldest of these initiatives is the decision of the global investment firm New Enterprise Associates to establish a multimillion-dollar "Experiment Fund" with Harvard University. The firm's partnership with Harvard's School of Engineering and Applied Sciences will use investment awards averaging $250,000 in an effort to seed startups in Boston's supercluster, focusing on the consumer Internet, enterprise software, alternative energy, and health care fields. "In youth, you tend to have really big ideas," said New Enterprise Associates' Patrick Chung. "This university has taught students how to solve major problems for 400 years."

In the quest for new models of technology transfer, incubation, and acceleration, understanding how incentives work and can be maximized both for discovery and commercial development is key. It is a delicate balance: Too much emphasis on discovery without the means or will to move promising ideas into the marketplace may mean that potentially valuable ideas sit around on university shelves; too much emphasis on commercialization can distort a university's basic research and educational mission. The balance can be tipped depending on the particular university and the state of the economy, including the local economy. When the global recession sent start-up investors into retreat, some universities bolstered their efforts to accelerate technology transfer. "We have to jump higher," said the founder of one university venture center. "Because of the recession, universities need to put more skin in the game." That "skin" includes gap funding through the "valley of death" between the research laboratory and a marketplace that kills many nascent companies before they get a chance to show their potential.

Some universities, at least with respect to their activity in the biosciences, are taking a different tack from that taken by venture centers, technology incubators, and technology accelerators to navigate the funding gap and the valley of death. A case in point is BioPontis Alliance headquartered in Research Triangle Park. BioPontis is building a network of university partners, contract research organizations, and pharmaceutical companies. The alliance exercises nonexclusive rights to review intellectual property in a university portfolio that the university chooses to make available. It then develops those assets its sees as promising to the proof-of-concept stage and makes them available to its pharmaceutical partners for further development. If the pharmaceutical company opts to develop the technology, the share that goes to the university partner depends on its stage of maturity when BioPontis licenses it. Under the BioPontis model, the goal is to bring the interests of all parties in the innovation stream into alignment.

As the science becomes ever more complex and its commercial potential ever more difficult to discern, that challenge is greater every day. "Everyone is dancing around each other, looking for new ways to work together" is the way the executive officer of Edinburgh BioQuarter described the effort to find new ways to create value from interorganizational cooperation. Tech transfer dances are yielding a steadily growing number of university spin-off startups. The problem is a dearth of financing for startups, particularly from venture capital. Given that reality, *Nature Biotechnology* business editor Brady Huggett counseled, universities should consider focusing less on startups and more on developing their own discovery validation capabilities, "getting them to the stage where they are attractive licensing options for the biotech and pharma industry."

Universities are now full-fledged players in the competitive game. The Federal Circuit Court of Appeals certified that reality with its 2002 ruling in *Madey v. Duke University*. The court held that universities are not automatically entitled to an experimental "limited use" exception for purely scientific research to study and understand a patented invention. Thus universities are competing not just for students and faculty and status but for the monetary rewards that their involvement with commerce can bring. No longer are they protected sanctuaries in a teeming social ecosystem. The ineluctable forces of politics, finance, demographics, and new media have taken care of that. Today, all institutions of higher education are obliged to pay attention to, as Adam Smith phrased it, "the current opinions of the world."

Since the founding of the University of Bologna during the High Middle Ages, universities have proven that they can change enough to deal with the realities and contingencies of the time. Natural ecosystems are compelled to adapt to flourish, as are economic and institutional ecosystems. As *The Economist* observed just before the turn of the millennium, no institution lasts nine centuries without adapting.

References

Listed sequentially based on their order in the chapter

Robert Williams, "Bologna and Emilia Romagna: A Model of Economic Democracy," Paper presented at the annual meeting of the Canadian Economics Association, University of Calgary, May/June 2002.

"Competitive Advantage from Co-operative Values: The Emilia Romagna Model," International Co-operative Alliance Annual Report 2006, Geneva, accessed January 22, 2010, http://www.ica.coop.

"Emilia-Romagna—Economy—The Combination of Traditional Products and Technological Innovation," Eurostat: Portrait of the Regions, March 2004, accessed January 22, 2010, http://circa.europa.eu/irc/dsis/regportraits/info/data/en/index.htm.

Michael Porter, *The Competitive Advantage of Nations* (New York: Free Press, 1990), 212.

Robert D. Putnam, Robert Leonardi, and Raffaella Y. Nanetti, *Making Democracy Work: Civic Traditions in Modern Italy* (Princeton, NJ: Princeton University Press, 1993), 97–103.

Philip Cooke, "Building a Twenty-First Century Regional Economy in Emilia-Romagna," *European Planning Studies* 4.1 (1996): 53–62.

"Emilia-Romagna—A Region of Excellence for Innovative Startups," PAXIS, the Pilot Action of Excellence in Innovative Startups, European Commission, March 27, 2006.

Nicola Baldini, Ricardo Fini, and Rosa Grimaldi, "The Transition Towards Entrepreneurial Universities: An Assessment of Academic Entrepreneurship in Italy," in *University of Chicago Handbook on Technology Transfer*, Albert Link, Donald S. Siegel, and Michael Wright, eds. (Chicago: University of Chicago Press, 2013).

ASTER: A Consortium Between Emilia Romagna Region, University, Research Bodies and Entrepreneurs Associations for Industrial Research, Technological Transfer and Innovation, October 16, 2003, accessed January 22, 2010, http://www.aster.it.

"Report of the MIT Taskforce on Innovation and Production," MIT Commission on Innovation, February 22, 2013, accessed February 25, 2013, http://web.mit.edu/newsoffice/2013/production-in-the-innovation-economy-report-mit-0222.html

Suzanne Berger and Richard M. Locke, "Il Caso Italiano and Globalization," in *European Industrial Restructuring in a Global Economy: Fragmentation and Relocation of Valve Chairs*, Michael Faust, Ulrich Voskamp, and Volker Witke, eds. (Gottingen: Soziologisches Forschungsinstitut, 2004).

Peter Dizikes, "Standing Up for Manufacturing," *Technology Review*, January/February 2012.

The Triple Helix Then and Now

Linus Pauling and Robert B. Corey, "A Proposed Structure for the Nucleic Acids," *Proceedings of the National Academy of Sciences USA* 39 (1953):84–97.

Henry Etzkowitz and Loet Leydesdorff, eds., *Universities and the Global Knowledge Economy: A Triple Helix of University-Industry-Government Relations* (London: Cassell Academic, 1997).

Leydesdorff, Loet, "The Mutual Information of University–Industry–Government Relations: An Indicator of the Triple Helix Dynamics," *Scientometrics* 58.2 (2003): 445–467.

Henry Etzkowitz and Loet Leydesdorff, "The Endless Transition: A 'Triple Helix' of University–Industry–Government Relations," *Minerva* 36.3 (1998): 203–208.

Henry Etzkowitz and Chunyan Zhou, "Building the Entrepreneurial University: A Global Perspective," *Science and Public Policy* 35.9 (2008): 627–635.

Adam Frank, "Kodak is in Bankruptcy, But Its Hometown Hasn't Lost its Sparkle," NPR.org, January 19, 2012, accessed January 30, 2012, http://www.npr.org.

Etzkowitz and Zhou, "Building the Entrepreneurial University."

Education and Knowledge Flow Early On

Elizabeth L. Eisenstein, *The Printing Press as an Agent of Change* (Cambridge, UK: Cambridge University Press, 1979), 520.

Elizabeth L. Eisenstein, *The Printing Revolution in Early Modern Europe* (Cambridge, UK: Cambridge University Press, 1983).

James Madison, "Address to the Agricultural Society of Albermarle, Virginia," May 12, 1818, in *Letters and Other Writings of James Madison*, III (Philadelphia, 1865, imprint of R. Worthington, NY, 1884), 66.

James Madison to W. T. Barry, August 4, 1822, in *Letters and Other Writings of James Madison*, III (Philadelphia, 1865, imprint of R. Worthington, NY, 1884).

James Madison, "Republican Distribution of Citizens," *The National Gazette,* March 5, 1792. Reprinted in *The Papers of James Madison*, William T. Hutchinson et al., eds. (Chicago: University of Chicago Press, 1962–1977), 244–246.

Roy Branson, "James Madison and the Scottish Enlightenment," *Journal of the History of Ideas* 40.2 (1979): 235–250.

Joel Mokyr, *The Gifts of Athena: Historical Origins of the Knowledge Economy* (Princeton, NJ: Princeton University Press, 2002), 221.

James Buchan, *The Authentic Adam Smith: His Life and Ideas* (New York: W.W. Norton, 2006).

Alexis de Tocqueville, *Democracy in America*, translated by Henry Reeve, Project Gutenberg, January 21, 2006, accessed February 6, 2012, http://www.gutenberg.org.

Adam Smith, *An Inquiry into the Nature and Causes of the Wealth of Nations* (London: W. Strahan and T. Cadell, 1776).

"The Land-Grant Tradition," Association of Public and Land-Grant Universities, Washington, DC, 2012, accessed November 20, 2013, http://www.aplu.org.

Tapping into the "Endless Frontier" for Innovation

Vannevar Bush, "Science, The Endless Frontier," A Report to the President by Vannevar Bush, Director of the Office of Scientific Research and Development, July 1945 (Washington, DC: US Government Printing Office, 1945).

G. Pascal Zachary, *Endless Frontier: Vannevar Bush, Engineering of the American Century* (New York: Free Press, 1997).

Elizabeth Popp Berman, *Creating the Market University: How Academic Science Became and Economic Engine* (Princeton, NJ: Princeton University Press, 2012).

David C. Mowery, "The Bayh–Dole Act and High-Technology Entrepreneurship in U.S. Universities: Chicken, Egg, or Something Else?" in *University Entrepreneurship and Technology Transfer: Process, Design, and Intellectual Property*, Gary B. Libecap, ed. (London: Elsevier, 2005), 39–68.

Vannevar Bush, *Frederick Gardner Cottrell 1877–1948* (Washington, DC: National Academy Press, 1952).

Spencer E. Ante, "The Prophet of Startups," *Harvard University Alumni Bulletin*, June 2008, accessed February 6, 2012, http://www.alumni.hbs.edu.

William D. Bygrave and Andrew Zackarakis, eds. *The Portable MBA in Entrepreneurship* (Hoboken, NJ: John Wiley & Sons, 2004), 27–42.

US Small Business Administration, FAQs, accessed January 25, 2012, http://www.sba.gov/aboutsba/sbaprograms/inv/faq/index.html.

"Financing High-Growth Firms: The Role of Angel Investors," Organisation for Economic Co-operation and Development, Paris, December 2011, accessed February 14, 2012, http://www.oecd.org.

Angus Loten, "Angel Investors Play Big Role for Startups, Think Tank Says," *Wall Street Journal*, January 24, 2012.

Wisconsin Alumni Research Foundation, accessed January 25, 2010, http://www.warf.org.

Mark Tatge, "Miracle in the Midwest: How Madison, Wis. Became a Hotbed of Capitalism," *Forbes*, May 24, 2004, accessed February 6, 2012, http://www.forbes.com.

Patent Rights and Patent Fights

Mowery, "The Bayh–Dole Act and High-Technology Entrepreneurship in U.S. Universities."

Harbridge House, Inc., *Government Patent Policy Study: Final Report*, Prepared for the Federal Council for Science and Technology, Committee on Government Patent Policy, Vols. 1–4 (Washington, DC: US Government Printing Office, 1968).

Norman J. Latker, Statement of Norman J. Latker, Patent Counsel, Department of Health, Education, and Welfare Before the Subcommittee on Science, Research, and Technology, House of Representatives, May 26, 1977.

Bayh–Dole on the Fast Track

William J. Broad, "Patent Bill Returns Bright Idea to Inventor," *Science* 205 (1979):473–476.

Berman, *Creating the Market University*.

Diamond v. Chakrabarty, 447 US 303 (1980).

Ashley J. Stevens, "The Enactment of Bayh–Dole," *Journal of Technology Transfer* 29.1 (2004): 93–99.

Association of University Technology Managers, accessed January 25, 2010, http://www.autm.net.

Berman, *Creating the Market University*.

Stevens, "The Enactment of Bayh–Dole."

Focusing on Innovation: Universities in the Balance

Madison to W. T. Barry, August 4, 1822.

Smith, *An Inquiry into the Nature and Causes of the Wealth of Nations*.

Dugald Stewart, *The Works of Adam Smith, L.L.D., with an Account of His Life and Writings*, Vol. IV. (London: T. Cadell and W. Davies, 1811), 169.

Peter David, "Universities: Inside the Knowledge Factory," *The Economist*, Special Insert, October 4, 1997, accessed February 6, 2012, http://www.economist.com.

"Innovation's Golden Goose," *The Economist*, December 12, 2002, accessed February 6, 2012, http://www.economist.com.

Mark Wriston Crowell, "Bayh–Dole at 25 Years: Where Have We Been, and What Can We Expect?" Plenary Speech I. Report on the International Patent Licensing Seminar, Tokyo, Japan, 2006.

Association of University Technology Managers, AUTM US Licensing Survey: FY 2004.

The Bologna Charter on SME Policies. Organisation for Economic Co-operation and Development, Paris. Adopted June 15, 2000, accessed January 30, 2012, http://www.oecd.org.

Brady Huggett, "Biotech's Wellspring: the Health of Private Biotech in 2012," *Nature Biotechnology* 31.5 (2013): 396–403.

Janet Bercovitz and Maryann Feldman, "Entrepreneurial Universities and Technology Transfer: A Conceptual Framework for Understanding Knowledge-Based Economic Development," *Journal of Technology Transfer* 31.1 (2005): 175–188.

Marcia Angell, "The Truth About Drug Companies," *New York Review of Books* 51.12 (July 15, 2004): 52–58, accessed February 6, 2012, http://www.nybooks.com.

Stevens, "The Enactment of Bayh–Dole," 93–99.

Michael Crichton, *Next* (New York: HarperCollins, 2007), 536.

Michael Crichton, Interview on the Charlie Rose Show, February 19, 2007, accessed September 22, 2012, http//www.charlierose,com.

Bernadette Tansey, "The Building of Biotech: 25 Years Later 1980 Bayh–Dole Act Honored as Foundation of an Industry," *San Francisco Chronicle*, June 21, 2005, February 6, 2012, http://www.sfgate.com.

Ed Silverman, "The Trouble With Tech Transfer," *The Scientist* 21.1 (January 1, 2007): 40, accessed February 6, 2012, http://classic.the-scientist.com.

"Bayhing for Blood or Doling Out Cash?" *The Economist*, December 24, 2005, accessed February 6, 2012, http://www.economist.com.

Jerry Thursby and Marie Thursby, "University Licensing: Harnessing or Tarnishing Faculty Research?" in *Innovation Policy and the Economy*, Vol. 10, Josh Lerner and Scott Stern, eds. (Chicago: University of Chicago Press, 2010), 159–189.

Mowery, "The Bayh–Dole Act and High-Technology Entrepreneurship in U.S. Universities."

Berman, *Creating the Market University*, 171, 176, 177.

Loet Leydesdorff and Martin Meyer, "The Decline of University Patenting and the End of the Bayh–Dole Effect," *Scientometrics* 83.2 (2010): 355–362.

Michael Nielsen, *Reinventing Discovery: The New Era of Networked Science* (Princeton, NJ: Princeton University Press, 2012).

Experimentation Underway in Knowledge Ecosystems

Wendy H. Schacht, "The Bayh–Dole Act: Selected Issues in Patent Policy and the Commercialization of Technology," Congressional Research Service (June 9, 2011), 23, accessed November 18, 2013, http://www.fas.org/sgp/crs/misc/RL32076.pdf.

Bhaven N. Sampat, "Lessons from Bayh–Dole," *Nature* 468 (2010):755–756.

Robert E. Litan and Lesa Mitchell, "A Faster Path from Lab to Market," in "The HBR List: Breakthrough Ideas for 2010," *Harvard Business Review* (January 2010): 52–53.

Arundeep S. Pradhan, "Defending the University Tech Transfer System," *Bloomberg BusinessWeek*, February 19, 2010, accessed February 6, 2012, http://www.businessweek.com.

Barack Obama, "Remarks by the President at Closing Session of Winning the Future Forum on Small Business in Cleveland, Ohio," February 22, 2011, accessed May 2, 2012, http://www.whitehouse.gov.

"National Bioeconomy Blueprint," The White House, April 2012, accessed May 2, 2012, http://www.whitehouse.gov.

United States Congress, S.493—SBIR/STTR Reauthorization Act of 2011, January 30, 2012, accessed January 30, 2012, http://www.opencongress.org.

"BIO Applauds Congress for Reauthorizing SBIR/STTR Programs," Biotechnology Industry Organization, December 16, 2011, accessed January 30, 2012, http://www.bio.org.

Krisztina "Z" Holly, "Universities in Innovation Networks: The Role and Future Promise of University Research in U.S. Science and Economic Policymaking," ScienceProgress.org, January 19, 2012, accessed February 1, 2012, http://www.scienceprogress.org.

Bob Tedeschi, "The Idea Incubator Goes to Campus," *The New York Times*, June 25, 2010.

Hans Zappe, "Innovation: Bridging the Market Gap," *Nature* 501 (September 18, 2013), accessed September 25, 2013.

Ken Walters, "Carnegie Mellon Launches 'Greenlighting Startups' Initiative." Carnegie Mellon University, June 1, 2011, accessed November 20, 2013, http://www.cmu.edu/news/archive/2011/June/june1_greenlightingstartups.shtml.

Wallace Loh, "Turning Research Into Jobs in Maryland," *The Baltimore Sun*, February 13, 2012.

Tedeschi, "The Idea Incubator Goes to Campus."

Lizette Chapman, "NEA Creates Harvard Seed Fund to Back Students' Big Ideas," *Wall Street Journal*, January 30, 2012, accessed January 30, 2012, http://www.wsj.com.

Nuala Moran, "New Models Emerge for Commercializing University Assets," *Nature Biotechnology* 29.9 (2011): 774–775.

Huggett, "Biotech's Wellspring."

Madey v. Duke University, 307 F.3d 1351 (2002).

6 }

Splicing and Dicing: Property, Information, and the DNA of Innovation

My Method of lessening the Consumption of Steam, and
consequently Fuel, in Fire Engines, consists of the following
Principles....
   —James Watt's patent application

Furthermore, by designing and manipulating genes in a synthetic
genome according to methods described herein, experimental studies
may be performed, e.g., to identify...features that are important to
impart 'life' to an organism, etc.
   —J. Craig Venter et al., US Patent Application 20070264688

The first patent issued by the United States was to Samuel Hopkins of Pittsford, Vermont, in 1790, two years after the nascent American Republic ratified its Constitution that established patent law "to promote the Progress of Science and useful Arts." The Hopkins patent, granting a fourteen-year temporary monopoly by the authority of President George Washington, Secretary of State Thomas Jefferson, and Attorney General Edmund Randolph who signed it, was for an improved method of making potash, potassium salts derived from tree ashes. Potash was used in making glass, saltpeter, which was a key ingredient in gunpowder, and a special cleansing soap for raw wool. It also was used in bleaching, mining, metallurgy, and fertilizer, which is its principal use today. Fruits and vegetables, rice, wheat, corn, soybeans, and cotton all benefit from the nutrient's growth and quality enhancing properties.

Potash was America's first industrial chemical. Hopkins' new and improved process for producing it doubled the domestic yield. Rather than running his competitors out of business, he came up with the idea of selling licenses to those who wanted to use his furnace process for five-year periods. To use his furnace process, licensees had the option of either giving him a down payment of $50 or a half-ton of potash, and then another $150 or a ton and a half of potash, over the five years. That led to the rapid rise of town-based asheries to meet the growing demand for potash-based soap used by woolen mills in the United States and Great Britain. Thus Hopkins developed both a novel industrial process and a business process that added value and distributed that value to the emerging American economy. Combined with his marketing acumen, these elements of innovation enabled the United States to dominate the potash industry until the 1860s, though in the end Hopkins failed as a businessman.

Putting aside the fact that procuring potash in those days meant destroying dense hardwood forests, it was an auspicious beginning for the American patent system. More than any other country, the United States promoted the ethos of material progress by securing in law the property rights of inventors. It fostered the democratic expansion of their ranks to help those inventors who lacked connections to financial resources or political power. In brief, the United States invested in the creativity of the individual operating freely in the marketplace of ideas. Philosopher John Locke may have developed his ideas about the primacy of property rights to individual freedom in his native England, but it was in American law where they took root and flourished. It is in America where Locke has achieved a nearly saintly status among libertarians and free-market thinkers.

More than two centuries and 8 million US patents later, Hopkins' name still resonates, including in the life sciences. Individual inventors apply for and are awarded patents, even when those inventors are employees of large organizations, corporations, or universities. The funding source for their research does not alter that reality, at least in current law. In 2010 the US Supreme Court ruled against Stanford University and for Roche Molecular Systems, Inc. in a patent rights dispute. The high court affirmed that the right to patent an invention ordinarily belongs to the person who creates the invention rather than to the inventor's employer, unless they agree otherwise. The fact that the invention was created with the assistance of federal funding and that patent rights were secured under the 1980 Bayh–Dole Act does not alter this fundamental rule. The primacy of the individual inventor supersedes such federal contracting arrangements. "Under that law [the Patent Act of 1790], the first patent was granted in 1790 to Samuel Hopkins, who had devised an improved method for making potash, America's first industrial chemical. U. S. Patent No. 1 (issued July 31, 1790)," wrote Chief Justice John Roberts, delivering the opinion of the Court. "Although much in intellectual property law has changed in the 220 years since the first Patent Act, the basic idea that inventors have the right to patent their inventions has not." In brief, inventors own the rights to their inventions regardless of their status as employees of federal contracting organizations under the Bayh–Dole Act. The National Institutes of Health (NIH) informs its contracting universities and institutes that they must obtain invention assignments because inventions vest with employees. The standard practice in research-intensive universities and corporations is that employees assign their patent rights to the parent organization as a condition of employment. However, contractors have not always secured such assignment from their inventor employees in a manner that is legally unambiguous.

Yet the net effect of the Court's decision in *Stanford v. Roche* was a sharing of patent rights to a polymerase chain reaction-based test for measuring human immunodeficiency virus (HIV) in a patient's blood. The patent rights were to be shared between the inventor's research institution, Stanford, where the inventor did his initial research on the project, and Cetus Corporation, the company where he developed the test as part of a Cetus-Stanford collaboration. Cetus, which pioneered polymerase chain reaction technology, was subsequently purchased by Roche, which commercialized the invention and reaped the financial rewards it brought. Stanford's claim to sole ownership of the patents involved was denied by the US Court of Appeals for the Federal Circuit, which also denied Roche's patent ownership counterclaim. The Supreme Court affirmed the Federal Circuit Court's judgment, which was a sharing of the patents ownership.

The contribution of the biosciences to future economic growth depends on a system that aligns incentives in a way that promotes and ideally maximizes the contribution of individuals and organizations all along the value chain. Intellectual property law is pivotal in the innovation paradigm. If court decisions like that in *Stanford v. Roche* fail to result in aligning incentives in such manner, regardless of legal precedent or historical circumstance, the actual consequence will be an innovation shortfall. In the *Stanford v. Roche* case, both parties clearly contributed to the development of a highly successful research-based commercial product, though Stanford did not share in the resulting royalties. Royalty sharing was not at issue in the case because no royalty agreement was ever established, but royalty sharing could become more important than it is today if optimal levels of social return from public research investment are to be achieved. Bringing the fruits of biological knowledge to patients and consumers is difficult given the increasingly expensive winner-take-all intellectual property system. Shared patent rights and royalty-sharing arrangements between publicly funded research institutions and companies, if properly designed, could constitute an innovative pathway to "derisking" drug discovery and development. Indeed, several major pharmaceutical companies have agreed to just such arrangements as part of a program, launched by the NIH's National Center for Advancing Translational Sciences, for repurposing compounds companies have tested for safety but then set aside.

The Hopkins legacy of the heroic solitary inventor experiencing his or her eureka moment in a garage seems singularly unsuited for the biosciences today. Biology is a world of abstruse genetic and biochemical pathways, cell-surface receptor complexes, vast cell-signaling networks, and computational models that remain underpowered for the task of accurately simulating molecular interactions in the cell or the whole organism. Congress recognized the reality of collaborative research and invention in 2004 when it passed the Cooperative Research and Enhancement Act. The CREATE Act provides that researchers working for different employers may collaborate without jeopardizing a future patent application, thus removing a barrier to joint research.

The Hopkins legacy is also under stress from globalization and the rapid economic rise of China, India, Brazil, Russia, Turkey, Mexico, South Africa, Indonesia, and other countries. American multinational corporations expanding into these markets will have to comply with the intellectual property law regimes established by these countries, regimes that reflect their own historical experience and self-interest. The reigning ethos of the early American Republic passed down in US patent law is not sacrosanct in such venues regardless of efforts of US negotiators to have it incorporated into trade agreements. India's Supreme Court rocked the pharmaceutical world in 2013 when it ruled that "evergreening," the incremental improvement of branded drugs, does not merit patent protection, opening the way for much cheaper generic drugs of which India is a major global producer. Many developed and developing countries, including the United States and China, view incremental improvement in drugs as meriting patent protection. Even when national governments acknowledge the legitimacy of intellectual property rights as defined by the World Trade Organization, their willingness to enforce them is often lacking.

Contemporary patent law creaks under the strain of trying to contain within one system at least two vastly different worlds of innovation. One is the world of infotech, software development, technical methods and processes, and systems and component design—together with their biological counterparts of genomics and synthetic biology. A second world is that of drug discovery, development, and delivery of, in the Food and Drug Administration's (FDA's) parlance, "new chemical entities" or "new molecular entities" or "therapeutic biological products" that require immense investment and regulatory approval before they can be marketed. Some patent law scholars contend that a well-designed unified system could reconcile these worlds, preferably through the courts rather than through legislation given how hostage lawmaking has become to shifting political winds.

But are these worlds really reconcilable in the long run? A few months after the death of Steve Jobs, best-selling author Malcolm Gladwell reflected on his legacy. Gladwell distinguished the classic inventor like James Watt from a "tweaker" like Jobs, the founder of Apple Computer, who "inherits things as they are, and has to push and pull them toward some more nearly perfect solution," which "is not a lesser task" than faced by the inventor but one that amplifies the economic value (and in Jobs' case, the beauty) of technology. That characterization was unfair to Jobs. Many if not most people would not agree with it, yet both Watt and Jobs are clearly identified with the first world above, the world of devices, design, systems, engineering, and the creative spark. Generally speaking, we do not tend to think of principals in the second world of molecular biology and pharmaceutical/biopharmaceutical research as inventors in the Watt, Hopkins, and Jobs tradition, even if these inventors owe more to others than is customarily acknowledged. Bioscience innovation, as with some other technologies, is achieved on the foundation laid by earlier innovators. Scientific co-authorship, regarded as a proxy for collaboration as opposed to individual achievement, has exploded in recent years. The efforts of Congress, the courts, the US Patent and Trademark Office (USPTO) and the intellectual property law establishment to have us meld these different worlds seem strained. To some observers, they have more to do with reaffirming a hallowed legacy than facing a new reality being forged by social, organizational, and technological change and geopolitics.

The America Invents Act: Needed Reform or Missed Opportunity?

Though he filed his patent from Philadelphia, Samuel Hopkins developed his potash processing technology in Pittsford, Vermont. Patrick Leahy, Vermont's longtime democratic senator and lead author of US Senate Bill S. 23, the America Invents Act (AIA), is quick to note that fact and that Vermont "is issued more patents per capita than any other state in the country." His six-year effort culminated in 2011 when the Senate passed by 89–9 a version of the bill that had previously passed US House of Representatives 304–117. President Barack Obama signed the bill into law September 16, 2011, at a ceremony at Thomas Jefferson High School for Science and Technology in Northern Virginia. Obama lauded the legislation as "the most significant reform of the Patent Act since 1952" and predicted that it "will speed up the patent process so that innovators and entrepreneurs can turn a new invention into a business as quickly as possible." The AIA was quickly embraced as the cornerstone of his administration's innovation policy.

But will the AIA serve to boost innovation in the biosciences? The property value of life was a latecomer to the American patent system. In 1873, eighty years after enacting its first patent statute, the US government granted patent US Patent 141,072 to Louis Pasteur with a claim to "yeast, free of organic germs of disease, as a living organism." It was the first instance in the history of patent law that a patent was granted on living matter, what science historian Daniel Kevles termed a "vital creature." The patent office considered the invention as an article of manufacture and did not deliberate concerning its unique characteristic as living matter, in which the office had no interest at the time. Not until 1930 with passage of the Plant Patent Act did Congress delve into the realm of living things, making eligible for patenting asexually reproduced plant varieties not found in nature but created by plant breeders like Luther Burbank who was posthumously granted a patent under the new act. As in other areas of law, Kevles observed, the patenting of life "has also been shaped by the play of economic interests operating in tandem with changes in science and technology."

Intellectual property law changes slowly. In the early days of the Industrial Revolution, the English judiciary struggled with its patent system. The English system, based largely on royal patronage, was designed for an economy resting on the inventions arising from, among others, artisans, craftsman, mechanics, millwrights, and amateur hobbyists. Soon the system faced the new technical and competitive conditions of industrialization. "Little agreement could be reached over what properly constituted a patentable invention, on what grounds it was entitled to this special protection and privilege, and how wide ranging the powers conferred by patent should be," wrote Christine MacLeod in her book *Inventing the Industrial Revolution*. The first legal treatise specifically on patents was not published until 1803, three years after the expiration of the patent granted by Parliament to entrepreneur Matthew Boulton and inventor James Watt for design of a steam engine with a separate condenser.

The patent system, in the words of the only American president to be awarded a patent, Abraham Lincoln, "added the fuel of interest to the fire of genius." But today many argue that the patent system is more about interest than genius, more responsible for producing an economy characterized by legal and entrepreneurial gridlock than by innovation. Indeed, patent history is being reexamined in the context of innovation. Experimentation with patent law to achieve optimal innovation in the economy is historically a rare species. Until the AIA passed in Congress and was signed into law, patent law in the United States was last recodified by the Patent Act of 1952. In that act the term *process* replaced the term *art* in the definition of patentable subject matter, which became "any new and useful process, machine, manufacture or composition of matter or any new and useful improvement thereof." Even then, a year before the discovery of the structure of DNA, an invention involving living matter as such rather than as an article of manufacture, as in the case of Pasteur's yeast, did not qualify. Not until the celebrated 1980 US Supreme Court case of *Diamond v. Chakrabarty* deciding that a live, human-modified microorganism is patentable subject matter, constituting a "manufacture" or "composition of matter" within the applicable statute, did the USPTO begin awarding patents on genetically modified organisms. In 1988, a US patent was awarded to Harvard University for "a transgenic non-human mammal whose germ cells and somatic cells contain a recombinant activated oncogene sequence introduced into said mammal," the OncoMouse. Less than three decades after *Chakrabarty*, genomics pioneer J. Craig Venter filed patent applications for making synthetic genomes, the basis for creating synthetic life.

If "process" replacing "art" was the key provision of the Patent Act of 1952, the "first-to-file" provision is the comparable feature of the AIA of 2011. Ever since Samuel Hopkins was assigned the first patent in 1790, the United States has operated under the "first-to-invent" system whereby one inventor can challenge the application of another if the former can present convincing evidence that he or she was first to conceive of the invention, then diligently reduced it to practice and then filed a patent application. Most other countries have adopted the first-to-file system as their norm. In the first-to-file system the patent is granted to the inventor who is the first to file a patent application, regardless of the date of invention.

The first-to-file provision is meant to streamline the patent application process, lessen the number of patent disputes and resulting litigation, and harmonize US patent practices with those followed in other countries. In brief, it is designed to promote innovation by making it unambiguous via assigning a firm date of patent application filing. Such practices are already common among many universities and biotech companies looking to protect their intellectual property in international venues. Whether adopting the first-to-file system will actually spur innovation is anyone's guess. The change may make it more difficult for inventors working inside smaller companies with less money, the source of much innovation, because they will no longer have as much time as before to assess the commercial value of an idea.

What is likely is that the transition from the first-to-invent to the first-to-file will produce litigation, probably lots of it, if for no other reason than the fact that the courts will need to rule on the parameters of first-to-file. In the short run, it is hard to see how such transitional litigation or any litigation resulting from implementation of the AIA will, to paraphrase President Obama, assist innovators and entrepreneurs in turning a new invention into a business quickly. The AIA is designed to reduce the time and expense involved in the patent application process. It is also intended to reduce litigation and make the entire process more transparent. The larger question is whether the new system will begin to do so once the transition phase, expected to last many years, is complete.

Although bioscience industry groups generally approved the legislation and were successful in removing provisions they opposed while it worked its way through Congress, critics of the AIA complained that the first-to-file system would favor large corporations over universities, small business and startups, and, importantly, the individual inventor. Patent law provides a one-year grace period for inventors to work on their inventions before filing a patent application. During the grace period, any patent applications or publications that disclose the invention cannot be used as a prior art claim against the inventors. Critics argue that the new law weakens the one-year grace period, a critical provision for invention disclosures by university faculty. Indeed, what the AIA introduced might be better characterized as a "first to file or disclose" system, a system that puts a greater burden on faculty inventors who publish their research or talk about it at scientific meetings to defend their claim to prior art against challengers.

The postgrant review process of a patent that takes place after a patent has been granted to an inventor is a novel provision of the new law that USPTO director David Kappos described as a postgrant "quality check." The process establishes procedures under which parties may challenge patents with the USPTO that have already been issued on the basis of prior art considerations or other faults. Postgrant review was adopted with the intent of reducing the frequency of lawsuits and lowering their impact on the validity of patents. Whether it will do so or whether it will put venture capital-backed start-up companies at a distinct disadvantage against big firms with deep pockets, as some critics contend, will become evident as the AIA is implemented. If entrepreneurial startups are substantially squeezed by implementation of the first-to-file system or postgrant review, or both, other provisions of the AIA that enhance patent application efficiencies and lower application fees may not matter so much in fostering innovation and economic growth tied to it. Indeed, given the fact that the USPTO approves a higher percentage of patent applications than patent offices in Europe and Japan and receives maintenance fees for those it approves, the AIA reforms may serve to provide an incentive for the agency to approve a larger number of patents of questionable validity.

Cries of a "patent crisis" in American law reverberated through recent decades and grew shrill enough for Congress to act beginning in the early years of the current century. Succeeding Congresses took up patent reform based largely on provisions in the Patent Reform Act of 2005 before the 112th Congress finally passed the Leahy-Smith AIA. But is Congress best equipped to institute substantial reform? Not everyone agrees that it is. The optimal approach may be for Congress "to provide statutory tools that allow courts to adapt patent law to changing needs of innovators in an ongoing, responsive manner," in the view of University of California Irvine patent law scholar Dan L. Burk testifying before the Judiciary Committee of the US House of Representatives.

Burk is a well-recognized authority on patent law and the life sciences industry, for which he advocates strong patent protections. Burk added his voice to concerns that the patent system is in part responsible for producing an economy characterized less by innovation and more by legal and entrepreneurial gridlock. "In order to foster innovation, our patent statutes must be dynamic, flexible, and capable of dealing with rapidly developing business needs that were never foreseen." While he advocates strong patent protections, Burk warns about the threat that upstream patents can pose to experimentation and downstream innovation, a threat economist Scott Stern and his colleagues found to be justified with respect to genetically engineered mice. Burk has written extensively about how the courts rather than legislatures or bureaucracies are best equipped to address the "patent crisis," a term used in the title of a book he coauthored with Stanford University's Mark Lemley. Litigation over the small percentage of patents that are both valuable and in dispute "elicits information from both patentees and competitors through the adversarial process, which is far superior to even the best-intentioned government bureaucracy [such as the USPTO] as a mechanism for finding truth," Lemley said following passage of the AIA.

Does the AIA disrupt the current patent system without implementing a system that will enhance innovation? The act encompasses many changes to settled patent law. "All will require review by the courts, implementation by the Patent Office and consultation between patent holders and their counsel," Burk argued shortly after the act was signed into law. If the AIA actually delivers the 200,000 jobs that the Obama administration promised it would, Burk wrote in the JURIST legal news and research service, "they seem likely to be jobs for patent attorneys, rather than jobs for technological innovators." Evidence for Burk's impression was soon reported by *Bloomberg News*: A surge reported in patent attorney demand "can be attributed in part to the AIA, the biggest overhaul to the US patent system in six decades. The legislation, which changes how patents are processed and reviewed, is spurring a race among law firms for star talent in a small pool of patent attorneys." It is a fairly safe bet that the 40,000 patent attorneys and agents registered with the USPTO will see their ranks swell and their business grow in coming years.

USPTO director David Kappos described the AIA's key first-inventor-to-file provision as bringing "certainty, clarity, and simplicity" to the "the world's only truly twenty-first century patent system that champions innovation." The patent attorney community, in contrast, appears to find in the AIA a complex law that represents opportunity, particularly for technically skilled patent lawyers. If the AIA, once fully implemented, turns out to be an organ of legal complexity rather than patent process clarity and simplification, it will give the United States more firepower in an arena where it already holds a significant international competitive advantage—intellectual property law. A competitive advantage in the number of patent attorneys does not necessarily translate into a competitive advantage in technological innovation.

Genes, Pathways, Methods and Schedules Go to Court

As the AIA was being legislated, approved, signed into law, and implemented, a series of high-profile court cases served to illustrate how difficult it will be for the US patent system to achieve the certainty, clarity, and simplicity that AIA was touted to provide. For the biosciences, the court cases and the judicial rulings the cases yielded seemed to augur greater complexity, uncertainty, and confusion and thus served as fertile ground for the field of intellectual property law, if not for innovation. Indeed, a provision of the AIA was inserted at the assistance of a member of Congress who was heavily lobbied by the plaintiff in one of the cases.

Section 27 of the AIA charges the director of the USPTO to conduct a study on "effective ways to provide independent, confirming genetic diagnostic test activity where gene patents and exclusive licensing for primary genetic diagnostic test exist" and deliver the report and its recommendations to Congress in nine months. In short, the agency was to recommend ways to solve the problem of gene patents, in particular the exclusive licenses that may serve to block a patient's access to independent confirming genetic diagnostic tests. For the first time, the USPTO was being asked to move beyond its focus of assessing the patentability of claims into the realm of the cutting-edge biomedical research, medical care quality, cost and access, and incentives to monetize patent rights, which are stronger in biotechnology than in any other industry.

Patient access to confirming or second-opinion genetic diagnostic tests led to challenging the validity of primary or initial diagnostic genetic tests that have been protected by patent. The issue was highlighted in the case *Association for Molecular Pathology v. US Patent and Trademark Office* in which the US Supreme Court ruled unanimously in 2013 that isolated DNA is not patent eligible. The decision was the culmination of a four-year court battle over gene patents—whether natural products can become patentable human inventions. Justice Clarence Thomas, writing for the Court, observed:

> We have "long held that this provision contains an important implicit exception[:] Laws of nature, natural phenomena, and abstract ideas are not patentable." *Mayo*, 566 U.S.... Rather, "they are the basic tools of scientific and technological work" that lie beyond the domain of patent protection.... As the Court has explained, without this exception, there would be considerable danger that the grant of patents would "tie up" the use of such tools and thereby "inhibit future innovation premised upon them.".... This would be at odds with the very point of patents, which exist to promote creation. *Diamond* v. *Chakrabarty*, 447 US 303, 309 (1980) (Products of nature are not created, and "'manifestations... of nature [are] free to all men and reserved exclusively to none'").

The four-year battle revealed how different courts in the American federal judicial system perceive the patent system and adjudicate patent disputes in different ways.

In 2009, organizations representing more than 150,000 researchers and pathologists, breast cancer and women's health groups, universities, genetic specialists, and patients—more than twenty plaintiffs in all supported by the America Civil Liberties Union—filed a lawsuit against the USPTO. The aim of the suit was to invalidate the patents the agency had granted for the BRCA1 and BRCA2 breast cancer genes. Myriad Genetics developed potentially life-saving predictive genetic tests from the genes, which are related to most cases of hereditary breast and ovarian cancers. These genes were discovered by analyzing families in which breast cancer developed over several generations. Myriad Genetics was granted an exclusive license to develop diagnostic tests based on the breast cancer genes by the University of Utah Foundation, which together with Myriad and others held the patents. The plaintiffs argued that tests were unconstitutional under the First Amendment and the Patent and Copyright Clause of the US Constitution. They insisted that the patents were unconstitutional and invalid because "human genes are products of nature, laws of nature and/or natural phenomena, and abstract ideas or basic human knowledge or thought" and thus are not patentable.

How widespread is the practice of patenting human genes? One brief study found that sequences from an estimated 20 percent of the 23,000 human protein-coding genes were patented as of 2005, most of them during a surge in the late 1990s and early 2000s when the NIH, universities, and private industry patented genetic sequences en masse. Law professor Christopher Holman called that conclusion "the 20 percent myth," that 20 percent of human genes are patented and thus sequencing an individual's whole genome would invariably result in the infringement of thousands of gene patents. Holman argues that from a practical standpoint the patent owners would be unlikely to assert their rights in the context of genetic testing.

US-based biotechnology and pharmaceutical companies that patented most of the human gene sequences were searching for therapeutic proteins and are not looking to block their use in genetic diagnostics, in his view. Yet human gene patenting can impede subsequent scientific research and product development for those genes, says economist Heidi Williams after studying human genomic sequences put into the public domain versus those held privately through the Human Genome Project.

Perhaps more troubling was Myriad Genetics' claim to any specific fifteen nucleotide sequence present in the BRCA genes. Patent claims drafting had run headlong into the mathematical realities of genetic sequence science. Researchers have found that fifteen nucleotide sequences from all known human genes match fifteen nucleotide sequences of at least one other gene, with the average gene possessing fifteen nucleotide sequences matching those present in several hundred other genes. BRCA1 has fifteen nucleotide sequences matching those of nearly 700 other genes. Looked at this way, the entire human genome plus animal and plant genomes have in effect already been patented by ownership claims to nucleotide sequences consisting of as few as fifteen DNA letters. A closer look showed that 8,000 approved US patents contain the type of composition-of-matter claims involving human, animal, plant, and microbial nucleotide sequences that the Supreme Court ruling in the Myriad Genetics case have put at risk.

While the Myriad case was being heard in Federal District Court in New York in 2010, a Department of Health and Human Services committee studying gene patents and licensing practices and their impact on patient access to genetic tests issued a draft report. The committee found that "a near perfect storm" was forming at the confluence of clinical practice and patent law. "The cost of genetic analysis is decreasing dramatically, while knowledge about the genetic foundations for health, illness, and responsiveness to medicine is growing exponentially." At the same time, trends in patent law are threatening the promise of these developments. Patenting has "moved upstream," covering not only commercial products but also making proprietary claims on research discovery platforms that enhance the possibility of future discoveries and are critical to medical research progress and innovation.

A New York federal district judge handed down a decision in 2010 that stunned the patent law community and the biotechnology industry. Judge Robert W. Sweet invalidated Myriad Genetics' BRCA1 and BRCA2 patents, ruling that they were improperly granted because they involved a "law of nature." Myriad's BRCA1 and BRCA2 DNA is not "markedly different" from native DNA, making both isolated native DNA and complementary DNA, based on its sequence, "unpatentable products of nature," he wrote in his decision. The US Court of Appeals for the Federal Circuit overturned Sweet's decision in 2011. A three-judge panel ruled 2–1 to reject his contention that isolated gene sequences should not be patentable. In its opinion, the court stated categorically that isolated DNA (as well as complementary DNA and engineered DNA) has a "markedly different structure to native DNA" and is thus patentable. But five of the company's broadest method claim patents were judged invalid because they were based on a way of comparing sequences to look for differences, in effect an abstract mental exercise that is not patentable. The paradox is that if a mathematical algorithm had been developed to do this comparison, that algorithm in all likelihood would be patentable if it were novel and nonobvious.

Appeals court Judge William C. Bryson dissented on the question of whether genes are patentable: "Extracting a gene is akin to snapping a leaf from a tree," he wrote. "Yet prematurely plucking the leaf would not turn it into a human-made invention." Bryson also questioned whether gene patents promote innovation: "Broad claims to genetic material present a significant obstacle to the next generation of innovation in genetic medicine—multiplex tests and whole-genome sequencing. New technologies are being developed to sequence many genes or even an entire human genome rapidly, but firms developing those technologies are encountering a thicket of patents." The fact that the USPTO had made it a practice for a decade to issue patents on genes in effect granted the agency "a lawmaking authority that Congress has not accorded it," Bryson argued. The comments the agency made in response to critics when it established its gene patenting guidelines in 2001 "do not reflect thorough consideration and study of the issue," he concluded, implicitly inviting the agency to conduct such a study, which the AIA obligated it to undertake.

Senator Patrick Leahy who authored the Senate bill insisted that Section 27 had "no bearing" on the Myriad Genetics case but acknowledged that the section "was championed by Ms. Wasserman Schultz" (Florida Democratic Representative Debbie Wasserman Schultz). The Association for Molecular Pathology, the lead plaintiff in the Myriad Genetics case, worked closely with Wasserman Schultz's staff to require that the legislation include a study designed to investigate the effects of gene patents on access to testing. Wasserman Schultz was diagnosed with breast cancer in 2007. Myriad's test showed that she carries the BRCA2 gene mutation. "Patients should not have to make life altering decisions based on the results of only one test," she said shortly after the AIA was signed into law, although many patients diagnosed with cancer make a decision based on a pathologist looking at a slide of a biopsy or surgical resection. In the weeks before the Supreme Court decision in the Myriad case, actress Angelina Jolie wrote in *The New York Times* that she had had both breasts removed after Mryiad's test showed she was at high risk for developing breast and ovarian cancer.

Striking the right balance between intellectual property protection and the free flow of information in the information age is difficult enough. Assigning property rights protection to the nature of life code, junctures in biochemical pathways, and immunization schedules, which is what courts are now dealing with, cannot bode well for the biosciences in the long run. It is a recipe for legal sclerosis and institutional confusion. Yet if protection of intellectual property rights is insufficient, for example in the codevelopment of drugs and companion diagnostic genetic tests for personalized medicine, the investment in start-up companies where so much innovation occurs could be and likely will be jeopardized.

In the Myriad Genetics case, as Judge Bryson of the Federal Circuit Court observed, entities in the executive branch of the federal government took different positions. Opposing the USPTO's implicit position and practice that isolated DNA is patent-eligible subject matter, the Department of Justice representing the Obama administration issued a brief in the case arguing that isolated but otherwise unmodified DNA segments are not patent eligible. Defenders of the Myriad Genetics gene patents such as the trade group BIO stressed that we are at the beginning of a paradigm shift in drug discovery and development brought about by genomics and personalized medicine. Strong intellectual property rights are essential for the nascent field to emerge and develop. Myriad Genetics reportedly burned through more than $500 million over seventeen years before finally turning a profit around the time the high-profile gene patent suit against it was launched. In fiscal year 2012 Myriad generated revenues of nearly $496 million, nearly all of it from its genetic diagnostic tests.

In 2012 the Supreme Court tied the fate of the Myriad Genetics case to another case on which it issued a unanimous ruling. *Mayo Collaborative Services v. Prometheus Laboratories, Inc.* highlighted claims to proprietary interest in a specific metabolite produced in a biochemical pathway—more generally whether the correlation between blood test results and patient health is patentable. Broadly at issue was whether such patents are valid. If the subject matter is patentable, it is because it meets the traditional legal threshold of novelty, inventiveness, and usefulness and because it constitutes a patentable business method or process, a standard affirmed unanimously by the Supreme Court in *Bilski v. Kappos* in 2010. If the subject matter is not patentable it is because the claim has to do with "laws of nature, natural phenomena, and abstract ideas," a disqualifying standard set by the Supreme Court in 1981.

In *Mayo v. Prometheus*, the patents granted to Prometheus Laboratories were disqualified. Writing for a unanimous Supreme Court, Justice Stephen Breyer put it succinctly: "If a law of nature is not patentable, then neither is a process reciting a law of nature, unless that process has additional features that provide practical assurance that the process is more than a drafting effort designed to monopolize the law of nature itself." Other steps are needed, and those steps must transform the process "into an inventive application of the formula." The decision was seen as a not-so-subtle correction of the way the Federal Circuit Court was interpreting certain aspects of patent law. A few days after its ruling the Supreme Court remanded the Myriad Genetics case to the Circuit Court for reconsideration in light of its *Mayo v. Prometheus* decision. After taking another look, a Circuit Court three-judge panel again ruled 2–1 in favor of Myriad, again with Judge Bryson dissenting on the validity of Myriad's patent claims. Bryson warned that the majority's decision, if upheld, "will likely have broad consequences, such as preempting methods for whole-genome sequencing."

In his *Mayo v. Prometheus* opinion, Breyer observed that patent protection "is, after all, a two-edged sword":

On the one hand, the promise of exclusive rights provides monetary incentives that lead to creation, invention, and discovery. On the other hand, that very exclusivity can impede the flow of information that might permit, indeed spur, invention, by, for example, raising the price of using the patented ideas once created, requiring potential users to conduct costly and time-consuming searches of existing patents and pending patent applications, and requiring the negotiation of complex licensing arrangements.

The "flow of information" was at play in *Classen Immunotherapies, Inc. v. Biogen Idec.*

The case involved claims to proprietary interest in an immunization schedule—a stepwise method of immunizing that makes vaccination safer for the patient. Claims under two of the three Classen patents were upheld 2–1 by a Federal Circuit panel in a decision prior to the Supreme Court's unanimous ruling in *Mayo v. Prometheus*.

These cases reveal the tension and the nuances between proprietary interests important for innovation and the practice of medicine and public health uninhibited by limits imposed by these interests. They also reveal a tension between the Federal Circuit Court and the Supreme Court about what is patent eligible and what approach best serves innovation. What is clear is that the courts have ventured into terrain from which no uniform and consistent position or test can be easily derived and applied. The Supreme Court's ruling in *Bilski v. Kappos* not to limit patent eligibility to whether business methods passed the "machine-or-transformation" test established by the Federal Circuit Court—that a claimed process must be "tied to a particular machine or apparatus" or must transform "a particular article into a different state or thing"—illustrates the quandary. "The machine-or-transformation test may well provide a sufficient basis for evaluating processes similar to those in the Industrial Age—for example, inventions grounded in a physical or other tangible form," wrote Justice Anthony Kennedy for the Court in *Bilski*. "But there are reasons to doubt whether the test should be the sole criterion for determining the patentability of inventions in the Information Age."

Advocates of broad patentability, including the bioscience and pharmaceutical industries, were greatly relieved by the Supreme Court's ruling in *Bilski v. Kappos* but troubled by its rulings in *Mayo v. Prometheus* and *Association for Molecular Pathology v. US Patent and Trademark Office*, which some of them say reflects a poor appreciation for the intricacies of patent law by the high court. In their view, restricting patents related to medicine, technology, and life sciences would stifle innovation because broad patents create the incentives essential for risk-taking and capital investment and the entrepreneurial efforts to develop a commercial product or process. The Supreme Court "has seen fit either to make continued viability of various categories and types of patents uncertain (reversing a generation of efforts by the Federal Circuit to increase certainty in patent law, pursuant to Congressional mandate) or to 'upset the settled expectations' of patent holders," wrote biotech patent attorney Kevin Noonan on his popular blog Patent Docs after the *Mayo v. Prometheus* unanimous decision. "Under these circumstances, investors can be expected to be less ready to choose already risky biotechnology inventions, where return on investment of even a successful product may be vitiated by the next decision of the Court." In the subsequent Myriad Genetics unanimous ruling, the Court approached the question of whether the claims at issue do or do not promote the progress of the useful arts "from a false premise, which colors and distorts the remainder of the decision," a decision that overall exhibits "glaring scientific and technological weaknesses."

In an amicus brief to the Federal Circuit Court reconsidering its earlier ruling in the Myriad Genetics case after the Supreme Court's unanimous decision in *Prometheus v. Mayo*, James D. Watson harkened back to his first presentation of the proposed Watson and Crick DNA double helix in June 1953. Leó Szilárd, the Hungarian physicist, asked him whether he would patent the structure. "That, of course, was out of the question," Watson wrote in a footnote. Watson was reflecting the scientific idealism that Horace Judson championed in his chronicle of the rise of molecular biology, *The Eighth Day of Creation*. Such idealism was reined in after passage of the Bayh–Dole Act in 1980, the establishment of Federal Circuit Court in 1982 to adjudicate intellectual property rights, and the rise of bioscience entrepreneurs within academia. Yet that idealism is not vanquished, and it is not incompatible with innovation, as the open innovation movement has shown.

Hospitals, health care professionals, and public interest advocates insist heightened standards for patent eligibility are essential for lowering costs and permitting broader use of novel processes beneficial to patients, a view shared by some free-market conservatives. Continuing down the path of permitting broad medical-diagnostic patents will only serve to further slow the economy, retard technological innovation, distort the free market, and place human health at risk, three prominent libertarian think-tanks argued in a court brief. In a brief he submitted in the Myriad case, genomics pioneer Eric Lander argued that Myriad's claims "erect an insurmountable barrier to studying these DNA sequences, with serious consequences to innovation in medicine." Yet it is a fundamental and historical tenet of the patent system that, without sufficient patent protection, high risk or venture capital will not step up to finance the development costs needed to bring these innovations to market. Is there anything in today's world that is shifting the foundation of that tenet?

The key is to pin down what is "sufficient" to spur entrepreneurial investment and avoid the legal complications and potential social loss that broad patentability invites. That presumably is what the courts are attempting to do—trying to find the right balance so that neither edge of Justice Breyer's two-edged sword of patent protection gains the advantage and thereby stifles innovation. They are deliberating at a time when the words *innovation* and *debt* are prominent in public discourse. If controlling the costs of health care is critical for both long-term debt reduction and economic growth, and if health technologies are key to lowering costs, then the pressure to tighten the patent-eligibility standards for medical diagnostic and therapeutic methods (in contrast to drug therapies or therapeutic devices) may become insurmountable. The Constitutional justification may reside in the "Progress Clause" that authorizes Congress "to promote the Progress of Science and useful Arts." An expansive interpretation of the Progress Clause, says science historian Daniel Kevles, recognizes the adverse consequences of monopoly control over human DNA, "a scientifically and medically essential substance for which there is no substitute."

Some 8,000 US patents with the type of composition-of-matter claims "to simple nucleic acid molecules with natural sequences" may be at risk following the Myriad Genetics ruling. Of them, less than half involve human genetic sequences. Most involve sequences from animals, plants, microbes, and other life forms, meaning that other industries besides health care, namely agriculture, have benefited from patenting isolated DNA. The lower courts and the Federal Circuit Court will be busy adjudicating challenges to these patents just as the patent holders and their attorneys will be resourceful in finding ways to distinguish their "proprietary" nucleotide sequences from their natural counterparts in the cell. At the same time, researchers note, for more than a decade patent applicants have been moving away from drafting claims to simple isolated DNA molecules. Now they will accelerate the practice of drafting composition-of-matter claims "more obliquely, within the context of sufficiently complex, nonnative genetic constructs, or with enhancing changes to the sequence, enough to make it a 'synthetic' or 'artificial' sequence, rather than a 'natural' sequence." Based on the high court's ruling in the Myriad Genetics case, such sequences would likely be patent eligible, although the distinction may be, as a leading scholar in the field sees it, "a lawyer's distinction, not a scientist's."

US patent law is at a crossroads. It will probably be at a crossroads after the AIA is fully implemented. Research advances in the biosciences will entail the consumption of more time, energy, and resources in pursuit of just resolution for disputes over DNA, genes, molecules, cells, proteins, metabolites, pathways, and self-replication. "Our holding today is limited—addressing the situation before us, rather than every one involving a self-replicating product," wrote Justice Elena Kagan for a unanimous Supreme Court in *Bowman v. Monsanto* upholding Monsanto's patent on its genetically engineered soybean seed against a defense of "patent exhaustion." Under the patent exhaustion doctrine, the sale of a patented product ends the patentee's monopoly right to control its use, but it also prohibits the buyer from copying it without the patentee's permission because that would "result in less incentive for innovation." Kagan acknowledged that such inventions "are becoming ever more prevalent, complex, and diverse" and that in a different case "the article's self-replication might occur outside the purchaser's control." The dance of biological product self-replication, social control, and economic incentives as understood by the Court is just getting started.

Sorting out legal claims for bioproducts and their methods and processes would be an enormous task under ordinary circumstances, but we are in extraordinary times. The information age "puts the possibility of innovation in the hands of more people and raises new difficulties for the patent law," wrote Justice Anthony Kennedy in *Bilski*. "With ever more people trying to innovate and thus seeking patent protections for their inventions, the patent law faces a great challenge in striking the balance between protecting inventors and not granting monopolies over procedures that others would discover by independent, creative application of general principles."

Indeed, methods and process patent applications are rife and growing in the life sciences and health care. But nothing in the court's opinion in *Bilski*, Kennedy went on, "should be read to take a position on where that balance ought to be struck." Kennedy's demurring on the question of where the balance should be struck is understandable. In truth, the scale of justice with respect to the values associated with intellectual property protection, innovation, patients, and the public interest, like so much in the information age, is in flux. Finding the proper weight to assign to these competing interests will be challenging and will evolve over time. On the international scene, it will be very difficult to harmonize US policy priorities with those of Europe and Asia with their diverse social, economic, and cultural make-ups and norms.

How incentives are structured to bring new technologies forward in the absence of a temporary monopoly and exclusive licensing practices is important. Those incentives have brought us new business methods, transaction services, supply chain management practices, and biological-pattern search algorithms largely without court battles over property rights. Activities like these constitute the nervous system of the information age. The digital networks, devices, and signals that surround us in our daily lives are its nerve fibers and transmitters. Their cumulative effect seems to have more in common with the "laws of nature, natural phenomena, and abstract ideas" than property rights and the legal institutions that adjudicate them. They constitute a kind of second economy, an economy that is expanding rapidly across the globe and insinuating its way into the larger economic organism. In the emerging networked economy, arguments that state-backed monopolies (patents) are essential to encourage the development of new business methods are less compelling. "In fact, we *want* people to copy the businesses of others, lowering prices as a result," asserts James Boyle, who teaches law and the public domain.

Traditional ideas about incentives upon which the patent system rests are under greater scrutiny than ever before thanks to new exploratory tools, new methods, and entirely new fields like neuroeconomics. Economics is at the start of a revolution that has its source in medical schools, specifically in departments of neuroscience, says Nobel economist Robert Shiller. Shiller told the Society for Neuroscience that the flow of information through a nerve is a useful metaphor for illustrating the workings of networks of computers and networks of people in the economy, so-called economic agents.

Why? Because understanding how the brain works is revealing how we make decisions. The role of incentives in that process is not what we thought. Since the beginning of capitalism, it has been accepted that rational utility-maximizing behavior is the driving force behind invention, entrepreneurship, and free markets. That assumption turns out to be wrong, Shiller contends, "or at least in need of fundamental revision." Human behavior expressing itself in the marketplace, as well as in society as a whole, is more complex than can be accounted for by the simple utility-maximizing proposition. A monetary incentive may be and usually is a motivating factor for inventors. Public recognition is another. There are others, including the innate satisfaction derived from interaction with like-minded individuals creatively designing solutions together.

Molecular biology, brain imaging, and computational modeling are combining to reveal how the brain's physical structures use neural networks to interact and produce a characteristic response to information from the environment and, indeed, the economy. Shiller sees a similarity in how the human brain, the computer, and the economy work. All are devices whose purpose is "to solve fundamental information problems in coordinating the activities of individual units—the neurons, the transistors, or individual people." Twentieth-century economics saw many new schools arise ranging from Keynesianism, to econometrics, to utility maximization, to rational expectations/efficient markets, to psychology-based behavioral economics. Now biology is being brought to bear on the mysteries of Adam Smith's "invisible hand," John Maynard Keynes' "animal spirits," and twenty-first-century innovation. Computers are being put to the task of exploring how the flow of information in the brain influences incentives, reward seeking, greed and fear versus cooperation and altruism in decision making, and what can be understood—and perhaps predicted—about collective human behavior in market economies.

This brings us to another, related revolution entirely consistent with what is being revealed by brain science. The world of human interaction is the proving ground for how incentives work and whether alternatives exist to the idea that the expression of individual self-interest is the exclusive avenue to marketplace innovation. Has the fact that the information age, which puts innovation in the hands of more people as US Supreme Court Justice Kennedy asserted, changed the incentives game? Cracks in that edifice appeared first in software production leading to the open-source software and licensing movements. The collaborative, transparent code-sharing culture of open-source software has now spread throughout the neural system of the economy, scaling rapidly to attach one sphere of human activity after another, including research and development (R&D) in the biosciences. Online tools amplify collective brainpower, spurring discovery and innovation through systems of distributed public–private and market-based incentives. The relationship between incentives and behavior, since the days of Adam Smith considered a settled matter in market economies and the legal institutions that undergird them, is being reexamined in light of what brain research and brains operating freely in networked communities are telling us. Smith himself believed that more was at work than purely self-interest for the butcher, the brewer, and the baker to give us our dinner. Brain science, advances in imaging, and network-based innovation appear to be proving him right.

Can Open Source Revolutionize Biotech?

Adam Smith railed against monopolies in his foundational document of free-market economics, *The Wealth of Nations*. But he does not use the term *patent* anywhere in the book, though he does discuss *invention*, particularly of labor-saving machines. Intellectual property laws were part and parcel of the first (1712–1830) and second (1870–1914) Industrial Revolutions. They were designed to set in motion a continuum of creative innovation and entrepreneurship that would constantly challenge established economic interests, seed entirely new industries, and contribute to economic growth. For all their wisdom, the designers of these laws did not foresee the era we are in, an era of rent-seeking patent aggregators or "trolls" that shift value creation from laboratories, manufacturers, and service providers to litigants and courtrooms. Within less than a decade patent trolls were operating on a scale that corroded the link between intellectual property protection and technological innovation. Their growth prompted the Obama administration to initiate rulemaking that makes it harder for trolls to hide their activities using shell companies. Yet the patent troll problem is really just a preview to the world of patent thickets, created by a decades-long patenting binge, that serves as a tax on innovative startups.

At its beginning, the biotech revolution was a quintessentially entrepreneurial activity, as the founding of its pioneering companies, Genentech, Amgen, Biogen, and Cetus demonstrated. It is still an entrepreneurial revolution, yet its dynamism is tempered by the hierarchical academic, legal, and regulatory systems with which it must contend. Bioscience innovation faces an increasingly complicated world with competing interests, a world of universities, research institutions, hospitals and clinics, venture capitalists, investment banking firms and hedge funds, multinational pharmaceutical companies, and regulatory authorities. The interaction of biotechnology entrepreneurs with these forces has yielded many useful products and services, some spectacularly so. But it has not produced the overall growth and profitability to the economy that was its promise more than three decades ago. Part of the problem may be linked to the "tragedy of the anticommons," a situation in which too many rights holders frustrate the goal of achieving a socially desirable outcome, which patent law is meant to achieve.

While R&D expenditures and patenting activity grew during the late 1990s and first decade of the twenty-first century, output in the way of FDA-approved drugs stalled or declined until it began to rebound in 2011. Is the trend related to patent law? Many industry observers think other factors are responsible, but a number of legal scholars who have studied the issue think there is a link between the overall disappointing number of drug approvals and an intellectual property system that can stifle innovation. "In sparking the biotech revolution, the federal government inadvertently created a property rights environment for basic medical research that can stymie collaboration and block the development of lifesaving drugs," says Michael Heller, who introduced the "tragedy of the anticommons" concept in a law review article. Heller teamed up with Rebecca Eisenberg in a provocative article titled "Can Patents Deter Innovation?" published in *Science* 1998. In it they argued that the anticommons effect was detrimental to biomedical innovation. A plethora of upstream patents and restrictive licensing practices is inhibiting downstream product development. Long-term research also can be inhibited by current practices. Universities do have the power under the Bayh–Dole Act to ensure that future research is not inhibited when they grant exclusive licensing rights under their patents, but they rarely exercise it. Heller and Eisenberg's conclusions are hotly disputed by BIO, the biotechnology industry's trade group and advocacy organization.

We are entering a new phase of biomedical research, one in which teams of researchers with complementary skills often from multiple institutions are engaged in a given research project. The team might include molecular biologists, synthetic biologists, bioinformatics specialists, population geneticists, epidemiologists, bioengineers, computer scientists, and clinicians all working in concert. The days of the solitary university scientist working with a couple of graduate students are waning if not largely gone. Large-scale team science embodies a research culture of knowledge sharing across disciplines, a culture enabled and enhanced by new communications technologies and devices. More and more, discovery takes place in what physicist and author Michael Nielsen has named "Networked Science," an international endeavor in Internet era. Thus research has progressed from the individual, the institutional, and the national to what Jonathan Adams describes in *Nature* as a "fourth age of research," driven by international collaborations between elite research groups where the most highly cited research now is done. "This will challenge the ability of nations to conserve their scientific wealth either as intellectual property or as research talent," Adams says. Establishing legal standards for patentability in the dizzying dynamic environment of the "Information Age," as Justice Anthony Kennedy labeled our time, will be a tall order as biology moves inexorably into the realm of information science distributed over networks.

While USPTO goes about implementing the AIA and the courts grapple with system of property rights burdened with ever-greater legal and scientific complexity, are there other avenues to spur innovation in the biosciences? "Can open source revolutionise biotech?" asked *The Economist* in a 2005 article, "The Triumph of the Commons." The computing industry has been transformed by the open-source software movement, upsetting business models throughout the economy and contributing to the digital transformation of entire industries like entertainment, journalism and publishing, real estate, retail, and advertising and marketing, not to mention politics. "Might the same happen in biotechnology?" *The Economist* wondered.

The key is the word *open*. Open-source innovation extends beyond computer programming to encompass an outside-in approach to solving problems that hold back many organizations. Open innovation propounds the idea that firms can and should use external ideas as well as internal ideas, and internal and external paths to market, as the firms look to advance their technology. As we have seen, open access includes unrestricted access via the Internet to articles published in technical journals and other activities where access to information has traditionally been restricted. In a networked world, the argument goes, companies are more likely to prosper by adopting a more "open" approach to innovation and knowledge distribution. They should move beyond their in-house research and instead collaborate to access manufacturing processes or inventions (i.e., patents) from other organizations. In addition, internal inventions not being used in a firm's business should be taken outside the company, for example, through licensing, joint ventures, or spin-offs, or placed in patent pools.

Many companies are now moving in this general direction. Open innovation collaboratives are also taking shape. An example is PIPRA, a nonprofit consortium of universities, research centers, and a pro bono attorney network that provides intellectual property rights and commercialization strategy services to drive innovation in developing countries. Launched by universities conducting agricultural research with the support of major foundations to manage intellectual property, today PIPRA's patent database has expanded beyond agricultural to health, water, and energy technologies.

The Economist's provocative question was prompted by a paper published in *Nature* in which a group of researchers described a way to transfer genes into plants that circumvents the standard technique that uses a bacterium called *Agrobacterium tumefaciens* to transfer genes. Such use of *A. tumefaciens* is protected by a thicket of hundreds of patents. The invention was a way of genetically modifying crops to improve yields. What caught *The Economist*'s attention was the involvement of Cambia, a nonprofit biotech research group in Australia, which made the invention free for anyone to use by obtaining an "open-source" license that stipulates users share any improvements they may make to the gene-transfer technology. In the open-source world the process is called "copyleft," a copyright licensing scheme in which an author surrenders some but not all rights under copyright law. Cambia's idea was "to make it easier for companies and researchers in poor countries to use agricultural gene-transfer technology, which today's patent-licensing approach impedes." Monsanto, the dominant patent holder in gene-transfer technologies for agricultural applications, told *The Economist* that the technology "seems to complement, not threaten, its business model."

Unlike open-source software that relies on user volunteers to constantly improve the product, Cambia relied on grants from foundations to develop the technology rather than on users volunteering their time to build the technology from the ground up. Richard A. Jefferson, an American-born molecular biologist, set up Cambia to increase innovation in the life sciences by applying software's open-source model to biotechnology. His goal is to change the global patent system, how people use intellectual property and break the grip that the big multinationals hold on the tools of innovation. Jefferson is not advocating the gene-transfer practices of his distant ancestor Thomas Jefferson, America's third president, who allegedly smuggled rice seeds sewn into the lining of his coat out of Italy in an attempt to introduce upland rice to cultivation in South Carolina. But Cambia's Jefferson *is* passionate about creating alternatives to the intellectual property system.

Cambia launched a technology development and sharing initiative called Biological Innovation for Open Society (BiOS) in 2005. The society enables scientists via the web to share the tools and "operating systems" of innovation. The online tool kit is designed to allow ag-bio startups to make genetic improvements to neglected crops without entering licensing or partnering agreements with the multinational companies that control most of the related intellectual property in the genetically modified crops and seeds business. That business is estimated to be worth more than $150 billion worldwide in annual revenues.

Cambia unveiled the Global Initiative for Open Innovation (IOI) in 2009, a program designed to, in the words of *The Economist*, "combine open-access software and sophisticated search features to make the confusing thicket of drug patents accessible to researchers the world over." In brief, the initiative is an attempt to make the world's patent system more transparent on behalf of global health, especially people in poor countries. With funding from the Bill and Melinda Gates Foundation, the Gordon and Betty Moore Foundation and others, Cambia developed an online IOI tool (Lens.org) that allows free patent searching of more than 80 million patent documents worldwide and more than 120 million DNA sequences and 10 million protein sequences. Jefferson and his colleagues say the *Myriad* gene patenting decision in the United States underscores the need for openly accessible data sets and analytical tools "to support biological innovation."

While open source is making inroads in agricultural biotechnology and global health, it is already commonplace in bioinformatics and genomics database sharing. It is also making its mark in a new, rapidly emerging scientific field with the potential to reshape a number of major industries. Synthetic biology combines science and engineering in order to design and build ("synthesize") novel biological systems from component parts. Synbio refers both to the design and fabrication of biological components and systems that do not already exist in the natural world and to the redesign and fabrication of existing biological systems. If its tools work as practitioners predict, "they could turn specialized molecules into tiny, self-contained factories, creating cheap drugs, clean fuels, and new organisms to siphon carbon dioxide from the atmosphere," wrote science journalist Michael Specter.

The field of synthetic biology, still very much in its infancy, is looked upon as a prime candidate for open-source innovation in the biological sciences because of similarities it shares with the computer industry. The most passionate advocates of the emerging field see a future in which decentralized networks of small firms and individual inventors can cheaply "hack" (i.e., improve) standardized DNA parts in the same way that they currently hack computer code, says legal scholar Arti Rai, who studies intellectual property and biological technologies. The availability of investment capital, particularly for start-up firms, may well depend on intellectual property protection. Patent applications in synthetic biology have been filed claiming novel methods used to create genetic components, whole genomes, and biochemical and metabolic pathways. Based on the unanimous Supreme Court decision in the Myriad Genetics case in which complementary and synthetic DNA were ruled to be patent eligible, synthetic biology should be in good standing with the courts as a creative rather than extractive enterprise. As with any emerging technology, however, patent protection that is too broad runs the risk of preventing potential innovators from bringing their ideas to market and thereby diminishing societal returns, whereas patent protection that is too narrowly construed might discourage investment and entrepreneurship at the early stages.

Nonproprietary as well as proprietary development and application of biological technologies will lay the foundation for a bioeconomy, says synthetic biologist and entrepreneur Robert Carlson. Carlson is perhaps best known for his "Carlson curves," time-series graphs showing the exponential growth of DNA sequencing and synthesis capabilities, sometimes called "DNA reading and writing." The medical and public health value of whole genome sequencing will only be realized when many people, hundreds of thousands, perhaps millions, have been sequenced. As relatively inexpensive whole genome sequencing becomes available, it will be our ability to organize and interpret the enormous data sets it produces that will constitute the bottleneck for genome-based personalized medicine and all that it promises. When that technical bottleneck is breached and patients are willing to pay for their whole-genome disease-risk profiles, it will be very difficult to deny them open access to their life source code on the basis of patent infringement. The question is, will people be willing to donate their source code to science?

An open-data project called the Portable Legal Consent for Common Genomics Research (WeConsent.us) aims to move beyond data privacy issues in the Genomic Exchange era by creating a large pool of openly available, user-donated data about health and genomics. The project is funded by, among others, the Ewing Marion Kauffman Foundation, the world's largest foundation devoted to entrepreneurship. The concern of the Kauffman Foundation and others is that informed consent and data-ownership issues are holding back research, discovery, and entrepreneurial activity as we enter the biodata era. On the corporate side, the DNA sequencing tools giant Illumina set up an open-source genomics version of Apple's Apps Store. BaseSpace Apps is described as a scalable cloud computing environment for all of Illumina's sequencing systems. Its mission is to encourage Illumina's software development corporate partners to find solutions to problems associated with analyzing genomic information, particularly the deluge of data expected from next-generation DNA sequencing.

Open Innovation in Drug Development

The open-source movement has established itself as a collaborative nonproprietary alternative to the proprietary system in bioinformatics, agricultural biotechnology, synthetic biology, genomics, and other fields where "platform" technologies play a role. In these fields, innovation can be "democratized" by many participating individuals, institutions, and small firms adding knowledge to the commons. The question asked by *The Economist* and other observers is, can open-source biotech also find its way into drug development, where the costs are much higher and potential profits much greater?

Drug development is indeed the acid test for the open approach movement in biotechnology. The biotechnology industry began mainly as a health care focused enterprise geared toward producing health care products. The rapidly growing global biopharmaceutical market in 2009 represented 17 percent of total pharmaceutical sales with revenues of $120 billion. But the cost of bringing a new biological drug to market today serves as a millstone to the entrepreneurial drive that characterized biotechnology in its early days and to innovation by means other than through today's rigid system that is heavily vested in the status quo. In 1979, in the industry's infancy, the cost to develop a new drug was $54 million, according to the Tufts Center for the Study of Drug Development. By 1991 the cost had climbed precipitously to $231 million, in 2001 to $802 million, and in 2006 the Tufts Center released its first comprehensive estimate of the average cost of developing a "new biotechnology product," pegging it at $1.2 billion. There is no clear evidence that the cost trajectory for new drug development will moderate any time soon.

The IMS Global Pharmaceutical Market and Therapy Forecast predicted in 2009 that global pharmaceutical sales would surpass $825 billion in 2010 (they reached $832 billion) and $975 by 2013, with more than a third of the pharmaceutical market in the United States, the world's largest. Sales are expected to continue to boom in emerging markets and slow in developed markets, particularly with the global economic slowdown. Indeed, by 2020 emerging markets are forecast to account for between one-third and one-half of total sales, though they will not be as profitable as developed markets often due to government-imposed price controls. As we saw in chapter 2, other forces drug companies are facing as they deal with shifting markets include specialist-driven products playing a larger role, blockbuster drugs going off patent, the rising rates of new drug failure in phase II and phase III clinical trials, and the rising influence of regulators and payers on health care decisions. Personalized medicine and the challenge of smaller market "niche" drugs are forcing a rethinking of the drug development paradigm and bringing new priorities to bear.

Pharmaceutical company executives frequently cite needed adaptations. They list filling pharmaceutical R&D pipelines with a broader array of products in addition to their patent-protected chemically based drugs and strategically employing incentives to boost entrepreneurship within the firms. They also advocate forming flexible business partnerships with small biotech firms or acquiring them and tapping into fast-growing international markets though discount pricing. Meanwhile, some observers have found big pharma showing some willingness to make riskier bets by in-licensing more drug candidates in the preclinical and early-phase clinical stages of development in an effort to replenish drug pipelines.

Since passage of the Bayh–Dole Act in 1980, biomedical research itself has become increasingly proprietary and secretive at universities and research institutions, leading to calls for open and collaborative science that overcomes access and licensing difficulties. Is there a place for open-source and open-innovation methods in drug development, an approach that engages the many versus the few knowledgeable players and democratizes the process? Is there room for a parallel system that provides for incentives and satisfactory rewards for investment while helping to accelerate a rather stalled system of innovation? If it takes more than a billion dollars to bring a new drug to market, what are the appropriate government policy and business models that ensure promising molecules move as expeditiously as possible through the drug R&D pipeline and that they are not dropped?

Besides making inroads in bioinformatics and research tools for drug discovery, R&D has spawned a variety of novel public–private partnerships that have adapted the open-source approach to create a new, low-cost business model—what one pharma R&D executive termed a "networked R&D model." Some of these partnerships have focused on new treatments for neglected diseases, including tropical diseases for which R&D funding is scarce. The "virtual pharma" approach engages governments and philanthropies to fund organizations that identify and help support drug development from promising private and academic research. Two of the best examples are the Institute for One World Health (Oneworldhealth.org), a nonprofit pharmaceutical company funded mainly through the Bill and Melinda Gates Foundation and other private sources, and the Drugs for Neglected Diseases Initiative (DNDi.org), an organization supported by public-sector agencies in seven countries plus the UN Development Programme, the World Bank, and the World Health Organization's Special Programme for Research and Training in Tropical Diseases. Rather than developing drugs itself, the Drugs for Neglected Diseases Initiative uses a cost-cutting model that capitalizes on existing, fragmented R&D capacity, especially in the developing world.

What about issues surrounding openness initiatives and intellectual property ownership in large pharmaceutical firms? Although there is a perception in these firms that such initiatives are hostile to patents and bent on putting discoveries in the public domain, the reality is more nuanced. Most open-innovation activities are concentrated at a precommercial R&D stage, before ideas and hypotheses take the kind of tangible shape that define inventions in patent law. They have been described as "an on-going scientific conversation that can be likened to a global instant-messaging system linking scientists interested in a topic" says pharma open-source advocate Bernard Munos. In that sense, open source is similar to other forms of scientific publishing and no more threatening. If a scientist who participates in such a forum comes up with an idea that can lead to a patentable invention, caution then kicks in with disclosures and the filing of patent applications, just as in a traditional research setting. The open communication that goes to the heart of open source promotes advancement of science but needs to be balanced by protection of the rights of inventors. Have changes in technology left the old way of innovating behind? Or is it simply that there is so much more information available and that the more we learn about biological systems the more complex and challenging they appear?

What is clear is that rigid institutional boundaries in hierarchical, tradition-bound universities and companies are inhibiting a more open flow of information to foster innovation in the pharmaceutical and biopharmaceutical arenas. The unanimous Supreme Court decision on gene patents did nothing to induce Myriad Genetics to share the information in its vast trade secret-protected patient database. The day the Court handed down its decision, a "Free the Data" effort (Free-the-data.org) was launched that encourages Myriad clients to make their genetic variant information publicly available via NIH's ClinVar database. Is it possible to imagine that horizontal public–private collaborative partnerships that incorporate proven mechanisms of information sharing could contribute to overcoming the current impasse? Are there elements of the open-source software movement and e-science that could be incorporated or adopted by the highly proprietary and monetized system of innovation as it now exists? If there are, could they accelerate the development of biomedical therapies for the diseases the various patient advocacy groups are interested in?

In the precompetitive space, big pharma appears to be more willing to take risks by experimenting with open-innovation business models in a number of ways. One way is to share research data on early-stage drugs to improve their drug development programs, as some companies are now doing. Another is to contribute proprietary compounds to open bioinformatics archives and by forming partnerships with nonprofit information collaborative platforms such as the Structural Genomics Consortium, the Biomarkers Consortium, Eli Lilly's Phenotypic Drug Discovery Initiative, and Europe's Innovative Medicines Initiative. Perhaps most notable among them is Sage Bionetworks, which is building a biology information commons for precompetitive drug development using open systems, incentives, and standards. Precompetitive cooperation brings companies with common interests and substantial resources together to develop technology platforms and derive mutual advantages from them. The experience of Boston-based Enlight Biosciences is a case in point.

Enlight Biosciences is creating drug discovery platforms that incorporate research in a variety of fields, including molecular imaging, drug design, synthesis, and formulation. It is tracking down technologies in academia and industry and building these platforms on a precompetitive basis. PureTech Ventures launched the startup in 2008. The firm initially brought in Eli Lilly, Johnson & Johnson, Merck, and Pfizer. Novartis and Abbott Laboratories joined the collaboration in 2009, raising the total investment in the venture at the time to nearly $80 million. All the drug companies participating with Enlight Biosciences have the option to use all the technology platforms once they are developed or choose among them. Its collaborative partnership model joins entrepreneurship to the pharma collaborative model. Knode, an Enlight company that has developed a search engine for scientists and innovators, formed a partnership with AstraZeneca. Another Enlight company, Entegra, is pioneering an oral delivery technology for the $100 billion injectable therapeutic protein market, a market in which Enlight's pharma partners participate.

Enlight's model addresses a critical structural impediment to funding innovation in the biopharma industry, namely "the fundamental tension between entrepreneurs and investors around how to apportion financial returns," as the professional services firm Ernst & Young puts it. Collaborating pharmaceutical companies are looking for strategic returns, not just zero-sum financial returns to be divvied up. They want to improve the chances of their drugs succeeding in preclinical testing and human trials. When organizations all along the value chain—government funding agencies, universities, entrepreneurial startups, pharmaceutical and biopharmaceutical firms— are more closely linked, argues biotech critic Gary Pisano, organizational learning can raise the odds of success for all concerned. This is particularly true for small firms and startups that today are too focused on monetization of intellectual property. Precompetitive collaboration is entirely in keeping with information-sharing communications technologies that are reshaping the culture of innovation, fueling new fields of science like genomics and synthetic biology, and undermining some of the more intransigent and counterproductive aspects of patent law.

Big pharma is also collaborating with universities to establish genomic and stem cell biobanks and with universities and disease advocacy nonprofits through translational acceleration programs. The combination of human physiological complexity stretching beyond the capacity of single firms to deal with plus the astronomical cost of bringing a new molecular entity to market—estimated to be more than $1.2 billion—is finally opening the way, if slowly, for more open approaches to drug target discovery and biomarker development, at least at the precompetitive stages of drug development. Companies appear to be willing to deemphasize intellectual property rights early in the process and share negative results with other firms so that knowledge of what works and what does not circulates more freely, thereby saving time, effort, and money.

With a change in federal statute, the FDA also could release data on failed drug development to companies and entrepreneurs without disclosing valuable proprietary information such as the biologic or pharmacological mechanism of action. For their part, research institutions have the opportunity to adjust their governance policies and licensing practices in a way that makes their research resources and products such as biological materials more widely available in research and innovation communities. To continue to protect collaborative research projects between universities and industry from the threat of patent litigation, the AIA incorporates a "continuity of intent under the CREATE Act" which permits the sharing of confidential information under a joint research agreement in the interest of spurring innovation.

As is the case with laboratory experiments, most business model experiments will likely fail. But some will succeed, and those successes could lead to a transformation in the structure of the life sciences industries and possibly radical new therapeutic advances. That is the view of Graham Dutfield, a leading thinker on intellectual property rights in the life sciences industries. We will continue to see business models of drug development "based on aggressive use of the patent system." But there will also be alternative strategies that incorporate open methods. The rise of R&D consortia in and of itself "should act as a Damascene conversion" for today's managers of bioscience companies, large and small, to look for new ways to grow their businesses, including ways that would have been unimaginable to their predecessors.

Despite the changes we see all around us, most companies will continue to maximize protection of their intellectual property. What matters is that, at a time of expanding neural networks that serve to bind research with production in unseen and sometimes autonomous ways, some firms are willing to remove their proprietary blinkers and experiment with new business models in an evolving economic ecosystem. Successful adaptation to economic swings and technological upheavals often leaves behind a series of experimental failures. They are the necessary markings along the trail to discovery and innovation, markings on our journey from painting on the walls of caves to discerning the code of life.

References

Listed sequentially based on their order in the chapter.

David W. Maxey, "Samuel Hopkins, the Holder of the First U.S. Patent: A Study of Failure," *The Pennsylvania Magazine of History and Biography* 122.1/2 (January–April, 1998): 3–37.

Stanford University v. Roche Molecular Systems, Inc. US Supreme Court, No. 09-1159. (Argued February 28, 2011—Decided June 6, 2011)

Public Law 96-517: Bayh–Dole Act (December 12, 1980).

Gardiner Harris, "Patent's Defeat in India Is Key Victory for Generic Drugs," *The New York Times*, April 1, 2013.

Malcolm Gladwell, "Steve Jobs's Real Genius," *New Yorker*, November 14, 2011.

Jonathan Adams, "The Rise of Research Networks," *Nature* 490 (2012):335–336.

Dan L. Burk and Mark A. Lemley, *The Patent Crisis and How the Courts Can Solve It.* (Chicago: University of Chicago Press, 2009).

The America Invents Act: Needed Reform or Missed Opportunity?

Public Law 11–29: Leahy–Smith America Invents Act (September 16, 2011).

Mark A. Lemley, "The Myth of the Sole Inventor," Stanford Public Law Working Paper No. 1856610, July 21, 2011, accessed June 20, 2012, http://papers.ssrn.com.

Daniel J. Kevles, "Ananda Chakrabarty Wins a Patent: Biotechnology, Law, and Society, 1972–1980," *Historical Studies in the Physical and Biological Sciences* 25.1 (1994): 111–135.

Plant Patent Act of 1930, enacted on June 17, 1930 as Title III of the Hawley-Smoot Tariff, ch. 497, 46 Stat. 703.

Christine MacLeod, *Inventing the Industrial Revolution: The English Patent System, 1600–1800.* (Cambridge, UK: Cambridge University Press, 1988), 62.

Diamond v. Chakrabarty, 447 US 303 (1980).

Utility Examination Guidelines. Docket No. 991027289-0263-02, US Patent and Trademark Office, Department of Commerce, December 29, 2000.

Philip Leder and Timothy Stewart, "Transgenic Non-Human Mammals," US Patent No. 4,736,866 (1988).

Patent Act of 1952, c. 950, 66 Stat. 792, *codified as* Title 35 of the United States Code, titled "Patents" (July 19, 1952).

Public Law 11–29.

Ashby Jones, "Inventors Race to File Patents," *Wall Street Journal*, March 14, 2013.

"New Patent Rules Substantially Weaken 'Grace Period,'" American Association of Medical Colleges, February 22, 2013, accessed December 2, 2013, http://www.aamc.org.

Thomas Cotter, University of Minnesota Law School, personal communication to authors, December 2, 2013.

Michele Boldrin and David K. Levine, *Against Intellectual Monopoly* (Cambridge, UK: Cambridge University Press, 2008).

James Bessen and Michael J. Meurer, *Patent Failure: How Judges, Bureaucrats, and Lawyers Put Innovators at Risk* (Princeton, NJ: Princeton University Press, 2008).

Zorina B. Khan, *The Democratization of Invention: Patents and Copyrights in America Economic Development, 1790–1920.* (Cambridge, UK: Cambridge University Press, 2005).

Eugene Samuel Reich, "US Legislation Aims to Simplify Rules for Inventor," *Nature* 472 (2011):149.

HR 2795, Patent Reform Act of 2005 (September 15, 2005).

Dan L. Burk, Statement before the United States House of Representatives Judiciary Subcommittee on Intellectual Property, Competition, and the Internet hearing on "Review of Recent Judicial Decisions on Patent Law," March 10, 2011.

Fiona Murray et al., "Of Mice and Academics: Examining the Effect of Openness on Innovation," NBER Working Papers Series, Working Paper 14819, March 2009, accessed February 10, 2012, http://www.nber.org.

Mark A. Lemley, "Fixing the Patent Office," Stanford Institute for Economic Policy Research Discussion Paper No. 11-014 (May 21, 2012), May 25, 2012, http://siepr.stanford.edu.

Dan Burk, "Disruptive Forces in Patent Law: Change Without Innovation," JURIST— Forum (October 4, 2011), accessed February 2, 2012, http://jurist.org/forum/2011/10/dan-burk-america-invents.php, accessed.

Burk and Lemley, *The Patent Crisis and How the Courts Can Solve It*.

Sophia Pearson, "Patent Lawyer Demand Rises Following U.S. Legislative Overhaul," *Bloomberg News*, October 8, 2011.

"Open For Questions: The America Invents Act with CTO Aneesh Chopra and USPTO Director David Kappos," White House video, September 16, 2011.

Genes, Pathways, Methods and Schedules Go to Court

Public Law 11–29.

Association for Molecular Pathology et al. v. United States Patent and Trademark Office et al. 569 US 12-398 (decided June 13, 2013).

Association for Molecular Pathology et al v. United States Patent and Trademark Office et al. 702 F. Supp. 2d at 231. 182. Id. at 201. 183 (filed May 12, 2009).

Kyle Jensen and Fiona Murray, "Intellectual Property Landscape of the Human Genome," *Science* 310 (2005):239–240.

Lori Pressman et al., "The Licensing of DNA Patents by US Academic Institutions: An Empirical Survey," *Nature Biotechnology* 24 (2006):31–39.

DNA Patent Database. http://dnapatents.georgetown.edu.

Christopher M. Holman, "Debunking the Myth that Whole-Genome Sequencing Infringes Thousands of Gene Patents," *Nature Biotechnology* 30 (2012):240–244.

Heidi Williams, "Intellectual Property Rights and Innovation: Evidence from the Human Genome," *Journal of Political Economy* 121.1 (2013):1–27.

Jeffrey Rosenfeld and Christopher E. Mason, "Pervasive Sequence Patents Cover the Entire Human Genome," *Genome Medicine* 5.27 (2013): 1–7.

Gregory D. Graff et al., "Not Quite a Myriad of Gene Patents," *Nature Biotechnology* 31.5 (2013): 404–410.

National Institutes of Health, "Impact of Gene Patents and Licensing Practices on Access to Genetic Testing," Compendium of Case Studies Commissioned for SACGHS by the Duke University Center for Genome Ethics, Law & Policy (Bethesda, MD: Office of Biotechnology Activities, 2009).

Association for Molecular Pathology et al v. United States Patent and Trademark Office et al. 09 Civ. 4515. US District Court, Southern District of New York (March 29, 2010).

Association for Molecular Pathology et al v. US Patent and Trademark Office et al., 35 USC § 103 (decided July 29, 2011).

Utility Examination Guidelines

Leahy-Smith America Invents Act. *Congressional Record* 157.132 (September 8, 2011).

Angelina Jolie, "My Medical Choice," *The New York Times*, May 14, 2013, accessed June 16, 2013, http://www.nytimes.com/2013/05/14/opinion/my-medical-choice.html.

Association for Molecular Pathology et al v. US Patent and Trademark Office et al., US Federal Circuit Court of Appeals, No. 2010-1046 (decided August 16, 2012).

Association for Molecular Pathology et al. v. United States Patent and Trademark Office et al. 09 Civ. 4515. US District Court, Southern District of New York (decided March 29, 2010).

Association for Molecular Pathology et al. v. United States Patent and Trademark Office et al. 09 Civ. 4515. On Writ of Certiorari to the US Court of Appeals for the Federal Circuit. Brief for the United States as *Amici Curiae* In Support of Neither Party (October 20, 2010).

Association For Molecular Pathology et al. v. United States Patent and Trademark Office et al. 09 Civ. 4515. US District Court, Southern District of New York. Brief of *Amici Curiae*. Biotechnology Industry Organization in Support of Defendants' Opposition to Plaintiffs' Motion for Summary Judgment (December 30, 2009).

J. J. Colao, "How a Breast Cancer Pioneer Finally Turned a Profit," *Forbes*, October 17, 2012.

"Myriad Genetics Reports Fourth Quarter and Fiscal Year 2012 Results," Myriad Genetics news release, August 14, 2012, accessed June 22, 2013, http://investor.myriad.com/releasedetail.cfm?ReleaseID=700461.

Mayo Collaborative Services v. Prometheus Laboratories, Inc. US Supreme Court, No. 10-1150 (March 21, 2011).

Bilski v. Kappos, 130 S. Ct. 3218 (2010).

Classen Immunotherapies, Inc. v. Biogen Idec. 130 S. Ct. 3541, Supreme Court 2010 (August 31, 2010).

Kevin E. Noonan, "Mayo Collaborative Services v. Prometheus Laboratories—What the Court's Decision Means," Patent Docs, March 22, 2012, accessed March 23, 2012, http://www.patentdocs.org.

Kevin E. Noonan, "*Association for Molecular Pathology v. Myriad Genetics, Inc.* (2013)," Patent Docs, June 13, 2013, accessed June 14, 2013, http://www.patentdocs.org.

Association for Molecular Pathology et al. v. United States Patent and Trademark Office et al., US Court of Appeals for the Federal Circuit, No. 2010-1406. Brief of *Amici Curiae*. James D. Watson In Support of Neither Party (June 15, 2012).

Mayo Collaborative Services v. Prometheus Laboratories, Inc. US Supreme Court, No. 10-1150. Brief of *Amici Curiae*. Cato Institute, Reason Foundation, and Competitive Enterprise Institute in Support of Petitioners (September 9, 2011).

Association for Molecular Pathology et al. v. United States Patent and Trademark Office et al. US Court of Appeals for the Federal Circuit, No. 2010-1406. Brief of *Amicus Curiae*. Eric S. Lander in Support of Neither Party (January 31, 2013).

Daniel J. Kevles, "Can They Patent Your Genes?" *New York Review of Books*, March 7, 2013, accessed February 17, 2013, http://www.nybooks.com.

Graff et al., "Not Quite a Myriad of Gene Patents."

Noah Feldman, "Supreme Court on Shaky Scientific Ground with Gene Patent Decision," *Bloomberg View*, June 13, 2013, accessed June 14, 2013, http://www.newsday.com/opinion/oped/feldman-supreme-court-on-shaky-scientific-ground-with-gene-patent-decision-1.5475664.

Bowman v. Monsanto Co. et al. US Supreme Court, No. 11-796 (decided May 13, 2013).

W. Brian Arthur, "The Second Economy," *McKinsey Quarterly*, October 2011.

James Boyle, *The Public Domain: Enclosing the Commons of the Mind* (New Haven, CT: Yale University Press, 2008).

Robert Shiller, "The Neuroeconomics Revolution," Project Syndicate, November 21, 2011, accessed March 23, 2012, http://www.project-syndicate.org/commentary/the-neuroeconomics-revolution.

Robert Shiller, "Animal Spirits: How Human Behavior Drives the Economy," Keynote address at Neuroscience 2011, November 19, 2011.

Debrah Meloso, Jernej Copic, and Peter Bossaerts, "Promoting Intellectual Discovery: Patents Versus Markets," *Science* 323 (2009):1335–1339.

Can Open Source Revolutionize Biotech?

Adam Smith, *An Inquiry into the Nature and Causes of the Wealth of Nations* (London: W. Strahan and T. Cadell, 1776).

Nathan Vardi, "President Obama Wallops The Patent Trolls," *Forbes*, June 4, 2013.

Michael Heller, *The Gridlock Economy: How Too Much Ownership Wrecks Markets, Stops Innovation, and Costs Lives* (New York: Basic Books, 2008).

Michael Heller, "The Tragedy of the Anticommons: Property in the Transition from Marx to Markets," *Harvard Law Review* 11.3 (1998):621–688.

Michael Heller and Rebecca Eisenberg, "Can Patents Deter Innovation? The Anticommons in Biomedical Research," *Science* 280 (1998):698–701.

Ted Buckley, "The Myth of the Anticommons." Biotechnology Industry Organization, May 21, 2007, accessed February 2, 2012, http://www.bio.org.

Michael Nielsen, *Reinventing Discovery: The New Era of Networked Science* (Princeton, NJ: Princeton University Press, 2012).

Jonathan Adams, "Collaborations: The Fourth Age of Research," *Nature* 497 (2013):557–560.

"The Triumph of the Commons: Can Open Source Revolutionise Biotech?" *The Economist*, February 10, 2005.

Richard C. Atkinson et al., "Public Sector Collaboration for Agricultural IP Management," *Science* 301 (2003):174–175.

"The Triumph of the Commons: Can Open Source Revolutionise Biotech?"

Cary Fowler, *Unnatural Selection: Technology, Politics, and Plant Evolution* (London: Routledge, 1994), 14.

"Drug Companies and Poor Countries: All Together Now," *The Economist*, July 16, 2009, accessed December 7, 2013, http://www.economist.com/node/14031424.

Osmat A. Jefferson et al., "Transparency Tools in Gene Patenting for Informing Policy and Practice," *Nature Biotechnology* (2013):1086–1093.

Michael Specter, "A Life of its Own," *New Yorker*, September 28, 2009.

Arti Rai, "The Paradigm Shift of Synthetic Biology: Tensions Between Innovation and Security," University of Minnesota's Consortium on Law and Values in Health, Environment and the Life Sciences and Joint Degree Program in Law, Health and the Life Sciences, November 4, 2008.

Cherise Fong, "Synthetic Biology Inches Toward the Mainstream," CNN.com, October 17, 2008, accessed February 3, 2012, http://www.cnn.com.

John Leguyader, "Patents and Synthetic Biology," US Patent and Trade Office, November 3, 2009, accessed February 2, 2012, http://www.synbioproject.org.

Robert H. Carlson, *Biology Is Technology:* The Promise, Peril, and New Business of Engineering Life (Cambridge, MA: Harvard University Press, 2010).

Erika Check Hayden, "Open-Data Project Aims to Ease the Way for Genomic Research," *Nature News*, April 25, 2012, accessed April 26, 2012, doi: 10.1038/nature.2012.10507.

Adrienne Burke, "There's an App Store Coming for Genomic Tools," *Forbes*, April 4, 2012.

Open Innovation in Drug Development

Research Milestones, Tufts Center for the Study of Drug Development, accessed February 10, 2012, http://csdd.tufts.edu.

"IMS Health Forecasts 4.5–5.5 Percent Growth for Global Pharmaceutical Market in 2009, Exceeding $820 Billion," IMS Health, October 29, 2009.

Hester Plumridge, "Rising Nations Are Not Remedy for Big Pharma," *Wall Street Journal*, May 25, 2010.

Jason Douglas, "Big Pharma Won't Wait in Rush for Biotech's Drugs," *Wall Street Journal*, August 4, 2010.

Matthew Herper, "Guest Post: Pfizer Research Boss On Why He's Working with the Cystic Fibrosis Foundation," *Forbes*, December 7, 2012.

"Malaria Disease Report," The *FasterCures* Philanthropy Advisory Service, *FasterCures / The Center for Accelerating Medical Solutions*, New York, NY, June 2009 accessed November 20, 2013, http://www.philanthropyadvisoryservice.org.

Bernard Munos, "Can Open-Source R&D Reinvigorate Drug Research?" *Nature Reviews Drug Discovery* 5.9 (2006): 723–729.

Stephen H. Friend and Thea C. Norman, "Metcalfe's Law and the Biology Information Commons," *Nature Biotechnology* 31.4 (2013): 297–303.

Ryan McBride, "Enlight Biosciences Forms Partnership with Abbott Labs," Xconomy.com, November 4, 2009, accessed February 16, 2010, http://www.xconomy.com.

Ryan McBride, "AstraZeneca Adopts Search Engine for Scientists," FierceBiotechIT.com, August 1, 2012, accessed December 12, 2012, http://www.fiercebiotechit.com.

Scott Kirsner, "Start-Ups Working on New Ways of Taking Medicine," *Boston Globe*, March 31, 2013, accessed April 25, 2013, http://www.bostonglobe.com.

Glen T. Giovannetti and Gautam Jaggi, "Beyond Borders: Global Biotechnology Report 2009," Ernst & Young, accessed February 16, 2010, http://www.ey.com.

Gary P. Pisano, *Science Business: The Promise, the Reality, and the Future of Biotech* (Cambridge, MA: Harvard Business School Press, 2006).

Daniel Carpenter, "Strengthen and Stabilize the FDA," *Nature* 485 (2012):169–170.

Jeffrey Furman, Fiona Murray, and Scott Stern, "More for the Research Dollar," *Nature* 468 (2010):757–758.

Graham Dutfield, *Intellectual Property Rights and the Life Science Industries: Past, Present and Future* (Hackensack, NJ: World Scientific Books, 2009).

Looking Ahead as an Industry Evolves

One can resist the invasion of armies; one cannot resist the
invasion of ideas.

  —Victor Hugo, *on the French Revolution*

Confident once more, like a high diver who had regained his nerve, he
launched himself across the light-years.

  —Arthur C. Clarke, *2001: A Space Odyssey*

Four and a half centuries after Leonardo da Vinci drew his anatomical masterpiece,
the "Fetus in the Womb," viewers of Stanley Kubrick's epic film *2001: A Space
Odyssey* saw a dying astronaut reborn as a fetus in a bubble-like womb. The film
ends with the fetus languidly looking down on the earth, completing a journey that
begins in the film with ape-like hominids fighting each other on the African savannah. In this manner, as one critic put it, the "cyclical evolution from ape to man to
spaceman to angel-starchild-superman is complete."

  The time between when Leonardo's drew what he called "the great mystery"
and the movie *2001* witnessed the scientific revolution, the French and Scottish
Enlightenments, two Industrial Revolutions, aviation, and the space program.
Since the fetal starchild cinematically hovered over the earth in 1968 when *2001*
opened in movie theaters, exploration has proceeded apace into the starry heavens,
planet Earth, and us. The Hubble space telescope has brought us images of the
birth of stars and galaxies in deep space. Molecular biology has brought us the
DNA sequence of the fetus in the womb discerned from the blood and saliva of its
creators.

  Empowered by technologies made possible by Moore's Law, we probe the natural world to reveal its workings, from the evolution of the universe through what
has been called cosmological natural selection to the evolution of our species, even
to tell us something about our future children while still encased in their maternal
space-bubble. Comedian Stephen Colbert joked that he was thrilled to have his
digitized DNA code sent to the International Space Station because "this brings

me one step closer to my lifelong dream of being the baby at the end of *2001.*" The stardust of the starchild is in all of us, as it is in all living things. Yet only we imagine launching ourselves "across the light years." Only we have moved from the man-ape Moonwatcher in the film's opening scene to the eighteenth-century Lunar Society to Apollo 11's lunar lander and astronaut Neil Armstrong's walk on the moon.

In the century that *2001* was fictionally commencing, innovation will be broadly framed by exploration of the ongoing technological and cultural evolution of *Homo sapiens.* We will see the evolution of institutions, organizations, and inno-vation networks; the evolution of public health agencies in response to pandemic influenza and other biothreats; and the evolution of the global middle class. These developments will occur in concert with the further evolution of urban-regions, the spatial crucibles where human interactions and the exchange of ideas is robust. The accelerating growth in the stock of knowledge about biology and biological technologies makes it likely that biology over time will account for a growing share of technological innovation. Within bioscience institutions and companies and outside of them—in schools and at science fairs, even in garages, basements, and kitchens—people will recruit biology to make interesting and useful things, gener-ate wealth, raise living standards, and preserve our biosphere.

The accelerating growth of biological knowledge is occurring during a period of accelerated human evolution. Physicist-turned-biologist Gregory Cochran and anthropologist Henry Harpending, authors of *The 10,000 Year Explosion,* find accelerated evolution in "the set of genetic changes that led to *an increased ability to* innovate." Each major innovation led to new selective pressures, which in turn led to more evolutionary change. That process had its most spectacular success with the development of agriculture. Early herding societies spurred the evolution and selection of genes for lactose tolerance, the ability as an adult to digest milk. The beer and wine producing and consuming peoples of early Europe altered the genet-ics of their alcohol dehydrogenase enzymes, enabling modern Europeans to tolerate alcohol better than Asians. Genetic adaptation generated by geography can now be tracked. Researchers have found that genes related to an ability to tolerate hypoxic stress from oxygen deprivation are under positive selective pressure in Tibetans who live at high elevations compared to Han Chinese from whom Tibetans split off as recently as 3,000 years ago. Evidence of similar genetic adaptation has also been found in Andean Altiplano and Ethiopian Highlander populations. Innovation itself is a product of genetic adaptation in the human brain. Looking deep into human prehistory, an international research team led by evolutionary geneticist Svante Pääbo has identified protein-coding genes conserved in monkeys, apes, Neanderthal, and Denisovan but that have changed in *Homo sapiens.* These genes affect axonal and dendritic growth and notably synaptic transmission in brain cells. Synaptic transmission is a feature of neural circuits thought to modulate learning and memory.

Cultural practices as well as environmental conditioning may be genetically adaptive over relatively short time spans measured in thousands of years. Evidence from converging scientific fields suggests that genes and culture evolved together and that such co-evolution is increasing and contributes to shaping the human genome. Cochran and Harpending base their acceleration hypothesis on genetic information accumulating at a rapid rate from the ongoing revolution in molecular biology along with evidence from paleontology, archaeology, and history. Human evolution has continued to accelerate down to the present day, particularly with ever larger and more concentrated populations and the dynamism such concentration engenders. Behavioral modernity is all about innovation and change.

Is there such a thing as accelerated evolution in economic development? Biologist and geographer Jared Diamond says there is and that technology serves as the accelerator, a key factor in the wealth of nations today. Ongoing national and cultural experiments are furnishing the evidence. Countries that were poor as recently as a half-century ago but developed complex institutions "caught up quickly, once they added advanced technology to their institutional advantages," with South Korea being a prime example.

The systematic practice of innovation in the industrial era—the application of ideas, inventions, and technology to markets, trade, and social systems—is now being joined with the code of life, the code that builds organisms, replicates them, packs craniums with neural networks, and over time creates new species. Today it is a code that can be read—and written—by desktop machines available on eBay. The bioscience future that the automated production of the code of life portends is, like the starchild in *2001*, in its embryonic state. That future will emerge from experiments with new bioscience tools, new business models, and new, innovative and transformational educational programs. For the rest of this chapter we will explore four factors that can help bring about transformational social change from the biosciences:

- Supporting the emergence of new biological technologies
- Exploring the nature of entrepreneurship and the "innovative brain"
- Sourcing talent wherever it exists
- Developing bioscience educational programs, particularly those that link discovery with innovation and production

Synthetic Biology: DNA, Cells, and the Bioeconomy

The apeman-turned-spaceman-turned-starchild framing of *2001: A Space Odyssey* brings to mind terms like *evolution* and *regeneration*, but the story is about our ceaseless search for new worlds. "And now, out among the stars, evolution was driving toward new goals," wrote the book's author Arthur C. Clarke. "The first explorers of Earth had long since come to the limits of flesh and blood; as soon as their machines were better than their bodies, it was time to move." One of those machines, the HAL 9000 computer, controls the systems of the Discovery One spacecraft.

HAL (Heuristically programmed **AL**gorithmic computer) is a vast mainframe with a long room full of memory banks, a sinister camera eye, and a mind of its own. The year the book was published and the movie released (1968), Moore's Law was just three years old. When the actual year 2001 rolled around, computing power had grown from 1,000 transistors per chip (1968) to 25 million transistors per chip, and the world was on the cusp of a revolution in mobile device communication. Today we carry HAL's theoretical computing power in our pockets.

Yet the extraordinary power of Moore's Law cannot keep pace with the exponential improvements in DNA sequencing and synthesis. "The rate at which this technology is now improving puts silicon to shame," wrote *The New York Times* technology reporter John Markoff after attending a short course on synthetic genetics given by Harvard geneticist George Church, founder of the open data Personal Genome Project (Personalgenomes.org), and genomics entrepreneur J. Craig Venter. Markoff, who has covered the computer industry for three decades, perceived the structure of the emerging synthetic genetics industry as "beginning to mirror that of the semiconductor and computer industries, which are based on modular components and design tools." He found a gathering consensus, at least in Silicon Valley, that the information age was giving way to the era of "synthetic genetics."

That consensus was reinforced just a year later when Venter's research group synthesized the genome of one species of bacteria, transferred the synthetic genome of more than a million DNA base-pairs to that of another species, and then "booted up" the synthetic genome in the recipient cell. The synthetic genome created by Venter's group reprogrammed the cell, producing the proper set of proteins based on its DNA code, and the cell replicated accordingly, ushering in what some called a new era in biology (Figure 7.1). Bioethicist Arthur Caplan said the dramatic accomplishment put an end to vitalism, the centuries-old theory that life cannot be created from nonliving parts. In brief, code generated by a digital computer generated a genome that recreated the genetic expression, operational machinery, and self-replicating capacity of a line of living cells. The recipe for success began with off-the-shelf chemical ingredients—adenine, thymine, cytosine, and guanine, the four DNA bases, the building blocks for the code of life.

The moment the first bacterium hosting a synthetic genome replicated successfully, the worlds of digital technology and biological evolution were irrevocably bridged, enabling a biorevolution. The programming language of the cell, nature's 3.5-billion-year-old search engine, will be used for building things, just as digitally programmed robots build things. Design is normally associated with fields like architecture and engineering, less so with fields like chemistry and biology. That is changing, and profoundly. Emerging research fields like genomics, synthetic biology, regenerative medicine, and 3D bioprinting employ design principles at the molecular level. Design has been key to the development of instruments that reveal to us the behavior of the invisible world of cells and microbes and their component parts. These parts have been painstakingly isolated, characterized, and catalogued

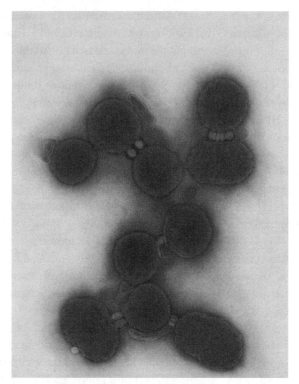

FIGURE 7.1 "Synthia" or *Mycoplasma laboratorium*, the name given to a bacterium whose genome was created synthetically from computer code based on the DNA sequence of another bacterial species.

Source: J. Craig Venter Institute. Negatively stained transmission electron micrographs of aggregated M. mycoides JCVI-syn1.0. Cells using 1% uranyl acetate on pure carbon substrate visualized using JEOL 1200EX transmission electron microscope at 80 keV. Electron micrographs provided by Tom Deerinck and Mark Ellisman of the National Center for Microscopy and Imaging Research at the University of California at San Diego. Reprinted with permission.

and are now available to anyone who wants to employ design principles to build useful things. Genetic and cellular components, integrated with computational platforms, are the biological successors of the standardized parts of the machines that launched the Industrial Revolution and the machines that have kept it going ever since.

The factory of tomorrow is the cell. Scientists, chemists, physicists, and engineers know enough about today's rendition of the cell that they can create circuits, toggle switches, and logic gates inside them and perform both digital and analog computing. We are forging ahead experimentally to see whether we can create novel forms of life from the chemicals of life. We have converted chemicals into synthetic DNA, the synthetic DNA into genes, and the genes into genomes, with the prospect of building completely new bioproductive organisms in the laboratory. We have reengineered the molecular machinery and control elements of cells to produce novel biosynthetic pathways for specific uses, such as the production of drugs, vaccines, biofuels, biocompatible materials, bioremediation enzymes for land and water conservation, and high-value biobased solvents. We are beginning to understand the cell as a chassis, a framework for the design, use, and importation of constructed genetic control elements, optimized genes, and functional genetic circuits that together can modulate the function of metabolic signaling pathways and control the cell, the basic unit of life.

Synthetic biology combines the specialized fields of DNA sequencing, synthesis, assembly, precise genomic editing, *in silico* design, metabolic engineering, microfluidics, microelectronics, miniaturization, and automation. It also employs directed evolution to deal with biological parts and signaling pathways, which are vastly more complex and unpredictable than mechanical and electrical parts and systems. Borrowing from nature's underlying principles and iterative tuning process, directed evolution allows synthetic biologists to design and test synthetic gene circuits with component parts to find optimal arrangements. Borrowing from platforms that allow computer devices to connect seamlessly for sharing data, scientists pursue a "plug and play" approach for constructing and modifying synthetic genetic networks. Simulations with the aid of computational tools are making significant headway in fields like functional genomics, nanotechnology, and cellular engineering. The successful creation and operation of transistor-like genetic logic gates (termed "Boolean integrase logic gates") inside living cells in 2013 established a proof of principle for cell-based biocomputing.

The biological equivalent of the "parts warehouse" at Matthew Boulton's Soho Manufactory at the dawn of the Industrial Age is the Registry of Standardized Biological Parts at Partsregistry.org. The open-source registry, which has been called a RadioShack for cells and their components, describes itself as "a collection of genetic parts that can be mixed and matched to build synthetic biology devices and systems." Founded in 2003 at the Massachusetts Institute of Technology (MIT), the registry is part of the synthetic biology community's efforts to make biology easier to engineer. Users are invited to browse parts by type (e.g., promoters, ribosome binding sites, protein-coding sequences, DNA, plasmids, primers, terminators), devices (protein generators, reporters, inverters, senders and receivers, and measurement devices), and chassis or host (*E. coli*, yeast, bacteriophage, *Bacillus subtilis*). These "BioBricks" come with a tutorial explaining how to inoculate or "streak" agar plates, grow cultures, prepare plasmid DNA, cut DNA at specific sites with restriction enzymes, insert and extract vectors from gels, transform ligation products into cells, and screen colonies for the correct plasmids.

The BioBricks Foundation (Biobricks.org) supports an open technical standards-setting process and promotes use of the registry, including by students in the annual International Genetically Engineered Machine (iGEM.org) competition launched in 2004 (Figure 7.2). Nearly 200 teams comprising more than 3,000 undergraduate students from thirty-four countries participated in iGEM 2012. A team from Germany won the grand prize for creating the Food Warden, a system for

FIGURE 7.2 iGEM 2010 [International Genetically Engineered Machines competition], Massachusetts Institute of Technology. Some thirteen hundred students comprising one hundred thirty teams from twenty-five countries participated in the synthetic biology competition. Students attempt to build simple biological systems from standard, interchangeable parts and operate them in living cells. *Source*: iGEM and Justin Knight, Wikimedia Commons.

detecting meat spoilage using *Bacillus subtilis* cells. A team from Slovenia was first runner-up for developing in situ production of biological drugs using microencapsulated engineered mammalian cells. In an interview after the competition, iGEM cofounder and the "father of synthetic biology" Tom Knight of MIT said that biology is a manufacturing capability. "We can have it build the things we want."

BioBricks, whether they are nucleotide sequences, switches, generators, regulators, promoters, primers, or any number of other parts and devices found in the cell, are the stuff of biological "hacking," in engineering parlance the creating of something useful or interesting out of a collection of parts. BioBricks is the brainchild of synthetic biologist Drew Endy when he was in the Department of Biological Engineering at MIT. In 2010 the National Science Foundation awarded seed money for bioengineers at Stanford University and the University of California (UC) Berkeley to ramp up efforts to characterize the thousands of control elements critical to the engineering of microbes through the establishment of Biofab, an open-source technology platform. Endy, who moved to Stanford in 2008, was appointed director. The term *biofab* borrows from Fab, the network of fabrication service foundries so important to the early semiconductor industry in Silicon Valley. Gene fabrication is becoming a large-scale industrial process with numerous commercial producers with names like Gen9 (cofounded by Endy, Harvard's George Church, and MIT's Joseph Jacobson), DNA 2.0, DNA Technologies, Ginkgo BioWorks, and the biotech tools giant Life Technologies.

Biotech development has occurred in three distinct technological stages stemming from three key scientific papers, as Endy sees it. First was the 1973 report showing the successful genetic engineering of bacterial plasmids, which led to the founding of Genentech and production of synthetic insulin in **E. coli**. Next was the 1985 paper reporting the cloning of the erythropoietin gene, the basis for Amgen's development of Epogen using mammalian cell culture. The third key scientific paper in biotech development was the 2006 article reporting a method to biosynthetically produce high yields of precursors of the anti-malarial drug artemisinin at levels matching those produced naturally by the wormwood plant but in only days compared to months in the natural system. In brief, scientists reconstituted a metabolic signaling pathway from yeast in *E. coli*, synthesized the drug precursor gene from sweet wormwood (*Artemisia annua*), then genetically optimized the pathway. The result was a multifold increase in the production of artemisinin precursor molecules. Scientists and bioengineers recreated simply and cheaply the chemical steps needed to synthesize the drug.

It was a coup for the field of metabolic pathway engineering. It put Amyris on the map and its founder, UC Berkeley professor of chemical engineering and bioengineering Jay Keasling, in the research celebrity spotlight. The Institute for OneWorld Health with $42 million from the Bill and Melinda Gates Foundation together with UC Berkeley, Amyris, and Sanofi-Aventis launched a collaboration to develop an affordable (cents a day rather than dollars a day) and reliable source of artemisinin for the developing world, large parts of which are ravaged by malaria. Sanofi's goal is to produce tons of semisynthetic artemisinin and meet the challenge of converting synthetically produced atemisinic acid to artemisinin. Keasling turned his attention to availability challenges by founding Zagaya (Zagaya.org), a nonprofit whose mission is to ensure access among affected populations to artemisinin-based therapies for treating malaria.

Amyris turned its novel signaling pathway bioengineering capability to producing biofuel from cane sugar using yeast as the "chassis" or production platform. It focused on making farnesene, a hydrocarbon that can be converted to a diesel-like fuel. The renewable fuel, which has been successfully tested in Brazilian buses and airliners, emits a fraction of the greenhouse gases that fossil-based fuels do. With investment dollars pouring in and partnerships established with oil and ethanol conglomerates, the company built an expansive scale-up production plant in Paraiso, Brazil, near large sugarcane plantations. But the plant failed to achieve its initial production goals, in part because of the difficultly of scale-up and the failure of yeast to digest cane sugar and yield farnesene at expected levels. The plant rebounded and shipped its first commercial order of farnesene early in 2013. Amyris's biofuels experience carried a cautionary note for synthetic biology. "When synthetic biologists announce they will treat microbes like tiny factories, investors and markets may be listening, but microbes are not," says Daniel Grushkin, who tracked the company's ups and downs for the business and technology magazine *Fast Company*. "Biology is not computing or engineering—at least not yet."

Commerce is in a hurry. Nature is not, as bioenergy firms using the tools of synthetic biology, firms like Synthetic Genomics, LS9, OPX Biotechnologies, KiOR, Gevo, and Joule, know from experience. Since the beginning of the Industrial Revolution the history of technology is littered with first-mover failure of promising discoveries and inventions—from propelling watercraft by steam power to building silicon semiconductors to social networking services. Choosing the right pathway from discovery to laboratory/bench scale process systems to pilot plant to industrial production scale-up, and choosing it at the right time under the right market conditions, can spell the difference between success and failure. That is especially true of biotechnology production process scale-up. Those who succeed are able to navigate between the treacherous shoals of cellular productivity and investor patience.

But it can be done, even in the biofuels space where scale-up challenges and a boom in oil and natural gas production in the United States would seem to lean against it. A case in point is the Bay Area firm Solazyme, whose algae makes specialty oils and biodiesel in closed fermentation tanks. The company uses what has been termed a synthetic nanobiology platform to select algal strains improved through classical methods or genetic engineering. In the case of the latter, microalgae are engineered to upregulate expression of a global regulator of fatty acid biosynthesis, thereby boosting lipid production.

Solazyme announced in 2008 that it had produced the world's first algal-based jet fuel. Four years later it achieved commercial-scale production at an Archer Daniels Midland plant in Iowa at four times the level of its own facility in Illinois. The carboniferous plant life that supports the world's energy infrastructure evolved from green algae. Solazyme's algal cells are proving to be the microbial factories touted by synthetic biologists and advocates of a bioeconomy. Is algae equipped to help replace the fossil fuels to which it helped to give birth? Is Solazyme poised to "disrupt the food industry" with its nonallergenic high-lipid whole algalin flour?

The challenge for the next decade will be to integrate molecular engineering and computing to make complex systems, says George Church. The development of engineering standards for biological parts "will permit computer-aided design at levels of abstraction from atomic to population scales." Church and others in the field of synthetic biology see a world in which biologists will have the tools that will allow them to arrange atoms to optimize biocatalysis or arrange populations of microorganisms to cooperate in making a chemical. Synthetic networks and biomolecules will be constructed in the predictive, self-configuring plug-and-play manner found in computing devices. Once scientists have the ability to program cells, in the view of MIT synthetic biologist Ron Weiss, "you don't have to be constrained by what the cells know how to do already. You can program them to do new things, in new patterns."

Synthetic biology has already spawned its own hacker culture. The biohacker culture is most clearly manifested in DIYbio (short for "do-it-yourself biology").

DIYbio is an organization of several thousand members whose stated aim is to help make biology a worthwhile pursuit for citizen scientists, amateur biologists, and DIY biological engineers who value openness and safety. As in the computer culture, hacking biology involves taking things apart and putting them back together in a new way that makes them better. It is reverse engineering with an eye toward reassembly with improved function. DIYbio cofounder Mackenzie Cowell wants DIYbio to become "the homebrew club" of biology, a reference to the early computer hobbyist group in Silicon Valley. "If you open up the science and allow diverse people to participate, it can really stimulate innovation," says molecular biologist Ellen Jorgenson, president of Genspace (Genspace.org), a community DIYbio in Brooklyn, New York. "There is something sacred about a space where you can work on a project and not have to justify to anyone that it's going to make a lot of money, or save the world."

DIYbio community labs similar to Genspace have sprouted up in other cities in the United States and around the world. They are proving grounds for whether biological technologies can draw talent from a large and vibrant hacker culture and benefit from the undeniable creativity of that culture. "You find the right people, crowdsource the genetic design, crowdfund the project and go to a community lab and do something that people on the street would consider science fiction," says Israeli synthetic biologist Omri Amirav-Drory, who is leading a project to grow bioluminescent plants. He is CEO and founder of Genome Compiler, software that helps users design DNA for certain functions. Amirav-Drory and his colleagues in the US plan to modify a DNA sequence of the mustard plant *Arabidopsis* so that it produces luciferase, the enzyme that makes fireflies glow. Their idea is to have Cambrian Genomics synthesize the sequence using its novel laser-printing technology. With the desired sequence in hand, they will transfer it into the plant in a community lab.

Perhaps first among the contemporary icons of hacker culture are Steve Jobs and Steve Wozniak, the founders of Apple Computer, but in the history of the American Republic the hacker mindset can be traced back to the tinkerer of note Benjamin Franklin. In Franklin's day Britain possessed the world's largest supply of highly skilled, mechanically able craftsmen thanks largely to its superb apprenticeship system. Ralf Meisenzahl and Joel Mokyr call them "tweakers," those who "improved and debugged an existing invention." Tweakers teamed up with "implementers" to translate invention into innovation and economic growth in the early decades of the Industrial Revolution. They gradually transformed a number of economic activities, including cotton textiles, ironmaking, and steam power but also food processing, brewing, glassmaking, papermaking, cement production, mining, and shipbuilding. They furnished the nitty-gritty brainpower for the new industrial economy, driven by reputation as much as financial reward.

Biological technology is entering a new era, one that will be characterized by economic disruption thus economic opportunity. Synthetic biology is a technological tsunami that enables practitioners to write life code and program cells. That is how the entrepreneur and investor Juan Enriquez describes the field. Enriquez and Venter launched Synthetic Genomics, Inc. in 2005 to commercialize genomic technologies. The synthetic biology tsunami, Enriquez says, will produce hundreds of open-source and private designs blossoming into thousands of applications and products. Synthetic biology is converging with a second tsunami, the field of stem cell science and regenerative medicine. This field can create cell therapies, bioartificial substitutes for diseased tissues and organs, and functioning mimics of complex physiological systems like the immune system. Cells can be genetically reprogrammed in vivo to a pluripotent state, a promising avenue for repairing diseased tissues and regenerating healthy tissues, including limbs and organs, from the inside. Tissues can be engineered to be compatible with promising nanomaterials like graphene for molecular-scale bioelectronic device interfaces. These fields, in Enriquez's view, are converging with a third technological tsunami, this one in robotics. Scientists, innovators, and biohackers now have the ability to manipulate genes, cells, and tissues with computers and automated devices and instruments such as inkjet printers (bioprinters) to bring new designs to biology and quite possibly to our own evolution.

Such empowerment is not without risk. Perhaps no development illustrates that risk more vividly than the dispute over whether and how research on the deadly H5N1 avian flu virus should proceed after Dutch and American scientists announced that they had created mutated versions that can be transmitted from ferret to ferret through the air. More than a year of governmental, regulatory, and academic wrangling ensued, some of it on the international stage. Publication of research results were delayed, and researchers ceased their experiments temporarily. With resumption of the research came a report that Chinese scientists succeeded in combining H5N1 with the highly contagious swine flu strain H1N1 and showed that the hybrid virus could be transmitted through the air among guinea pigs.

These H5N1 gain-of-function studies prompted a call for an Asilomar-like conference to address the risks posed by viral engineering. The Asilomar meeting, convened in 1975 to deal with risks and public fear surrounding recombinant DNA technology, set standards for geneticists so that they could pursue their research without endangering public health. The risks today, however, are not limited to use of a single emerging technology (viral engineering/synthesis) but are represented in a menu of converging and expanding technologies. They include DNA sequencing and synthesis, cellular and tissue engineering, genetic editing, and automation of human immune system function, all empowered by and sometimes melding with digital technology. Increasingly they are geotechnologies, automated bioanalytical and biosynthesis instruments and systems linked to data networks. Some are flowing into the global plug-and-play biohacking culture. Finding agreement on what constitutes the appropriate use of these biological technologies will be one of the great challenges of twenty-first-century global governance. The health and well-being of living systems and the environment will depend on how that challenge is met.

Entrepreneurs: What Makes Them Tick?

"Every generation needs a new revolution," Thomas Jefferson wrote near the end of his life. By then (1826), the word *entrepreneur* was only a century old. It was introduced to general use in Jefferson's day by French economist Jean-Baptiste Say and translated into *adventurer* for English readers of Say's work. Adam Smith never used the term, nor did Smith find a place for disruptive adventurers in his understanding of how a free-market economy works. But in today's technology-based economy, the entrepreneur is key.

The new revolution of our generation is the entrepreneurial revolution, says Adrian Wooldridge of *The Economist,* who writes a column named for economist Joseph Schumpeter of "creative destruction" fame. The entrepreneurial revolution has spread from America and Britain to the rest of the world and from the private to the public sector. It is a revolution that "is doing something remarkable: applying more brainpower, in more countries and in more creative ways, to the art of raising productivity and solving social problems." Today Schumpeter's entrepreneurs take basic science discoveries from government-funded research and give them technological legs. They are "global heroes," the title of a special report on entrepreneurship by Wooldridge published in the months after the financial crisis of 2008. They have gone mainstream in the economy itself, in the realm of economic ideas, whether Paul Romer's new growth theory, Michael Porter's clusters of innovation, or Brian Arthur's insights into how technology evolves. The strategic plan for the future bioeconomy released by the White House in 2012 elevated entrepreneurship to a level perhaps not seen before in any comparable document. At the federal level, the Startup America program, the National Science Foundation's I-Corps program, the National Institutes of Health's SBIR technology transfer program, and the Food and Drug Administration's entrepreneurs-in-residence pilot program all are designed to spur entrepreneurship. The Challenge.gov prize competition is a public-sector initiative in the spirit of contests offered by the nonprofit X-Prize Foundation to promote technological development. In the White House plan, academic institutions are encouraged to provide incentives "to enhance entrepreneurship and restructure training programs [that] would better prepare the future bioeconomy workforce."

As is often said, entrepreneurs set out to change the world, not rule it. They are making lots of headway. When it began, capitalism was largely the province of white men. Today, entrepreneurial capitalism is the province of men—and women—of all ethnicities and backgrounds, unleashing proportionately many times the human brainpower than that available for the Industrial Revolution. Entrepreneurs are less exquisitely tuned to the price of risk as are, say, financiers and corporate managers. That is why they receive their reward at the end of the process, after their ideas have been accepted by the market. Only after the hard work of evangelizing the idea and giving it a serious road test can the dream of entrepreneurial success be realized.

In the biosciences and other knowledge-based industries, that road test can be long, grueling, and expensive, which is why the translation of science and technology into products and services stands in stark contrast to the pursuit of "easy money" and "get rich quick" schemes typical of Wall Street. The trains of science-based entrepreneurship and financial arbitrage run on separate tracks. Moreover, the project to bridge the gap between scientific knowledge and practical knowledge launched by the Lunar Society at the dawn of the Industrial Age remains a challenge. It is not helped by the probability that future biotech startups will have to do more with less, which is what global venture capital industry analysts expect. Nor is the trend toward downsizing core employment and outsourcing laboratory science to contract research—so-called virtual biotechnology—a proven model for bridging the gap, particularly as contract researchers do not have their own "skin in the game." Such outsourcing forgoes the imagination and entrepreneurial drive employee scientists can bring to the task.

Managerial capitalism embodied in giant, bureaucratic post–World War II companies gradually gave way to what some describe as financial-market capitalism that spurred growth in entrepreneurial start-up activity for three decades beginning in the 1980s. Financial-market capitalism shifted value from labor to capital and, as technology advanced, from blue-collar workers to workers with college degrees, especially graduate technology degrees. A productive system of entrepreneurial capitalism requires a stable, well-regulated, and well-run financial system that can make risk capital available for risky ventures with long idea-to-product timelines. Entrepreneurship is the proven vehicle for converting new knowledge-intensive fields like biotechnology, nanotechnology, robotics, green technology, and clean energy—fields that will drive innovation in this century—into fields of dreams, dreams of commercial and societal as well as personal success.

At the peak of the dot.com boom, *The Economist* observed that innovation had become the "industrial religion" of the late twentieth century. The religion of innovation is preached louder from the business pulpit today than it was in the 1990s. Yet the individual and organizational appetite for actual risk-taking, which is where the rubber meets the road, has diminished in the face of massive economic dislocations in the industrialized world. That needs to change. The human imagination must be freed from fear and take flight, as it did the day James Watt took a walk across Glasgow's College Green in 1765. In a flash of insight Watt realized that the efficiency of Thomas Newcomen's steam engine could be greatly improved if the steam could be drawn off and condensed back into water without cooling the steam piston cylinder. Thus was born the idea of a steam engine with a separate condenser, the iconic machine that launched the Industrial Revolution with the aid of the entrepreneur Matthew Boulton. If the future refutes economist Robert J. Gordon's pessimistic outlook of painfully slow economic growth and embraces Juan Enriquez's optimistic outlook of technological tsunamis just over the horizon, entrepreneurship in the Genomic Exchange era will be one reason why.

For more than three decades, innovation has been the mantra of business and public policy worlds, with scientific research institutions willing and cooperative partners in the economic enterprise. Entrepreneurship is the principal driver of innovation in a capitalist system of production, and entrepreneurs are the creative destroyers of entrenched interests in Schumpeter's scheme of how innovation occurs—which is why research on entrepreneurship and how to teach and encourage it are thriving. What relative contributions do nature and nurture make to building an entrepreneur? What configurations of neural circuits or expressions of neurotransmitters trigger novel ideas in the heads of entrepreneurs? What in their brains or backgrounds is responsible for their irrepressible drive to put these ideas to work in the human social system we call the economy?

The professional services firm Ernst & Young set about "decoding the DNA of the entrepreneur" to find an answer to the nature versus nurture question. From a survey of nearly 700 entrepreneurs from more than thirty countries and twenty-five industry sectors, Ernst & Young analysts concluded that

- Entrepreneurs are made, not born. Education and a period of corporate employment are key.
- Serial entrepreneurship is the norm, one-off entrepreneurship the exception.
- Funding, people, and know-how are the biggest barriers to entrepreneurial success.
- Though there is no "entrepreneurship gene," entrepreneurs share common traits, including an innate belief that an individual's action can shape events and an ability to see opportunity where others do not.
- Traditional companies can learn from entrepreneurial leaders.

Twin studies have long been the gold standard for sorting out nature versus nurture questions. A genetic study of nearly 5,000 identical (monozygotic) and fraternal (dizygotic) twins in the United States and the United Kingdom revealed that although phenotypic correlations between certain personality characteristics (extraversion, openness to experience) and the tendency to be an entrepreneur were small in size, genetic factors accounted for most of them. The authors suggested that the DRD4 gene, a dopamine receptor subtype previously found to be associated with openness to experience, would be "a good candidate gene to test for association with the tendency to be an entrepreneur." They cautioned that the patterns they identified are "merely predispositions" and definitely not deterministic relationships, though confirmation of their findings would be of value in encouraging people to become entrepreneurs. Personality traits with a strong environmental component rather than those with a strong genetic component could be the focus of entrepreneurial education to expand the ranks of future entrepreneurs.

Cognitive enhancement is part and parcel of "The Mental Wealth of Nations," the title of a commentary published by British scientists in *Nature*. If countries are to prosper economically and socially in the new century, they need to take measure of their collective brainpower and find ways to shore it up. The commentary focused on findings from the United Kingdom's Foresight Project on mental capacity and well-being. Countries need to boost brainpower in the young and old, monitor biomarkers that can reveal learning difficulties as well as mental disorders early in life, promote life-long learning, and attend to employee mental well-being in a rapidly changing workplace. It is as if Adam Smith's *The Wealth of Nations* is to be viewed through the lens of his earlier work, *A Theory on Moral Sentiments*, in which Smith provided the ethical, philosophical, and psychological underpinnings for all his writings on political economy and trade. How a nation develops and uses its mental capital, the British scientists say, "not only has a significant effect on its economic competitiveness and prosperity, it is also important for mental health and well-being and social cohesion and inclusion." Interventions, including behavioral, educational, social, and pharmaceutical, are entirely justified under the appropriate circumstances to keep a country mentally healthy and competitive.

Today's wealth of nations ethos is extending Adam Smith's ideas of specialization and division of labor to take into account knowledge of how people are neurologically wired to excel at certain tasks but not others. The division of labor mobilizes human neurodiversity in powerful ways economists are only beginning to understand, says economist, author, and blogger Tyler Cowen, noting the unique perspectives and high achievements of many people with autism. A high percentage of successful entrepreneurs are dyslexic, business magnate Richard Branson, discount brokerage pioneer Charles Schwab, and Cisco Systems CEO John Chambers among them. Sharp human resources departments and bioscience startups can tap the power of neurodiversity in their efforts to find the right fit for prospective employees.

Neuroscience, social science, genetics, and other disciplines relevant to the nuts and bolts of human behavior are scouring the entrepreneurial brain. Scientists in these fields want to find out what makes it tick under normal circumstances and how neural signaling pathways related to entrepreneurial risk-taking could be stimulated in ways that are safe and effective in healthy adults. A successful entrepreneur possesses the characteristics of vision, passion, determination, tenacity, perseverance, ingenuity, mastery of the knowledge field and, of course, a healthy dose of luck. Entrepreneurs also have a tendency to behave like "juvenile delinquents" by acting on impulse. Indeed, "functional impulsivity" is exactly how the authors of "The Innovative Brain" characterize the risk-taking behavior of entrepreneurs in a cognitive study they undertook with matched groups of entrepreneurs and conventional managers who served as controls. They pulled together sixteen serial entrepreneurs who helped to make Silicon Fen, the Cambridge (UK) high-tech cluster, a success and seventeen business managers. They tested these subjects using a computerized neurocognitive assessment and functional neuroimaging and compared their abilities to make certain kinds of decisions, specifically "cold" and "hot" reasoning processes that imaging has shown to be associated with different areas of the brain.

Entrepreneurs displayed significantly riskier behavior than their managerial counterparts. They scored high on personality impulsiveness and cognitive flexibility.

Neurobiology is throwing some light on the molecular constituents of marketplace success. Paul Zak and other neuroeconomists have shown that the neurotransmitter oxytocin is key to the trust that is essential for markets to work. Risk-taking is associated with the neurotransmitter dopamine. Entrepreneurs are much more willing to place a high-risk bet than their managerial counterparts, a behavior that neuroimaging has revealed is tied to activity in the midbrain, the seat of emotions and site of dopaminergic neurons. Brain researchers have shown that the density of the brain's dopamine receptor D2 is associated with divergent or "out of the box" thinking and creativity. They speculate that D2 is a sort of gatekeeper for the flow of information, with increased D2 densities enabling greater flow. Maybe what hit James Watt as he walked across the Glasgow Green that fine Sabbath afternoon in 1765 was in fact a D2-mediated midbrain information surge.

If risk-taking and cognitive flexibility are indeed linked to levels of the neurotransmitter dopamine, which both neuroimaging of human subjects and behavioral pharmacological studies of laboratory animals appear to suggest, it would raise the question of whether one could enhance entrepreneurship through neurochemistry. The practice of pharmacologically enhancing traits such as those associated with memory, focus, self-control, or willpower is likely to come with trade-offs, say psychologists Thomas Hills and Ralph Hertwig. Broadly speaking, evolution appears to prefer symmetry among many related traits. In that context, use of cognitive-enhancing drugs to boost certain performance functions may well have detrimental cognitive side effects on other important functions such as originality or the ability to interact with others, keys to collective technological progress.

One of the goals of the Human Connectome Project, the effort to map the brain's connections, is to decipher how the pattern of electrical signals generate thoughts, feelings, and behaviors and how they interrelate. The project, with National Institutes of Health funding, is also exploring how abnormal "wiring" during brain development can result in neurological disorders such as schizophrenia and autism spectrum disorders. Are the estimated 85 billion human brain neurons connected in a way that distinguishes the innovative brain from others? However that question is answered in the decades ahead, "connectomics" itself is a bold exercise in innovation. Scientists from neuroscience plus nanotechnology and genetic engineering are teaming up in a federally funded effort called the BRAIN Initiative (**B**rain **R**esearch through **A**dvancing **I**nnovative **N**eurotechnologies). Like the Human Genome Project, the BRAIN Initiative is a "big science" public–private partnership designed to accelerate the invention of new tools that will help reveal how brain function is linked to disease but also behavior and learning. Meanwhile, Canadian and German scientists sliced a human brain into thousands of sections, which were stained and digitized to create an ultrahigh-resolution 3D brain atlas. The open-access reference brain, named "BigBrain," is part of the European Union's Human Brain Project. It reveals the brain's organization in exquisite detail and how its various regions are connected. Could stained and digitized brain slices reveal patterns that can be linked to human behavior, even a propensity to take chances?

Entrepreneurial training that teaches risk tolerance in behavior and personality as well as the "know-how" of entrepreneurship could help would-be entrepreneurs to reframe their decisions, mitigating the negative perception of risk in starting a new venture. Geography also could play a role: "One of the beneficial effects of entrepreneurial clusters in regions such as Silicon Fen may be that the increased networking and contact amongst the entrepreneurs works to create a culture that normalizes a more risk-tolerant type of decision-making," says Cambridge, UK, investor Hermann Hauser, a fellow of the Royal Society who pioneered Europe's venture capital industry. An entrepreneur's natural ability is founded on the inter-action of genes and environment, Hauser says. In 1997 only 17 percent of entre-preneurs in the portfolio of Amadeus Capital, his firm, were serial entrepreneurs. By 2009 about 70 percent were serial entrepreneurs, contributing substantially to the more than 1,000 high-tech companies in the Cambridge region. "Know-how is transmitted 'in the air' within these high-technology clusters," Hauser observed, bor-rowing a phrase from the nineteenth-century Cambridge economic historian Alfred Marshall, who used it to describe the seemingly spontaneous growth of industrial districts. "They [entrepreneurs] are also valuable in tolerating risk and possible failure well." Hauser cited the bioscience entrepreneur Andy Richards, who once characterized Cambridge as "a low-risk area for doing high-risk things." Indeed, entrepreneurial risk-taking is increasingly concentrated in regional hotbeds, which separates them from the risk-averse mood of advanced economies as a whole.

As we saw earlier, the coffee houses and tea rooms of eighteenth-century London, Edinburgh, Amsterdam, Paris, Venice, and other major trading cities provided the venues for development of modern capitalism. There was something "in the air" in these locales. But there was more. Ideas of trade, private property and legal contracts, the chartered joint-stock companies, the accounting systems, the deal making, Josiah Wedgwood's ceramics cluster—all were formulated and exercised with the aid of the neurochemical agent caffeine. Nor should it be forgot-ten that Adam Smith himself worked on revisions of *The Wealth of Nations* in a London coffee house, the Starbucks of his day, a day when no one ordered decaf. Today few would dispute that the association of free-market capitalism, entrepre-neurial neurochemistry, networks, and risk-taking behavior is a key contributor to the wealth of nations and the health and well-being of their peoples.

Sourcing Talent for Successful Innovation

The international search for talent and technology is at least as old as the Industrial Revolution. Take the case of George Parkinson, an English weaver who worked at America's first planned industrial park at the Great Falls of the Passaic River in

Paterson, New Jersey. The Society for Establishing Useful Manufactures, the brain-child of Treasury Secretary Alexander Hamilton and his deputy secretary Tench Coxe, hired Parkinson as a foreman in 1791. Parkinson had been granted a British patent for making incremental improvements in Richard Arkwright's flax-spinning machine. In contravention of British law, Secretary of State Thomas Jefferson approved Parkinson's US patent application and helped arrange the transport of his family to America. Whereas Jefferson had some misgivings about technology piracy, Hamilton had none, granting Parkinson a federal subsidy to cover his living expenses. But technology piracy in itself was not enough. America was woefully lacking in skilled mechanical labor. Hamilton saw that luring British mechanics and machine operators to the United States was a strategic necessity.

Today the United States has earned the reputation of welcoming the best and brightest from around the world. Foreign students come to the United States to receive a world-class education, yet many are then forced to leave, taking what they have learned with them. The strategic plan for the future bioeconomy released by the White House in 2012 elevates entrepreneurship but does not mention foreign-born talent or visas, glaring omissions in an increasingly international enterprise. "The competition for top scientific talent is now global, and we need to recognize it as such," the Milken Institute argued in its 2011 report on accelerating innovation in the bioscience revolution. "The U.S. needs to make it easier for talented graduate students to not only study here, but stay and conduct research here." Lawmakers should fix the nation's broken visa system "so the world's best students can stay in the U.S. throughout their careers."

Silicon Valley and its neighboring universities brought what UC Berkeley's AnnaLee Saxenian calls the "New Argonauts" to America: immigrant entrepre-neurs from China, India, Taiwan, Israel, and other countries. These immigrants founded more than half of Silicon Valley's startups from 1995 to 2005 and 44 per-cent from 2006 to 2012. Immigrant entrepreneurs founded nearly a quarter of bio-science companies in the United States between 2006 and 2012, based on a study by the Kauffman Foundation. Immigrant founders start their businesses where inno-vation thrives, exhibiting "a high pattern of clustering" in the immigrant gateways of California, Massachusetts, Florida, Texas, New Jersey, and New York where bio-technology and pharmaceutical companies are also concentrated. The decline in the percentage of immigrant-founded startups occurred as "wait times" for immigrants to be granted permanent residence visas or green cards averaged about a decade. Some immigrant entrepreneurs went home in frustration or were lured home by opportunity that the United States could no longer provide. The bipartisan push for immigration reform in the US Congress in 2013 included a proposed doubling in the number of H-1B guest worker visas available to skilled foreign-born work-ers, including those with advanced degrees from US universities. In draft legisla-tion supporters acknowledged that "it makes no sense to educate the world's future innovators and entrepreneurs only to ultimately force them to leave our country at the moment they are most able to contribute to our economy."

The year the epic futuristic film *2001: A Space Odyssey* was released, 1968, a Hungarian immigrant entrepreneur named Andy Grove joined the newly established firm Intel Corporation in what became known as Silicon Valley. After Grove became Intel's chief executive officer in 1987 he transformed the company into one of the world's dominant producers of microprocessors. Reflecting on his immigrant experience, Grove said that America "must be vigilant as a nation to have a tolerance for differences, a tolerance for new people." The "tolerance" for differences and new people has been upgraded to a strategic "quest" for many forward-looking organizations, and with good reason.

No women appeared in a prominent role in *2001: A Space Odyssey*. By the year 2001 thirty-seven women from seven countries had flown in space, most of them in the 1990s aboard NASA's space shuttle. That year the Institute of Medicine published a report exploring the contributions of biological research to human health and answering the question whether gender makes a difference: "Sex does matter. It matters in ways that we did not expect. Undoubtedly, it matters in ways that we have not yet begun to imagine." Since then neuroscience has revealed differences in brain anatomy, chemistry, and function between the sexes that may help explain differences in how men and women make decisions and solve problems. If incorporating these differences within top management and adapting to them is valuable for next-generation organizational performance—and leading thinkers believe that it is—countries and industries, including biobased industries, would be well advised to take note and take action. Nordic countries, which typically rank at or near the top in international competitiveness and innovation indices, also rank at the top in gender equality. All in all, editorialized *Nature* in a special issue on women in science published in 2013, there is "a huge amount of talent going to waste" given their low numbers with full professorships (20 percent in the United States and Europe) and their marginal representation on the scientific advisory boards of start-up companies.

Higher education is laying the foundation for future innovation in the biosciences. The statistics are telling. About half of recipients of degrees in the biological sciences in the United States and Europe are women. The number of women earning science and engineering PhD and MD degrees in the United States today has reached parity with men, although women remain well behind men in their representation among tenured faculty in US universities, as medical school department chairs and deans, as managers of biotechnology and pharmaceutical companies, and in board rooms. Women now account for nearly 60 percent of students attending institutions of higher education overall and 60 percent of students who graduate. More than 53 percent of doctorate (PhD, EdD) and professional degrees (MD, DDS, law) in 2009–2010 were awarded to women. Such a profound shift in higher education "is likely to have major implications for future changes in the gender gap in average earnings, the fraction of heads of business that are women, and other measures of gender differences in achievement," says Nobel economist Gary Becker.

Since Genentech launched the biotechnology industry more than three decades ago, women have been underrepresented in top leadership positions in biotech policymaking, research, and management. So began a commentary published in *Nature Biotechnology* calling for more women in leadership roles. The authors, members of a joint European Commission–US Task Force on Biotechnology Research that convened a workshop to look at women's leadership issues, averred that without equal input from both men and women, optimal solutions to intractable problems and global challenges are unlikely to be found. A "female perspective" would broaden and expand the types of novel approaches that could be brought to the task. Collaborative network organizations, which are on the rise, "are more conducive to women's leadership in biotech than are rigid, less transparent hierarchies" that are characteristic of the biotechnology and pharmaceutical industries as well as of academia.

Half a century has passed since the biologist Rachel Carson published *Silent Spring*, thereby launching the modern environmental movement. As we observed in our introduction, when Carson was writing and publishing her book describing how manufactured chemicals were damaging natural ecosystems, Marshall Nirenberg and Har Gobind Khorana and other geneticists and biochemists were deciphering the genetic code and protein synthesis and explaining the chemistry of life. Today the tools of bioscience and genetics are showing that healthy ecosystems are critical for economic development. To be successful, future innovation ecosystems will need to be both economically productive and environmentally sound. If we are in fact entering the Anthropocene era, a new story about the earth's natural systems being altered by human beings, Rachel Carson wrote the introduction.

Women entrepreneurs are proving themselves all over the world—at the Biotech Park for Women Society in Chennai, India, as well as at Women in Bio (Womeninbio.org) chapters in the Baltimore–Washington–Northern Virginia region, the San Francisco Bay Area, Boston, Seattle, Pittsburgh, Chicago, Atlanta, and Research Triangle Park. If gender does indeed matter in ways we have not yet begun to imagine, the cultivation of talented women in the biosciences around the world would seem to constitute a prudent investment for our economic as well as our ecological future.

Education and Brainpower for a New Century

The spirit that drives us to learn, take what we learn and build new things and better ways of doing things, that spirit needs to be rekindled and widely shared. With a global population projected to reach 9 billion people in the decades ahead when the planet is struggling to sustain the population it already has, countries best able to summon the imagination and creativity of their citizens are the countries that will be best able to innovate. That cannot happen, at least not in the biosciences, without a solid education in programs ranging from specialized degrees in the sciences to interdisciplinary degrees that prepare students for work in areas such as product development and regulatory affairs. The "silent Sputnik" challenge for the United States is to pay more attention to its educational system, says former National Science Foundation director Rita Colwell, the first woman and the first biologist to head the agency.

Innovation in education is key. Colwell champions professional science master's (PSM) degree programs. These two-year programs "aim to engage students with professional goals and help them become scientists uniquely suited to the twenty-first century workplace." The America COMPETES Act (Public Law 110-69) passed by Congress in 2007 noted the importance of the PSM degree to the nation's overall competitiveness, a point reiterated by the National Research Council of the National Academies, which urged expansion of the PSM degree in a 2008 report "Master's Education for a Competitive World." With seed money from the Alfred P. Sloan Foundation and other foundations and an initial $15 million from National Science Foundation provided through the American Recovery and Reinvestment Act of 2009, more than 100 PSM programs were established. Currently some 300 PSM programs have been set up at nearly 130 affiliated universities located in more than half the states (Sciencemasters.com). PSM programs enrolled 5,500 students in 2011 and awarded 1,600 degrees

Two-thirds of PSM programs are in the biosciences (including environmental science). Colwell believes that the capacity to innovate depends not only on scientific discovery, typically the bailiwick of PhDs and MDs in the biosciences, "but also on the ability to translate new knowledge into products and services." This is where graduates of professional science master's degree programs come in. They can have a "major impact" because their education by its very nature requires strong university–industry linkages. Such linkages, particularly those formed between universities and the sector-specific firms that cluster around them, hold the key to both regional economic innovation and global competitiveness. They foster growth cultures, rich ecosystems teeming with ideas. It is disproportionately within these ecosystems where technology evolves.

Graduate-level professional programs in the biosciences are a valuable complement to doctoral degree programs. PhD programs furnish most of the scientific talent for the biotechnology and pharmaceutical industries as well as for academic bioscience. The United States is the global leader in terms of both the number of PhD programs and the annual number of graduates from them. The dominance of these programs is due in no small part to the growing percentage of bright and driven international students, many of whom upon graduation enter the US workforce and make a major contribution to the domestic economy and standard of living. Foreign-born PhD students have become a major force in US higher education over the past four decades. In 1966, US universities awarded 23 percent of science and engineering PhDs to foreign-born students. In 2006, that figure had risen to 48 percent. Foreign-born students earned one-third of all doctoral degrees granted in the life sciences by US educational institutions in 2012. Nearly two-thirds of them who study the life sciences at US universities want to stay in the United States following graduation.

The growth in applications of international students to attend American universities began to decline early in this century as competition for these students picked up from Europe, Asia, Australia, and other nations. Asian countries such as China, India, and South Korea have succeeded in keeping more of their brightest students at home for graduate study and to do their innovation in their own economies. Those students who do spend time in the United States in training or education are more likely to seek opportunity back home, particularly since the severe recession increased the competition for American jobs. It is no small irony that among those who return home are graduates of US medical schools. Some of them perform surgery on medical tourists from the United States.

It is often said that the rapid growth of institutions of higher education in Asian countries is a good thing for the global economy and for America. The stagnation of its own system of public high-school education is not. The scientific literacy of high-school students in the United States has declined, and their performance in international comparative testing in science and math has lagged behind that of many other nations including Japan, Singapore, and South Korea. In international educational assessment tests conducted in 2009 and 2012, the average US mathematics literacy score for fifteen-year-old students was below the average score of the thirty-four Organisation for Economic Co-operation and Development member countries; the average score for science literacy was about the average. East and Southeast Asian students dominated these subjects. In the United States, student achievement in reading and mathematics has been essentially flat since the 1970s. Is it possible that the baby-boom generation, the generation that produced the scientists, engineers, venture capitalists, managers, marketers, and technicians that launched the biotechnology industry in the 1970s and 1980s, received better primary and secondary education than Generations X and Y?

A state-by-state analysis of bioscience education prepared by Battelle, BIO, and the Biotechnology Institute in 2009 found national data on life sciences achievement "spotty" but good enough to conclude that the nation is "falling short," particularly in preparing students for careers in science. Only 28 percent of the high school students taking the American College Test (ACT) scored well enough to indicate readiness for college biology. No state, including Massachusetts, California, Maryland, North Carolina, and Pennsylvania which host a large number of life sciences institutions and companies, reached the 50-percent mark in student preparedness for college biology, based on ACT scores. All of the states were found to have schools with a focus on science, technology, engineering, and mathematics (STEM) education. At least half of them had at least one school with a bioscience focus. Of those, a number of states have multiple schools with such focus. Many of these programs work well but "need to be replicated and states need to commit resources to them."

As the Battelle report was released, however, state coffers were beginning to bleed dollars by the millions and even billions, with no relief in sight for years to come. In its 2013 budget the Obama administration proposed to invest $260 million in STEM programs and set the ambitious goal of preparing 100,000 high-quality STEM teachers over the next decade. STEM education efforts "are expected to contribute broadly to the development of a bioeconomy workforce for the 21st century," according to the administration's bioeconomy blueprint report. STEM degree program graduates, including foreign nationals, are entering the labor force in record numbers. About 5 percent of all jobs in the United States are considered STEM positions. But with the projected squeeze on federal discretionary spending, the future of federal support for STEM programs and other innovative educational initiatives is uncertain.

Critics of international science and math test results rankings point out that the percentage of students who perform at the top are most likely to be tomorrow's innovators, and on that score the United States does very well. Indeed, the United States produces the lion's share of the world's best students, wrote Hal Salzman and Lindsay Lowell in "Making the Grade" published in *Nature*. The averages are less important than the "tails." The United States should pay more attention to the circumstances of its own low-performing students in the context of how its best schools educate top performers, they contend, not the test scores of students from very different countries and cultures. Average test scores "are largely irrelevant as a measure of economic potential" because it is the high-performing students who are critical for future innovation, and for now the United States has the largest share. But will the United States continue to produce the largest share of high-performing students? Does evidence that a small number of high-impact high achievers generates a great deal of economic dynamism, what has been termed "cognitive capitalism," mean that policymakers should not worry too much about international test scores?

Student mean testing scores and innovation do not march in lock step. Countries whose cultures discourage creativity, individual achievement, and risk-taking may not benefit regardless of whether their students score highly in international science and math testing. A culture of innovation "requires the encouragement of conflict within the larger culture of transparency and trust, placing a premium on cross-cultural competence," says the academic and foreign policy analyst Anne-Marie Slaughter about America's edge in the networked economy. It is a culture for which Americans "are ideally suited by both temperament and history."

In an age of borderless communication, growing international education and travel, and the shifting focus of capital to emerging markets, can the seeds of creativity find fertile soil in cultures and countries not known for nurturing them and begin to grow? If they can, having lots of students who test well is likely to become a competitive economic advantage, their absence a disadvantage. Besides, even in a country like China saddled by centuries of economic stagnation and state bureaucracy, Chinese-born entrepreneurs created an entirely new and highly competitive semiconductor industry, first in Taiwan and today on the Chinese mainland. There is nothing to prevent a repeat of such successful entrepreneurship in the biosciences, particularly as many Chinese "sea-turtle" students and professionals in the field are returning home from the United States coupled with the Chinese government's investment of hundreds of billions of dollars in the field.

When entrepreneurship in parts of the world where it has not flourished begins to grow while talent is not fully used or is wasted in places where entrepreneurial talent has historically flourished, that could be a game changer. There is plenty of evidence that society benefits more when talented people become entrepreneurs who start companies and create real innovation than when they go into rent-seeking and redistributionist activities. Indeed, America's economic future may hinge on the extent to which the financial crisis and its aftermath serves to shunt young talent away from Wall Street and toward creative and productive enterprise in fields like the biosciences, engineering, medical technology, nanotechnology, communications, robotics, clean energy, education, design, and many others. These are the wealth-producing enterprises of the future in an evolving ecosystem of innovation. Their creators do not see themselves as global heroes, the entrepreneurs who hold the key to the future of the global economy. But they are in fact the indispensable economic adventurers that Jean-Baptiste Say first described more than two centuries ago when the modern economy was in its neonatal stage, when the word *innovation* began to appear in publications reflecting the discussions of the day. It is the entrepreneurs who seize an idea, imagine its possibilities, and in their own way, in the words of Arthur C. Clarke, launch themselves across the light years.

References

Listed sequentially based on their order in the chapter

Tim Dirks, "*2001: A Space Odyssey*," The Greatest Films of All Time, accessed December 10, 2012, http://www.filmsite.org/twot.html.

H. Christina Fan et al., "Non-invasive Prenatal Measurement of the Fetal Genome," *Nature* 487 (2012):320–324.

Chris Gaylord, "Stephen Colbert's DNA Heads Toward Final Frontier," *Christian Science Monitor*, September 8, 2008.

Gregory Cochran and Henry Harpending, *The 10,000 Year Explosion: How Civilization Accelerated Human Evolution* (New York: Basic Books, 2009), 23.

Tatum S. Simonson et al., "Genetic Evidence for High-Altitude Adaptation in Tibet," *Science* 329 (2010):72–75.

Xin Yi et al., "Sequencing of 50 Human Exomes Reveals Adaptation to High Altitude," *Science* 329 (2010):75–78.

Laura B Scheinfeldt et al., "Genetic Adaptation to High Altitude in the Ethiopian Highlands," *Genome Biology* 13.1 (2012): R1, accessed December 11, 2012, doi:10.1186/gb-2012-13-1-r1.

A. Bingham et al., "Identifying Signatures of Natural Selection in Tibetan and Andean Populations Using Dense Genome Scan Data," *PLoS Genetics* 6.9 (2010):e1001116, accessed November 30, 2013, doi: 10.1371/journal.pgen.1001116.

Svante Pääbo, Presentation at "The Genetic Revolution and Its Impact on Society," Nobel Week Dialogue, December 9, 2012, accessed December 12, 2012, http://www.nobelweek-dialogue.org.

Matthias Meyer et al., "A High Coverage Genome Sequence From an Archaic Denisovan Individual," *Science* 338 (2012):222–226.

Kevin N. Laland, John Odling-Smee, and Sean Myles, "How Culture Shaped the Human Genome: Bringing Genetics and the Human Sciences Together," *Nature Genetic Reviews* 11 (2010):137–148.

Simon E. Fisher and Matt Ridley, "Culture, Genes, and the Human Revolution," *Science* 340 (2013): 929–930.

Cochrane and Harpending, *The 10,000 Year Explosion.*

Jared Diamond, "The Wealth of Nations," *Nature* 429 (2004):616–617.

Synthetic Biology: DNA, Cells, and the Bioeconomy

Arthur C. Clarke, *2001: A Space Odyssey* (New York: New American Library, 1968).

John Markoff, "Synthetic Life," Tierney Lab blog, *The New York Times*, August 3, 2009.

Moore's Law Timeline, Intel Corp., accessed December 13, 2012, http://www.intel.com.

Daniel G. Gibson et al., "Creation of a Bacterial Cell Controlled by a Chemically Synthesized Genome," *Science* 329 (2010):52–56.

Arthur Caplan, "The End of Vitalism," *Nature* 465 (2010):423.

Mike May, "Engineering a New Business," *Nature Biotechnology* 27.12 (2009): 1112–1120.

Synthetic Biology Website, accessed November 21, 2013, http://syntheticbiology.org.

Adrian L. Slusarczyk, Allen Lin, and Ron Weiss, "Foundations for the Design and Implementation of Synthetic Genetic Circuits," *Nature Reviews Genetics* 13 (2012):406–420.

Kevin D. Litcofsky, "Iterative Plug-and-Play Methodology for Constructing and Modifying Synthetic Gene Networks," *Nature Methods* 9 (2012):1077–1080.

Jerome Bonnet et al., "Amplifying Genetic Logic Gates," *Science* 340 (2013):599–603.

International Genetically Engineered Machine Competition Website, accessed December 13, 2010, http://igem.org.

Andy Coghlan, "Biology Is a Manufacturing Capability," *New Scientist*, December 11, 2012, accessed December 14, 2012, http://www.newscientist.com.

Ariene Weintraub, "Stealthy Gen9 Rolls Out BioFab for Large-Scale Gene Manufacturing," Xconomy.com, July 17, 2012, accessed December 14, 2012, http://www.xconomy.com.

Drew Endy, "Dialogue and Notes on Synthetic Biology," Consortium on Law and Values in Health, Environment, and the Life Sciences, University of Minnesota, April 16, 2009, accessed February 11, 2010, http://www.lifesci.consortium.umn.edu.

"OneWorld Health, Amyris Biotechnologies and Sanofi-Aventis Announce Development Agreement for Semisynthetic Artemisinin," Institute for OneWorld Health, March 3, 2008, accessed February 11, 2010, http://www.oneworldhealth.org.

"Stable Supply of Artemisinin—Important Step Toward Making Malaria a Disease of the Past," Institute for OneWorld Health, April 25, 2012.

Daniel Grushkin, "The Rise and Fall of the Company That Was Going to Have Us All Using Biofuels," *Fast Company*, August 8, 2012, accessed December 14, 2012, http://www.fastcompany.com.

Donald E. Trimbur et al., "Lipid Pathway Modification in Oil-Bearing Microorganisms," US Patent Application 20090061493, March 5, 2009, accessed December 15, 2012, http://patft.uspto.gov.

"Solazyme Pre-IPO Report: Synthetic Nanobiology Comes of Age," Seeking Alpha, May 19, 2011, accessed December 15, 2012, http://www.seekingalpha.com.

"Are You Ready To Eat Algae? Solazyme Knows You Are," Seeking Alpha, January 26, 2013, accessed January 27, 2012, http://www.seekingalpha.com.

"Solazyme Bags U.S. Navy Contract for Green Jet Fuel," Reuters, September 24, 2009.

"Solazyme Shares Soar on Production News," Associated Press, December 14, 2012.

Frederick Leliaert, "Phylogeny and Molecular Evolution of Green Algae," *Critical Reviews in Plant Sciences* 31 (2012):1–46.

Tom Ellis, Xiao Wang, and James J. Collins, "Diversity-based, Model-guided Construction of Synthetic Gene Networks with Predicted Functions," *Nature Biotechnology* 27 (2009):465–471.

Ron Weiss, quoted in Ray Kurzweil, *The Singularity is Near: When Humans Transcend Biology* (New York: Viking Press, 2005), 221.

Carolyn Y. Johnson, "Accessible Science: Hackers Aim to Make Biology Household Practice," *Boston Globe*, September 15, 2008.

Ben Lillie, "Do-It-Yourself Biotech: Ellen Jorgensen at TEDGlobal 2012," TED blog, June 26, 2012, http://blog.ted.com, accessed December 17, 2012.

Liat Clark, "Glowing Trees Could Pave the Way for Solving World Problems with Biology," *Wired*, May 9, 2013, accessed May 12, 2013, http://www.wired.co.uk/news/archive/2013-05/9/glowing-plants-kickstarter.

Ralf Meisenzahl and Joel Mokyr, "The Rate and Direction of Invention in the British Industrial Revolution: Incentives and Institutions," NBER Working Paper 16993, April 2011, accessed December 16, 2012, http://www.nber.org/papers/w16993.

Juan Enriquez, "The Next Big Thing: A New You," *Foreign Policy*, April 15, 2009, accessed December 17, 2012, http://www.foreignpolicy.com.

Leo Furcht and William Hoffman, *The Stem Cell Dilemma: The Scientific Breakthroughs, Ethical Concerns, Political Tensions, and Hope Surrounding Stem Cell Research*, 2nd ed. (New York: Arcade, 2011).

Maria Abad et al., "Reprogramming *In Vivo* Produces Teratomas and iPS cells with Totipotency Features," *Nature* 502.7471 (2013): 340–345.

Tzahi Cohen-Karni, Robert Langer, and Daniel S. Kohane, "The Smartest Materials: The Future of Nanoelectronics in Medicine," *ACS Nano* 6.8 (2012): 6541–6545.

Sander Herfst et al., "Airborne Transmission of Influenza A/H5N1 Virus Between Ferrets," *Science* 336 (2012):1533–1541.

M. Imai et al., "Experimental Adaptation of an Influenza H5 HA Confers Respiratory Droplet Transmission to a Reassortant H5 HA/H1N1 Virus in Ferrets," *Nature* 486 (2012):420–428.

Y. Zhang et al., "H5N1 Hybrid Viruses Bearing 2009/H1N1 Virus Genes Transmit in Guinea Pigs by Respiratory Droplet," *Science* 340 (2013):1459–1463.

Simon Wain-Hobson, "H5N1 Viral Engineering Dangers Will Not Go Away," *Nature* 495 (2013):411.

William Hoffman, "Sword and Shield: The Dual Uses of Pathogen Research," ScienceProgress. org (Center for American Progress), January 5, 2012, accessed April 8, 2012, http://scienceprogress.org/2012/01/sword-and-shield-the-dual-uses-of-pathogen-research/.

William Hoffman, "The Shifting Currents of Bioscience Innovation," *Global Policy* 5.1 (February 2014).

Entrepreneurs: What Makes Them Tick?

Jean-Baptiste Say, *A Treatise on Political Economy or the Production, Distribution, and Consumption of Wealth* (Philadelphia: Grigg & Elliot, 1846), 78.

Adrian Wooldridge, "Global Heroes: A Special Report on Entrepreneurship," *The Economist*, March 12, 2009, accessed December 18, 2012, http://www.economist.com/node/13216025.

Adrian Wooldridge, *Masters of Management* (New York: HarperBusiness, 2011).

"National Bioeconomy Blueprint," The White House, April 2012, accessed January 24, 2012, http://www.whitehouse.gov.

"Globalizing Venture Capital: Global Venture Capital Insights and Trends Report 2011," Ernst & Young, April 17, 2012, accessed May 6, 2013, http://www.ey.com.

Heidi Ledford, "Virtual Reality," *Nature* 498 (2013):127–129.

Michael Lind, *Land of Promise: An Economic History of the United States* (New York: HarperCollins, 2012), 364.

"Industry Gets Religion," Special Report, *The Economist*, February 18, 1999, accessed December 19, 2012, http://www.economist.com/node/186620.

Robert J. Gordon, "Is US Economic Growth Over? Faltering Innovation Confronts the Six Headwinds," Policy Insight No. 63, Center for Economic Policy Research, September 2012, accessed December 23, 2012, http://www.cepr.org.

"Nature or Nurture: Decoding the DNA of the Entrepreneur," Ernst & Young, 2011, accessed December 26, 2012, http://www.ey.com.

Scott Shane et al., "Genetics, the Big Five, and the Tendency to be Self-Employed," *Journal of Applied Psychology* 95.6 (2010): 1154–1162.

John Beddington et al., "The Mental Wealth of Nations," *Nature* 455 (2009):1057–1060.

Adam Smith, *The Theory of Moral Sentiments* (Cambridge UK: Cambridge University Press, 2002).

Tyler Cowen, "An Economic and Rational Choice Approach to the Autism Spectrum and Human Neurodiversity," Social Science Research Network, December 22, 2011, accessed December 26, 2012, http://papers.ssrn.com/sol3/papers.cfm?abstract_id=1975809.

Thomas Armstrong, *The Power of Neurodiversity: Unleashing the Advantages of Your Differently Wired Brain* (Cambridge, MA: DaCapo Press, 2010).

Andrew Lawrence et al., "The Innovative Brain," *Nature* 456 (2008):168–169.

Michael Kosfeld et al., "Oxytocin Increases Trust in Humans," *Nature* 435 (2005):673–676.

Örjan de Manzano et al., "Thinking Outside a Less Intact Box: Thalamic Dopamine D2 Receptor Densities Are Negatively Related to Psychometric Creativity in Healthy Individuals," *PLoS ONE* 5.1 (2010): e10670, accessed December 18, 2012, doi: 10.1371/journal.pone.0010670.

Nicholas W. Simon et al., "Dopaminergic Modulation of Risky Decision-Making," *Journal of Neuroscience* 31.48 (2011): 17460–17470.

Thomas Hills and Ralph Hartwig, "Why Aren't We Smarter Already? Evolutionary Trade-Offs and Cognitive Enhancements," *Current Directions in Psychological Science* 20.6 (2011): 373–377.

Human Connectome Project, accessed February 5, 2013, http://www.humanconnectomeproject.org/.

Francis Collins, "The Symphony Inside Your Brain," National Institutes of Health Director's blog, November 5, 2012, accessed February 5, 2013, http://directorsblog.nih.gov/the-symphony-inside-your-brain/.

John Markoff, "Obama Seeking to Boost Study of Human Brain," *The New York Times*, February 18, 2013.

Barack Obama, "Remarks by the President on the BRAIN Initiative and American Innovation," April 2, 2013. Accessed May 5, 2013, http://www.whitehouse.gov/briefing-room/speeches-and-remarks.

Katrin Amunts et al., "BigBrain: An Ultrahigh-Resolution 3D Human Brain Model," *Science* 340 (2013):1472–1475.

Hermann Hauser, "Right Environment Can Enhance 'Innate' Entrepreneurial Skills," *Nature* 456 (2008):700.

Ben Casselman, "Risk-Averse Culture Infects U.S. Workers, Entrepreneurs," *Wall Street Journal*, June 2, 2013.

Sourcing Talent for Successful Innovation

Doron Ben-Atar, "Alexander Hamilton's Alternative: Technology Piracy and the Report on Manufactures," *William & Mary Quarterly* 52.3 (1995): 389–414.

Doron S. Ben-Atar, *Trade Secrets: Intellectual Piracy and the Origins of American Industrial Power* (New Haven, CT: Yale University Press, 2004).

"Accelerating Innovation in the Bioscience Revolution," Milken Institute, April 2012, accessed January 24, 2011, http://www.milken.org.

AnnaLee Saxenian, *The New Argonauts: Regional Advantage in a Global Economy* (Cambridge MA: Harvard University Press, 2006).

Vivek Wadhwa, AnnaLee Saxenian, and F. Daniel Siciliano, "America's New Immigrant Entrepreneurs: Then and Now," Kauffman Foundation, October 2012, accessed January 24, 2013, http://www.kauffman.org.

Jennifer Martinez, "Bipartisan Group of Senators to Introduce High-Skilled Immigration Bill," *The Hill* (Washington, DC), January 24, 2013, accessed January 28, 2013, http://www.thehill.com.

Charles E. Schumer et al., "Bipartisan Framework for Comprehensive Immigration Reform," *The New York Times*, January 28, 2013.

Mike Sager, "What I've Learned: Andy Grove," *Esquire*, May 1, 2000.

List of Female Astronauts, Wikipedia, accessed December 30, 2012, http://en.wikipedia.org/wiki/List_of_female_astronauts.

Theresa M. Wizemann and Mary-Lou Pardue, eds. *Exploring the Biological Contributions to Human Health: Does Sex Matter?* Committee on Understanding the Biology of Sex and Gender Differences, Board on Health Policy, Institute of Medicine (Washington, DC: National Academy Press, 2001), x.

Larry Cahill, "Why Sex Matters for Neuroscience," *Nature Reviews Neuroscience* 7 (2006):477–484.

"Global Competitiveness Report 2012–2013," World Economic Forum, September 5, 2012, accessed February 6, 2013, http://www.weforum.org/issues/global-competitiveness.

Ricardo Hausmann, Laura D. Tyson, and Saadia Zahidi, "The Global Gender Gap Report 2012," World Economic Forum, October 24, 2012, accessed February 6, 2013, http://www3.weforum.org/docs/WEF_GenderGap_Report_2012.pdf.

"Science for All," *Nature* 495 (2013):5.

"Fast Facts," National Center for Educational Statistics, US Department of Education, accessed December 29, 2012, http://nces.ed.gov/fastfacts/display.asp?id=72.

Gary Becker, "The New Gender Gap in Education," The Becker-Posner blog, March 2, 2008, accessed February 17, 2010, http://www.becker-posner-blog.com.

Laurell Smith-Doerr et al., "A Global Need for Women's Biotech Leadership," *Nature Biotechnology* 29.10 (2011): 948–949.

Education and Brainpower for a New Century

Rita R. Colwell, "Professional Science Master's Programs Merit Wider Support," *Science* 323 (2009):1676–1677.

Public Law 110-69: America COMPETES Act, August 9, 2007. US Government Printing Office, accessed December 31, 2012, http://www.gpo.gov/fdsys/pkg/PLAW-110publ69/content-detail.html.

Science Professionals: Master's Education for a Competitive World, National Research Council (Washington, DC: National Academies Press, 2008), accessed December 31, http://www.nap.edu/catalog.php?record_id=12064.

Carol Lynch, "The Case for Professional Science Master's Degrees," *BioScience* 62.8 (2012): 705–706.

Richard Freeman, "What Does the Global Expansion of Higher Education Mean for the U.S.?" NBER Working Paper 14962, May 2009.

Science and Engineering Indicators Digest 2012, National Science Board, National Science Foundation, January 2012, accessed January 25, 2013, http://www.nsf.gov/statistics/digest12/nsb1202.pdf

Christine M. Matthews, "Foreign Science and Engineering Presence in U.S. Institutions and the Labor Force," Congressional Research Service, October 28, 2010, accessed December 31, 2012, https://opencrs.com/document/97-746/2010-10-28/.

NAEP 2008: Trends in Academic Progress, National Center for Educational Statistics, accessed February 19, 2012, http://nces.ed.gov.

Performance of U.S. 15-Year-Old Students in Mathematics, Science, and Reading Literacy in an International Context–First Look at PISA 2012 (NCES 2014024), National Center for Educational Statistics, December 3, 2013, accessed December 3, 2013, http://nces.ed.gov.

"Taking the Pulse of Bioscience Education in America: A State-by-State Analysis," Report Prepared by Battelle in Cooperation with Biotechnology Industry Organization and the Biotechnology Institute, May 2009, accessed February 19, 2010, http://www.batelle.org.

Budget of the US Government for Fiscal Year 2013, Summary and Background Information, The White House, p. 22, accessed December 31, 2012, http://www.whitehouse.gov/omb/budget.

"National Bioeconomy Blueprint."

Ruth Ellen Wasem, "Immigration of Foreign Nationals with Science, Technology, Engineering, and Mathematics (STEM) Degrees," Congressional Research Service, November 26, 2012, accessed January 25, 2013, http://www.fas.org/sgp/crs/misc/R42530. pdf

Hal Salzman and Lindsey Lowell, "Making the Grade," *Nature* 453 (2008):28–30.

Heiner Rindermann and James Thompson, "Cognitive Capitalism: The Effect of Cognitive Ability on Wealth, as Mediated Through Scientific Achievement and Economic Freedom," *Psychological* Science (2011) 22.6:754–763.

Anne-Marie Slaughter, "America's Edge: Power in the Networked Economy," *Foreign Affairs* (January/February 2009): 94–113.

George Baeder and Michael Zielenziger, "China, the Life Sciences Leader of 2020," The Monitor Group, November 17, 2010, accessed November 30, 2010, http://www.monitor. com.

John Carroll, "China to Spend $308B, Gain 1M New Jobs in 5-year Biotech Plan," FierceBiotech.com, accessed December 31, 2012, http://www.fiercebiotech.com.

Conclusion: Seeding the Future

At the end of the last millennium, the scientist and Nobel economist Robert W. Fogel was astonished that so soon after Kitty Hawk a man was standing on the moon. Fogel used the example in his presidential address to the American Economic Association to illustrate the challenge the economics profession faces in trying to keep up with the accelerating pace of technological change. "We are slow in pondering such grand questions as the implications of the Human Genome Project, which is now nearing completion, and the emergence of molecular medicine for the future of economic life," said Fogel, the world's leading scholar on the effects of agricultural productivity and technological evolution on human physiology since 1700. We have entered an era in which purposeful intervention in evolutionary processes is passing beyond plant and animal breeding, he said. "The new growth economics needs to incorporate at least some aspects of directed, rapid human evolution."

Fogel created a much-cited timeline showing world population growth alongside some major events in the history of technology from the beginning of agriculture to the Human Genome Project. We have modified Fogel's timeline, first published in 1999, focusing on discoveries and developments in the biosciences and their convergence with those in digital technology and networks (Figure C.1). Biological technologies like genomics, synthetic biology, regenerative medicine, 3D bioprinting and others are in their infancy, perhaps where computer technology was in the 1950s and 1960s. A self-replicating microbe created entirely from computer code has yet to debut. If it does, life itself will take on new meaning. Meanwhile, as tracked by Fogel over three centuries, technological innovation's contribution to human health, nutrition, longevity, and well-being grows apace.

Yet we remain some distance from knowing what a field like genomics means, in Fogel's words, "for the future of economic life." We do know that biological knowledge is moving out of the laboratory, into the economy, and into all aspects of our lives. It is happening more subtly than the technologies that launched the Industrial Revolution symbolized by James Watt's separate condenser steam engine. The naturalist Erasmus Darwin, Watt's Lunar Society friend, called the engine "the most ingenious of human inventions." Darwin subscribed to a then-fashionable theory of evolution—that characteristics acquired during life could be passed on

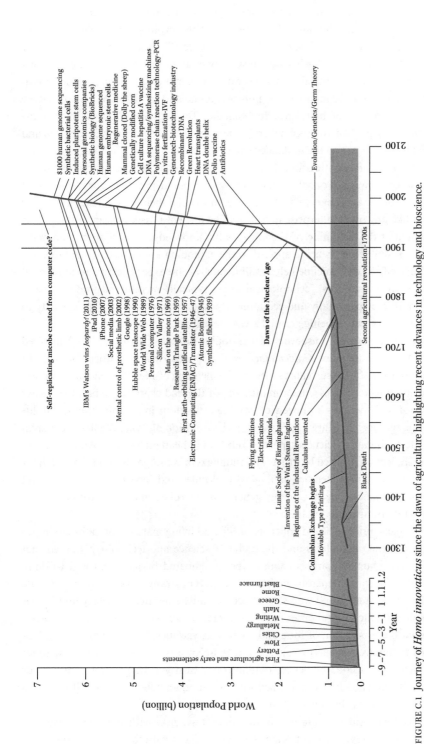

FIGURE C.1 Journey of *Homo innovaticus* since the dawn of agriculture highlighting recent advances in technology and bioscience.

Source: The authors with the assistance of James Hudak. Modified from Figure 1 of Robert W. Fogel, "Catching Up With the Economy," *American Economic Review* 89/1 (1999): 2, with permission.

to offspring—that his much more famous grandson Charles contradicted with his theory of evolution by natural selection. But the evolution of technology does reflect the inheritance of acquired characteristics. The tools and instruments of Watt's day—products of what was then known about physics, mechanics, chemistry, electricity, optics, metallurgy, and craft—have evolved through countless iterations, alterations, recombinations, improvements, and miniaturization to become today's basic tools of bioscience and genetics. These tools are poised to alter the evolutionary path of the toolmakers' creative power, their economic fortunes and social institutions, and the toolmakers themselves, us.

The ethos of relentless innovation that arose in the eighteenth century today gives us the power not only to shape how we live, and in a profound way, but indeed to alter who we are. That is why the sciences of life and their material manifestations are unlike any other enterprise in history. The ancient Greek concept of *techne*, as we noted at the beginning of the book, is the knowledge of how to do and make things. Today *techne* is being applied to the exploding knowledge of the processes of life itself. As our timeline helps to illustrate, we are experiencing a technological surge in the biosciences. Technological surges typically precede the production and distribution of personally and economically valuable innovations. Our capacity to imagine and create is being granted a vast new arena.

Watson and Crick's discovery of the double-helical structure of DNA inspired Salvador Dali to paint *Galacidalacidesoxyribonucleicacid* (subtitled Homage to Crick and Watson), a surrealist's view of the meaning of life. The genetic revolution gave film director James Cameron the idea for the lead character in his blockbuster movie *Avatar*, perhaps a preview of coming attractions in neurotechnology. In literature, writers like Margaret Atwood in the tradition of Mary Shelley draw upon the latest laboratory science to portray our Promethean pretensions. Amid exercises in self-reflection induced by artists and filmmakers, we forge ahead in the search for life-saving and life-extending biological molecules and bioartificial organs, brain function-enhancing agents, energy-generating microbes, and food crops equipped with superior nitrogen uptake traits.

The basic principles and concepts of life and living matter prompt broad inquiries about the possibilities and the limits of science and technology and whether the creative human spirit is being properly tempered by judgment and wisdom. The questions are numerous and compelling. They range from when human life begins, when it ends and who should decide, to whether our planetary footprint is stomping on life's moral order in the inexorable march of progress. Yes, bioscience is about the production of food and medicines, vaccines, green chemicals, and bio-based materials and energy sources. But bioscience is also about our story, about our ability to decode ourselves, peer into our own genetic past and ponder our genetic future, about how and when we reproduce, and about whether it is wise to take steps to alter the course of our own evolution.

It is been said that biology is the biggest science, with the most scientists, the most funding, the most scientific results, the most ethical significance, and where we have the most to learn given its billions of years of experimental results. Biology's domain spans our living space from the invisible to the panoramic, from biomolecules to the biosphere. Bioscience encompasses the study of archaea buried deep in ocean vents and the earth's crust to the weighty quadrupeds that walk the planet to the bacteria that hover in clouds over it to speculation about the possibility of life forms on other planets or in other galaxies. But biology as a technology capable of making major contributions to sustainable economic growth and development used to be limited and slow. Its tools and methods were largely in the hands of specialists. Those limitations no longer exist as they once did. Powerful new tools of biological exploration are now in the hands of college students, high schoolers, hobbyists, and hackers as well as bench scientists with academic degrees.

Because the health of living systems is critical to our future, basic education in biology is migrating beyond the basic principles of the cell and the development of single-cell and multicellular organisms to the ecosystems in which they flourish and fade. Because living systems possess unparalleled creative powers, it is only natural that our uniquely creative species will draw on those powers to improve our lot in life as well as understand the world around us. The tools of bioscience give us the power to delve into our brains to visualize the seeds of imagination and creativity, the neural networks that bring art, music, literature, design, and those very tools into existence. They bring us full circle as "strange loops," self-referential beings embarked on an endless journey of our own creation, a journey in which we build new tools not only to monitor but to sustain and revitalize the health and well-being of the living systems that support us.

In ancient Athens, Athena was the patron god of the *polis*, the urban space where people and ideas come together to create culture, technology, and commerce. The spirit of Athena bridged art and *technê* before they went their separate ways. She embodied practical intelligence, which is what successful innovators and entrepreneurs possess. They dream what might be possible based on what is known and push back the unknown and the impossible with their enterprise. The sciences of life are just beginning to seed these dreams, all over the world. The rest of the human journey that began on the African savannah will be marked indelibly by the creative energy these dreams set free.

References

Robert W. Fogel, "Catching Up with the Economy," *American Economic Review* 89.1 (1999): 1–21.

SELECTED BIBLIOGRAPHY

Adams, Jonathan. "The Rise of Research Networks." *Nature* 490 (2012):335–336.

Appleby, Joyce. *Relentless Revolution: A History of Capitalism* (New York: W.W. Norton, 2010).

Armstrong, Karen. *A Short History of Myth* (Edinburgh: Canongate Books, 2005).

Arthur, W. Brian. "'Silicon Valley' Locational Clusters: When Do Increasing Returns Imply Monopoly?" *Mathematical Social Sciences* 19.3 (1990): 235–235.

Arthur, W. Brian. *The Nature of Technology: What It Is and How It Evolves* (New York: Free Press, 2009).

Association for Molecular Pathology et al. v. *United States Patent and Trademark Office*, 569 US 12-398 (decided June 13, 2013).

Baeder, George, and Michael Zielenziger. "China, the Life Sciences Leader of 2020." The Monitor Group, November 17, 2010, accessed November 30, 2010, http://www.monitor.com.

Beddington, John et al. "The Mental Wealth of Nations." *Nature* 455 (2009):1057–1060.

Berman, Elizabeth Popp. *Creating the Market University: How Academic Science Became an Economic Engine* (Princeton, NJ: Princeton University Press, 2012).

Berra, Rajendra K. "The Story of the Cohen-Boyer Patents." *Current Science* 96.6 (2009): 761–762.

Bilski v. Kappos, 130 S. Ct. 3218 (2010).

Boycott, Kym M., Megan R. Vanstone, Dennis E. Bulman, and Alex E. MacKenzie. "Rare-disease Genetics in the Era of Next-generation Sequencing: Discovery to Translation." *Nature Reviews Genetics* 14 (2013):681–691.

Broad, William J. "Patent Bill Returns Bright Idea to Inventor." *Science* 205 (1979):473–476.

Buchan, James. *Crowded with Genius: The Scottish Enlightenment: Edinburgh's Moment of the Mind* (New York: HarperCollins, 2003).

Bud, Robert. *The Uses of Life: A History of Biotechnology* (Cambridge, UK: Cambridge University Press, 1993).

Burk, Dan L., and Mark A. Lemley. *The Patent Crisis and How the Courts Can Solve It.* (Chicago: University of Chicago Press, 2009).

Burrill, G. Steven. "The Global Transformation: The International Nature of Research and Development Is Changing the Life Sciences in Radical Ways." *The Burrill Report*, May 16, 2007, accessed September 16, 2012, http://www.burrillreport.com/printer_article-198.html.

Bush, Vannevar. "As We May Think." *The Atlantic Monthly* (July 1945):101–108.

Bush, Vannevar. "Science, The Endless Frontier." A Report to the President by Vannevar Bush, Director of the Office of Scientific Research and Development, July 1945. Washington, DC: US Government Printing Office, 1945.

Cahill, Larry. "Why Sex Matters for Neuroscience." *Nature Reviews Neuroscience* 7 (2006):77–484.

Carlson, Robert. "Biodesic 2011 Bioeconomy Update." August 2011, accessed November 18, 2013, http://www.biodesic.com.

Carlson, Robert H. *Biology Is Technology: The Promise, Peril, and New Business of Engineering Life* (Cambridge, MA: Harvard University Press, 2010).

Chakma, Justin, Stephen M. Sammul, and Ajay Agrawal. "Life Sciences Venture Capital in Emerging Markets." *Nature Biotechnology* 31.3 (2013):195–201.

Chatterjee, Anusuya, and Ross DeVol. "Estimating Long-term Economic Returns of NIH Funding on Output in the Biosciences." The Milken Institute, August 31, 2012, accessed January 31, 2012, http://www.milkeninstitute.org/pdf/RossandAnuNIHpaper.pdf.

Christensen, Clayton, Jerome H. Grossman, and Jason Hwang. *The Innovator's Prescription* (New York: McGraw-Hill, 2008).

Church, George M., Yuan Gao, and Sriram Kosuri. "Next-Generation Digital Information Storage in DNA." *Science* 337 (2012):1628.

Clarke, Arthur C. *2001: A Space Odyssey* (New York: New American Library, 1968).

Cochran, Gregory, and Henry Harpending. *10,000 Year Explosion: How Civilization Accelerated Human Evolution* (New York: Basic Books, 2009).

Cockburn, Iain M., and Matthew J. Slaughter. "The Global Location of Biopharmaceutical Knowledge Activity: New Findings, New Questions." In Josh Lerner and Scott Stern, eds., *Innovation Policy and the Economy*, Vol. 10 (Chicago: University of Chicago Press, 2010).

Cohen, Stanley N., et al. "Construction of Biologically Functional Bacterial Plasmids *In Vitro*." *Proceedings of the National Academy of Sciences USA* 70.11 (1973):3240–3244.

Collins, Francis S., Michael Morgan, and Aristides Patrinos. "The Human Genome Project: Lessons from Large-Scale Biology." *Science* 300 (2003):286–290.

Colwell, Rita R. "Professional Science Master's Programs Merit Wider Support." *Science* 323 (2009):1676–1677.

Cooke, Philip E. "Global Bioregional Networks: A New Economic Geography of Bioscientific Knowledge." *European Planning Studies* 14 (2006):1265–1285.

Cooke, Philip. *Growth Cultures: The Global Bioeconomy and its Bioregions* (London: Routledge, 2007).

Cookson, Clive. "The Lucrative Allure of the Double Helix." *Financial Times*, April 19, 2013, accessed April 23, 2013, http://www.ft.com/intl/cms/s/0/4269758c-a776-11e2-9fbe-00144feabdc0.html.

Cowen, Tyler. *The Great Stagnation: How America Ate All The Low-Hanging Fruit of Modern History, Got Sick, and Will (Eventually) Feel Better* (New York: Dutton Adult, 2011).

Crosby, Alfred W. *The Columbian Exchange: Biological and Cultural Consequences of 1492* (Westport, CT: Greenwood Press, 1973).

David, Peter. "Universities: Inside the Knowledge Factory." *The Economist*, Special Insert, October 4, 1997, accessed February 6, 2012, http://www.economist.com.

De Chadarevian, Soraya. *Designs for Life: Molecular Biology After World War II* (Cambridge, UK: Cambridge University Press: 2002).

Delbecco, Renato. "A Turning Point in Cancer Research: Sequencing the Genome." *Science* 231 (1986):1055–1056.

Diamond v. Chakrabarty, 447 US 303 (1980).

Diamond, Jared. *Guns, Germs and Steel: The Fates of Human Societies* (New York: Norton, 1997).

Diamond, Jared. "The Wealth of Nations." *Nature* 429 (2004):616–617.

DiMasi, J. A., L. Feldman, A. Seckler, and A. Wilson. "Trends in Risks Associated with New Drug Development: Success Rates for Investigational Drugs." *Clinical Pharmacology & Therapeutics* 87.3 (2010):272–277.

Dobzhansky, Theodosius. "Nothing in Biology Makes Sense Except in the Light of Evolution." *American Biology Teacher* 35 (1973):125–129.

Dodgson, Mark,and David Gann. *Innovation: A Very Short History* (Oxford: Oxford University Press, 2010).

Dutfield, Graham. *Intellectual Property Rights & the Life Science Industries: Past, Present & Future* (Hackensack, NJ: World Scientific Books, 2009).

"Economic Impact of the Human Genome Project." Battelle Technology Partnership Practice, Battelle Memorial Institute, May 2011, accessed January 9, 2012, http://www.battelle.org.

Eisenstein, Elizabeth L. *The Printing Press as an Agent of Change* (Cambridge, UK: Cambridge University Press, 1979).

Ellis, Aytoun. *Penny Universities: A History of the Coffee-Houses* (London: Secker &Warburg, 1956).

ENCODE Project Consortium. "An Integrated Encyclopedia of DNA Elements in the Human Genome." *Nature* 489.57 (2012):57–74.

Enriquez, Juan. "The Next Big Thing: A New You." *Foreign Policy*, April 15, 2009, accessed December 17, 2012, http://www.foreignpolicy.com.

Etzkowitz, Henry, and Loet Leydesdorff, Loet, eds. *Universities and the Global Knowledge Economy: A Triple Helix of University-Industry-Government Relations* (London: Cassell Academic, 1997).

Fedoroff, N. V., et al. "Radically Rethinking Agriculture for the 21st Century." *Science* 327 (2010):327–328.

Feldman, Maryann. "Place Matters." ScienceProgress.org, January 21, 2009, accessed August 31, 2012, http://www.scienceprogress.org/2009/01/place-matters.

Fisher, R. A. "Has Mendel's Work Been Rediscovered?" *Annals of Science* 1 (1936):115–137.

Florida, Richard. "The World Is Spiky." *Atlantic Monthly* (October 2005):48–51.

Floud, Roderick, Robert W. Fogel, Bernard Harris, and Sok Chui Hong. *The Changing Body: Health, Nutrition, and Human Development in the Western World since 1700* (Cambridge, UK: Cambridge University Press, 2011).

Fogel, Robert W. "Catching Up with the Economy." *American Economic Review* 89.1 (1999):1–21.

Fogel, Robert W. *The Escape from Hunger and Premature Death, 1700–2100* (Cambridge, UK: Cambridge University Press, 2004).

Foley, J., et al. "Solutions for a Cultivated Planet." *Nature* 478 (2011):337–342.

Furcht, Leo and William Hoffman. *The Stem Cell Dilemma: The Scientific Breakthroughs, Ethical Concerns, Political Tensions, and Hope Surrounding Stem Cell Research*, 2nd ed. (New York: Arcade, 2011).

Genome Resources Website. National Center for Biotechnology Information, National Institutes of Health, accessed August 19, 2012, http://www.ncbi.nlm.nih.gov/genome.

Gibson, D. G., et al. "Creation of a Bacterial Cell Controlled by a Chemically Synthesized Genome." *Science* 329 (2010):52–56.

Glaeser, Edward L., Hedi Kallal, Jose A. Scheinkman, and Andrei Shleifer, "Growth in Cities." *Journal of Political Economy* 100.6 (1992):1126–1152.

"Global Trends 2030: Alternative Worlds." U.S. National Intelligence Council, Office of the Director of National Intelligence, December 2012, http://www.dni.gov/index.php/about/organization/national-intelligence-council-global-trends, accessed December 10, 2012.

Gonzaga-Jauregui, Claudia, James R. Lupksi, and Richard A. Gibbs. "Human Genome Sequencing in Health and Disease." *Annual Review of Medicine* 63 (2012):35–61.

Gordon, Robert J. "Is U.S. Economic Growth Over? Faltering Innovation Confronts the Six Headwinds." NBER Working Paper No. 18315 (Cambridge, MA: National Bureau of Economic Research, 2012), accessed December 5, 2012, http://www.nber.org/papers/w18315.

Green, Eric. "Welcome and Opening Remarks: HGP10: The Genomics Landscape a Decade after the Human Genome Project." National Human Genome Institute, April 25, 2013, accessed May 21, 2013, http://www.genome.gov/27552257.

Harrison, Charlotte. "Dangling from the Patent Cliff." *Nature Reviews Drug Discovery* 12 (2013):14–15.

Heller, Michael and Rebecca Eisenberg, "Can Patents Deter Innovation? The Anticommons in Biomedical Research." *Science* 280 (1998):698–701.

Hermans, Raine, Alicia Löffler, and Scott Stern. "Biotechnology." In *Innovation in Global Industries: U.S. Firms Competing in a New World* (Washington, DC: National Research Council, National Academy Press), 231–272.

Hicks, Justin, and Robert D. Atkinson. "Eroding Our Foundation: Sequestration, R&D, Innovation and U.S. Economic Growth." Information Technology & Innovation Foundation, September 2012, accessed March 4, 2013, http://www2.itif.org/2012-eroding-foundation.pdf.

Hoffman, William. "The Shifting Currents of Bioscience Innovation." *Global Policy* 5.1 (February 2014).

Human Microbiome Project, National Institutes of Health, accessed January 15, 2013, http://commonfund.nih.gov/hmp/.

"Innovation's Golden Goose." *The Economist*, December 12, 2002, accessed February 6, 2012, http://www.economist.com.

International Human Genome Sequencing Consortium. "Initial Sequencing and Analysis of the Human Genome." *Nature* 401 (2001):860–921.

Jackson, David A., Robert H. Symons, and Paul Berg. "Biochemical Method for Inserting New Genetic Information into DNA of Simian Virus 40: Circular SV40 DNA Molecules Containing Lambda Phage Genes and the Galactose Operon of *Escherichia coli.*" *Proceedings of the National Academy of Sciences USA* 69.10 (1972):2904–2909.

Judson, Horace. *The Eighth Day of Creation: Makers of the Revolution in Biology* (New York: Simon & Schuster, 1979).

Kamb, Alexander, Sean Harper, and Kari Stefansson. "Human Genetics as a Foundation for Innovative Drug Development." *Nature Biotechnology* 31 (2013):975–978.

Kelley, Brian. "Industrialization of mAb Production Technology." *mAbs* 1.5 (2009):443–452.

Kevles, Daniel J. "Can They Patent Your Genes?" *New York Review of Books*, March 7, 2013, accessed February 17, 2013, http://www.nybooks.com.

Kneller, Robert. "The Importance of New Companies for Drug Discovery: Origins of a Decade of New Drugs." *Nature Reviews Drug Discovery* 9 (2010):867–882.

Kumar, Dhavendra, ed. *Genomics and Health in the Developing World*, Oxford Monographs on Medical Genetics (New York: Oxford University Press, 2012).

Laland, Kevin N., John Odling-Smee, and Sean Myles. "How Culture Shaped the Human Genome: Bringing Genetics and the Human Sciences Together." *Nature Genetic Reviews* 11 (2010): 137–148.

Landes, David. *The Wealth and Poverty of Nations: Why Some are So Rich and Others So Poor* (New York: W.W. Norton, 1998).

Lawrence, Andrew, et al. "The Innovative Brain." *Nature* 456 (2008):168–169.

Lessig, Lawrence. *Remix: Making Art and Commerce Thrive in the Hybrid Economy* (New York: Penguin Press HC, 2008).

Lin, F.-K., et al. "Cloning and Expression of the Human Erythropoietin Gene." *Proceedings of the National Academy of Sciences USA* 82 (1985):7580–7584.

Lind, Michael. *Land of Promise: An Economic History of the United States* (New York: Har perCollins, 2012).

Litan, Robert E., and Lesa Mitchell. "A Faster Path from Lab to Market." In "The HBR List: Breakthrough Ideas for 2010," *Harvard Business Review* (January 2010):52–53.

Litcofsky, Kevin D. "Iterative Plug-and-Play Methodology for Constructing and Modifying Synthetic Gene Networks." *Nature Methods* 9 (2012):1077–1080.

Loman, Nicholas J., and James Hadfield. "World Map of High-Throughput Sequencers." accessed November 1, 2012, http://omicsmaps.com.

Madison, James. "Address to the Agricultural Society of Albermarle, Virginia," May 12, 1818. In *Letters and Other Writings of James Madison*, III (Philadelphia, 1865, imprint of R. Worthington, NY, 1884).

Malakoff, David and Robert F. Service. "Genomania Meets the Bottom Line." *Science* 291 (2001):1193–1203.

Mandel, Michael. "The Failed Promise of Innovation in the U.S." *Bloomberg BusinessWeek,* June 3, 2009, accessed November 18, 2013, http://www.businessweek.com.

Mann, Charles C. *1493: Uncovering the New World Columbus Created* (New York: Knopf, 2011).

Marshall, Alfred. *Principles of Economics* (London: Macmillan and Co., 1890).

Matthews, Christine M. "Foreign Science and Engineering Presence in U.S. Institutions and the Labor Force." Congressional Research Service, October 28, 2010, accessed December 31, 2012, https://opencrs.com/document/97-746/2010-10-28.

Mayo Collaborative Services v. Prometheus Laboratories, Inc. US Supreme Court, No. 10-1150 (March 21, 2011).

Mayr, Ernst. *The Growth of Biological Thought: Diversity, Evolution, and Inheritance* (Cambridge, MA: Harvard University Press, 1982).

McNeil, Maureen. *Under the Banner of Science: Erasmus Darwin and His Age* (Manchester, UK: Manchester University Press, 1987).

Mokyr, Joel. "Punctuated Equilibria and Technological Progress." *American Economic Review* 80.2 (1990): 350–354.

Mokyr, Joel. *The Gifts of Athena: Historical Origins of the Knowledge Economy* (Princeton, NJ: Princeton University Press, 2002).

Mokyr, Joel. "The Contribution of Economic History to the Study of Innovation and Technical Change: 1750–1914." In Bronwyn H. Hall and Nathan Rosenberg, eds., *Handbook of the Economics of Innovation*, Vol. 1 (Holland: Elsevier, 2010).

Moore's Law Timeline. Intel Corp., accessed December 13, 2012, http://www.intel.com.

Mowery, David C. "The Bayh-Dole Act and High-Technology Entrepreneurship in U.S. Universities: Chicken, Egg, or Something Else?" In Gary B. Libecap, ed., *University*

Entrepreneurship and Technology Transfer: Process, Design, and Intellectual Property (London: Elsevier, 2005), 39–68.

Munos, Bernard. "Can Open-Source R&D Reinvigorate Drug Research?" *Nature Reviews Drug Discovery* 5.9 (2006):723–729.

Munos, Bernard. "Lessons from 60 Years of Pharmaceutical Innovation." *Nature Reviews Drug Discovery* 8 (2009):959–968.

"National Bioeconomy Blueprint." The White House, April 2012, accessed May 2, 2012, http://www.whitehouse.gov.

National Research Council. *Toward Precision Medicine: Building a Knowledge Network for Biomedical Research and a New Taxonomy of Disease* (Washington, DC: National Academies Press, 2012).

Nielsen, Michael. *Reinventing Discovery: The New Era of Networked Science* (Princeton, NJ: Princeton University Press, 2012).

1000 Genomes Project Consortium. "An Integrated Map of Genetic Variation From 1,092 Human Genomes." *Nature* 491 (2012):58–65.

Pääbo, Svante. Presentation at "The Genetic Revolution and Its Impact on Society," Nobel Week Dialogue, December 9, 2012, accessed December 12, 2012, http://www.nobel-weekdialogue.org.

Pisano, Gary P. *Science Business: the Promise, the Reality, and the Future of Biotech* (Cambridge, MA: Harvard Business School Press, 2006).

Pisano, Gary P., and Willy C. Shih, "Restoring American Competitiveness." *Harvard Business Review* (July/August 2009):114–125.

Porter, Michael E. "The Adam Smith Address: Location, Clusters and the 'New' Microeconomics of Competition." *Business Economics* 33.1 (1998):7–13.

Porter, Michael E. *The Competitive Advantage of Nations* (New York: Free Press, 1990).

Powell, Walter W. et al. "The Spatial Clustering of Science and Capital: Accounting for Biotech Firm-Venture Capital Relationships." *Regional Studies* 36 (2002):291–305.

Presidential Commission for the Study of Bioethical Issues, "Privacy and Progress in Whole Genome Sequencing." October 2012, accessed November 2, 2012, http://www.bioethics.gov/cms/node/764.

Public Law 11–29: Leahy–Smith America Invents Act (September 16, 2011).

Public Law 96-517: Bayh–Dole Act (December 12, 1980).

Public Law 110-69: America COMPETES Act (August 9, 2007).

Rathman, George B., Chairman, CEO and President of Amgen, 1980-1988, Oral History Transcript / George B. Rathmann, BANC MSS 2005/108, The Bancroft Library, University of California, Berkeley, California.

Report of the MIT Taskforce on Innovation and Production. MIT Commission on Innovation, February 22, 2013, accessed February 25, 2013, http://web.mit.edu/newsoffice/2013/production-in-the-innovation-economy-report-mit-0222.html.

"Report to the President: Transformation and Opportunity: The Future of the U.S. Research Enterprise." President's Council of Advisors on Science and Technology, The White House, November 2012, December 5, 2012, http://www.whitehouse.gov/administration/eop/ostp/pcast/docsreports.

Ridley, Matt. *The Rational Optimist: How Prosperity Evolves* (New York: HarperCollins, 2010).

Rinaldi, Andrea. "More Than the Sum of Their Parts." *EMBO Reports* 7.2 (2006):133–136.

Ro, Dae-Kyun, et al. "Production of the Antimalarial Drug Precursor Artemisinic Acid in Engineered Yeast." *Nature* 440 (2006):940–943.

Roach, J. C., et al. "Analysis of Genetic Inheritance in a Family Quartet by Whole-Genome Sequencing." *Science* 328 (2010):636–639.

Roberts, N. J., J. T. Vogelstein, G. Parmigiani, K. W. Kinzler, B. Vogelstein, and V. E. Velculescu. "The Predictive Capacity of Human Genome Sequencing." *Science Translational Medicine* 4.133 (2012):133ra58, doi: 10.1126/scitranslmed.3003380.

Roberts, Richard J. "Perspective: How Restriction Enzymes Became the Workhorses of Molecular Biology." *Proceedings of the National Academy of Sciences USA* 102.17 (2005): 5905–5008.

Romer, Paul M. "Endogenous Technological Change." *Journal of Political Economy* 98.5 (1990):S71–S102.

Romer, Paul M. "Economic Growth." In *The Concise Encyclopedia of Economics* (New York: Warner Books, 1993), accessed February 2010, http://www.econlib.org.

Saiki, R., et al. "Primer-directed Enzymatic Amplification of DNA with a Thermostable DNA Polymerase." *Science* 239 (1988):487–491.

Salzman, Hal, and Lindsey Lowell. "Making the Grade." *Nature* 453 (2008):28–30.

Samways, Michael. "Translocating Fauna to Foreign Lands: Here Comes the Homogenocene." *Journal of Insect Conservation* 3.2 (1999): 65–66.

Scannell, J. W., A. Blanckley, H. Boldon, and B. Warrington. "Diagnosing the Decline in Pharmaceutical R&D Efficiency." *Nature Reviews Drug Discovery* 11 (2012):191–200.

Schacht, Wendy H. "The Bayh-Dole Act: Selected Issues in Patent Policy and the Commercialization of Technology." *Congressional Research Service* (June 9, 2011), accessed November 18, 2013, http://www.fas.org/sgp/crs/misc/RL32076.pdf.

Schumpeter, Joseph A. *The Theory of Economic Development: An Inquiry into Profits, Capital, Credit, Interest, and the Business Cycle* (Cambridge, MA: Harvard University Press, 1934).

Shiller, Robert. "The Neuroeconomics Revolution." Project Syndicate, November 21, 2011, accessed March 23, 2012, http://www.project-syndicate.org/commentary/the-neuroeconomics-revolution.

Sinsheimer, Robert. "A Database of Genomes." Interview for DNA Interactive (DNAi), Dolan DNA Learning Center, Cold Spring Harbor, New York, accessed January 21, 2012, http://www.dnalc.org/view/15334-A-database-of-genomes-Robert-Sinsheimer.html.

Smil, Vaclav. "Harvesting the Biosphere: The Human Impact." *Population and Development Review* 37.4 (2011):613–636.

Smith, Adam. *An Inquiry into the Nature and Causes of the Wealth of Nations* (London: W. Strahan and T. Cadell, 1776).

Smith-Doerr, L., et al. "A Global Need for Women's Biotech Leadership." *Nature Biotechnology* 29.10 (2011):948–949.

Smýkal, P., et al. "Pea (*Pisum sativum* L.) in the Genomic Era." *Agronomy* 2 (2012):74–115.

Solow, Robert M. "Technical Change and the Aggregate Production Function." *Review of Economics and Statistics* 39.3 (1957):312–320.

Specter, Michael. "A Life of Its Own." *The New Yorker*, September 28, 2009.

Stanford University v. Roche Molecular Systems, Inc. US Supreme Court, No. 09-1159. (Argued February 28, 2011—Decided June 6, 2011.)

Stern, Nicholas. *The Economics of Climate Change: The Stern Review* (Cambridge, UK: Cambridge University Press, 2007).

Stevens, Ashley J. "The Enactment of Bayh-Dole." *Journal of Technology Transfer* 29.1 (2004):93–99.

"The Global Use of Medicines: Outlook Through 2016." IMS Institute for Healthcare Informatics, July 2012, accessed November 16, 2012, http://www.imshealth.com.

"The Impact of Genomics on the U.S. Economy." Battelle Technology Partnership Practice, Battelle Memorial Institute, June 2013. Prepared for United for Medical Research, accessed June 12, 2013, http://www.unitedformedicalresearch.com.

"The Triumph of the Commons: Can Open Source Revolutionise Biotech?" *The Economist*, February 10, 2005.

Thiers, F. A., A. J. Sinskey, and E. R. Berndt. "Trends in the Globalization of Clinical Trials." *Nature Reviews Drug Discovery* 7 (2008):13–14.

Thorsteinsdóttir, Halla, et al. "Conclusions: Promoting Biotechnology Innovation in Developing Countries." *Nature Biotechnology* 22 (Suppl.) (2004):DC48–DC52.

Townsend, Anthony, Alex Soojung-Kim Pang, and Rick Weddie. "Future Knowledge Ecosystems: The Next Twenty Years of Technology-Led Economic Development." Institute for the Future, accessed February 2, 2010, http://www.iftf.org.

Uglow, Jenny. *The Lunar Men: Five Friends Whose Curiosity Changed the World* (New York: Farrar, Straus and Giroux, 2002).

Van Dongen, J., et al. "The Continuing Value of Twin Studies in the Omics Era." *Nature Reviews Genetics* 13 (2012):640–653.

Varmus, Harold. "The DNA of a New Industry." *The New York Times*, September 24, 2002.

Venter, J. C., et al., "The Sequence of the Human Genome." *Science* 291 (2001):1304–1351.

Vogelstein, Bert, et al., "Cancer Genome Landscapes." *Science* 339 (2013):1546–1559.

von Hippel, Eric, and Michael Schrage. "Users Are Transforming Innovation." *Financial Times*, July 11, 2007.

Wadhwa, Vivek, AnnaLee Saxenian, and F. Daniel Siciliano. "America's New Immigrant Entrepreneurs: Then and Now." Kauffman Foundation, October 2012, accessed January 24, 2013, http://www.kauffman.org.

Wasem, Ruth Ellen. "Immigration of Foreign Nationals with Science, Technology, Engineering, and Mathematics (STEM) Degrees." Congressional Research Service, November 26, 2012, accessed January 25, 2013, http://www.fas.org/sgp/crs/misc/R42530.pdf.

Watson, James D. *The Double Helix: A Personal Account of the Discovery of the Structure of DNA* (New York: Scribner, 1968).

Wizemann, Theresa M., and Mary-Lou Pardue, eds. *Exploring the Biological Contributions to Human Health: Does Sex Matter?* Committee on Understanding the Biology of Sex and Gender Differences, Board on Health Policy, Institute on Medicine (Washington, DC: National Academy Press, 2001).

Wooldridge, Adrian. "Global Heroes: A Special Report on Entrepreneurship." *The Economist*, March 12, 2009, accessed December 18, 2012, http://www.economist.com/node/13216025.

Zucker, Lynn G., Michael R. Darby, and Marilyn B. Brewer. "Intellectual Human Capital and the Birth of U.S. Biotechnology Enterprises." *American Economic Review* 88.1 (1998): 290–306.

INDEX